NONLINEAR OPTICS

NELINEINAYA OPTIKA

НЕЛИНЕЙНАЯ ОПТИКА

The Lebedev Physics Institute Series

Editor: Academician D. V. Skobel'tsyn

Director, P. N. Lebedev Physics Institute, Academy of Sciences of the USSR

Volume 25	Optical Methods of Investigating Solid Bodies
Volume 26	Cosmic Rays
Volume 27	Research in Molecular Spectroscopy
Volume 28	Radio Telescopes
Volume 29	Quantum Field Theory and Hydrodynamics
Volume 30	Physical Optics
Volume 31	Quantum Electronics in Lasers and Masers
Volume 32	Plasma Physics
Volume 33	Studies of Nuclear Reactions
Volume 34	Photomesonic and Photonuclear Processes
Volume 35	Electronic and Vibrational Spectra of Molecules
Volume 36	Photodisintegration of Nuclei in the Giant Resonance Region
Volume 37	Electrical and Optical Properties of Semiconductors
Volume 38	Wideband Cruciform Radio Telescope Research
Volume 39	Optical Studies in Liquids and Solids
Volume 40	Experimental Physics: Methods and Apparatus
Volume 41	The Nucleon Compton Effect at Low and Medium Energies
Volume 42	Electronics in Experimental Physics
Volume 43	Nonlinear Optics

In preparation

Volume 44	Nuclear Physics and Interaction of Particles with Matter
Volume 45	Experimental Physics: Programming and Computer Techniques
Volume 46	Cosmic Rays and Interaction of High-Energy Particles
Volume 47	Radio Astronomy Instruments and Observations
Volume 48	Surface Properties of Semiconductors and Dynamics of Ionic Crystals
Volume 49	Quantum Electronics and Paramagnetic Resonance
Volume 50	Electroluminescence

Proceedings (Trudy) of the P. N. Lebedev Physics Institute

Volume 43

NONLINEAR OPTICS

Edited by
Academician D. V. Skobel'tsyn
Director, P. N. Lebedev Physics Institute
Academy of Sciences of the USSR, Moscow

Translated from Russian by
Frank L. Sinclair

Springer Science+Business Media, LLC
1970

The Russian text was published by Nauka Press in Moscow in 1968 for the Academy of Sciences of the USSR as Volume 43 of the Proceedings (Trudy) of the P. N. Lebedev Physics Institute. The present translation is published under an agreement with Mezhdunarodnaya Kniga, the Soviet book export agency.

Труды Ордена Ленина
Физического института
им. П. Н. Лебедева
Том XLIII

Library of Congress Catalog Card Number 72-107530
SBN 306-10840-2

ISBN 978-1-4615-7521-4 ISBN 978-1-4615-7519-1 (eBook)
DOI 10.1007/978-1-4615-7519-1

Originally published by Consultants Bureau, New York in 1970

CONTENTS

SOME QUESTIONS OF GAS LASER THEORY

S. G. Rautian

Introduction

§ 1. Optical Masers (Lasers)

Optical masers (lasers) owe their existence to a combination of three ideas originating in different branches of physics. The first fundamental idea was formulated as long ago as the beginning of the twentieth century by Einstein [1], who predicted the process of induced emission. The second idea consisted essentially in the application of thermodynamically unstable and temporally unsteady systems. In such systems amplification (and not absorption) of electromagnetic waves can occur [2, 3]. Finally, the third, purely electronic idea consisted in the utilization of positive feedback to convert the amplifying system to a self-oscillating one, i.e., to a generator producing coherent electromagnetic waves [4-6].

The first masers operated in the SHF range [5-7]. The most interesting and important results, however, were obtained when the fundamental ideas were extended to the optical region of the spectrum [3-11].

The special features of a laser as a radiation source are due to the spatial and temporal coherence of the waves emitted in macroscopically different parts of the laser. The energy characteristics of lasers are illustrative in this respect. In some characteristics lasers are quite ordinary: Pulse generators emit energy of the order of a joule or less during one pulse; continuous generators have a mean power of 10^{-6}-1 W. What is quite unusual is the ability of lasers to concentrate energy – to concentrate it in time, in space, in the direction of emission, and in a narrow spectral interval. The fundamental limitation on the time interval in which a macroscopic system can give up the energy stored in it in the form of a coherent electromagnetic field pulse is the finiteness of the velocity of light – the pulse length is determined simply by the time of flight of the photon through the laser and is 10^{-9} sec. Hence, when the total energy is of the order of 1 joule the instantaneous value of the power may reach 10^9 W. Owing to spatial coherence the directionality of laser emission is limited only by diffraction effects, i.e., depends on the ratio λ/D, where λ is the wavelength, D is the characteristic dimension of the wave front, which depends on the geometry of the laser. Hence, for a given total flux the brightness of laser emission is $4\pi(D/\lambda)^2$ times greater than that of noncoherent sources, which radiate isotropically. By means of simple optical systems all the laser emission can be concentrated

*Dissertation presented for the Degree of Doctor of Physicomathematical Sciences, 1966.

on an area of the order of the square of the wavelength, which leads to light intensities of the order of 10^{15}-10^{17} W/cm^2 or to fields comparable with intraatomic fields ($e/a_0^2 = 1.7 \cdot 10^7$ CGSE units $= 5 \cdot 10^9$ W/cm). The width of the spectral interval within which the laser emission is concentrated is determined by the temporal instabilities of the system as a whole. The fundamental limitations here are statistical effects – spontaneous emission and Brownian motion of the macroscopic parts of the system. The smallest width ($\Delta\nu \sim 1$ Hz, $\Delta\nu/\nu \sim 10^{-14}$) which has been attained so far [12, 13] is no smaller probably due to Brownian motion of the mirrors.

The remarkable features of laser emission have given physicists a new tool for experimental work. The colossal power of the emission lends itself to the most varied applications. The standard phrase from textbooks on electrodynamics "owing to the weakness of the electromagnetic field Maxwell's material equations can be regarded as linear" has become a thing of the past. The simplest nonlinear effect is the redistribution of the populations of the levels and the associated nonlinear field dependence of the polarization. This effect was discovered in the SHF range, has been subjected to thorough investigation in that range, and is called the "saturation effect" (see [14], for instance). A large class of nonlinear effects is due to electrostrictive forces, which were previously ignored in optical problems (self-focusing of beam [15], induced Mandel'shtam–Brillouin scattering [16], shock waves produced in the medium when the emission pulse passes through it, and so on).

Of considerable interest (especially from the viewpoint of use of laser radiation for information transmission) is the generation of sum, difference, and multiple harmonics arising at field strengths comparable with intraatomic fields [17]. We can mention, finally, the numerous effects of "nonelectrodynamic" origin due to the great power of the emission (stepwise ionization [18] of a sample at the focus of the beam, evaporation of solids and liquids [20], and so on).

Gas lasers occupy a special position. Their energetic characteristics are relatively modest (10^{-6}-10^{-2} W in continuous operation and 10^2-10^4 W in pulsed operation with a pulse energy of 10^5-10^{-3} J), although very recently lasers with outstanding parameters have been designed (a continuous generator producing ~10 W with an efficiency of 4% [21] and a pulse generator using photodissociation to produce population inversion [22-24]). However, the main fascination of gas lasers is of quite a different nature. As in other regions of physics, gas systems have the indisputable advantage of simplicity; it is much easier to distinguish in them the relatively small number of decisive factors and to avoid subsidiary effects of no interest from the physical viewpoint. Hence, gas lasers can be used for a much more thorough investigation of the elementary processes in the active medium of the laser and the interaction of the electromagnetic field with atoms. In particular, the properties of lasers as nonlinear self-oscillating devices have been investigated mainly in gas systems [25-30]. In several investigations various processes associated with atom–atom and atom–electron collisions have been investigated and their cross sections have been determined (see [11, 27, 31-34]). An analysis of the spontaneous and induced emission of lasers reveals new opportunities for measurement of spontaneous emission probabilities and the lifetime of levels (see [27, 33, 35-37]). We note, finally, new methods of investigating hyperfine line structure by using nonlinear effects [38, 39]. Such spectroscopic applications of gas lasers are of great interest and promise.

Thus, the invention of lasers enables investigation of the interaction of an electromagnetic field with matter in new and quite unusual conditions, the specific nature of which is due to the great power and high monochromaticity of the field. There are so many special effects that we can speak of an interesting new field of physics.

§2. Content of Work

This dissertation summarizes the author's investigations on the theory of lasers (mainly gas lasers) carried out in 1960-1965. In the earliest investigations different variants of the optical method of producing population inversion were dealt with [23, 24, 43, 44]. Some of these methods (those discussed in [22, 41]) were later realized in practice [23, 24, 43, 44]. Subsequently, however, the author's attention was concentrated on the investigation of the interaction of atoms with an electromagnetic field in operating lasers, and on the relaxation of excited states. The excitation processes leading to the inverse population were assumed to be prescribed in these investigations. Research of this nature provided the basis for the dissertation [35, 37, 45-55].

Laser theory is fundamentally nonlinear (see [5], for instance) and in most cases it is sufficient to consider the saturation effect, i.e., the change in populations of the levels due to induced transitions. This kind of effect must be taken into account in the analysis of thermal radiation — the changeover from Planck's formula for equilibrium radiation to the Rayleigh–Jeans formula means that induced transitions are more probable than spontaneous transitions, i.e., induced transitions mainly determine the distribution of atoms in the levels. In this case, however, it is a question of the interaction of an atom with radiation possessing a broad spectrum. The specificity of the effects taking place in lasers is due to the fact that the width of the radiation spectrum is much less than the widths of the levels. In these conditions the kinetics of induced transitions is of a very special nature and this is responsible for the nature of the many nonlinear effects.

It is known [56] that in the first approximation of perturbation theory the excited states of an atom relax (owing to interaction with zero oscillations of the field) according to an exponential law. If the external field is sufficiently strong the time dependence of the probability of finding an atom in excited states is greatly altered. Oscillations with a frequency which depend on the field intensity may appear on the decay curve, the rate of decay may change, and so on. This group of questions is discussed in Chapter I and from the vast diversity of situations we choose those which are typical of lasers.

According to the known results of radiation theory (see [56], for instance), the spontaneous emission line of the isolated atom has a dispersion form with a width equal to the sum of the decay constants of the upper and lower states. This line shape is a consequence of the exponential decay law. Since the kinetics of the decay of the states of an atom situated in a strong field is greatly altered, one can expect that the spectral composition of the spontaneous emission will also be different.

Cahtper II gives an account of the quasi-classical theory of spontaneous emission of an atom in an external field (§ 7) and discusses some particular cases typical of lasers. In the case of gas systems, where the Doppler effect must be taken into account, we consider another nonlinear effect — the change in the velocity distribution of the atoms.

In many problems of the theory of lasers and amplifiers it is necessary to calculate the probability of processes induced by a "weak" field (a field which has practically no effect on the population) in the presence of a strong one. Here we use the same apparatus as in the problem of spontaneous emission. The main result is that the frequency dependence of the amplification (absorption) factor of the weak field is quite different from the form of the spontaneous emission line. In particular, this factor may change sign within the natural line width. In other words, in some regions of the spectrum the weak field may be amplified, and in others it may be absorbed. This obviously means that in this case we cannot introduce the spectral density of the second Einstein coefficient (in contrast to the first approximation of perturbation theory; see [57], for instance).

In Chapters I-III the theory is developed on the assumption that the decay of the excited states of the atom is due entirely to spontaneous transitions. It is found that the result of the interaction of the strong field with the atom depends significantly on the relationship between the decay rates of the upper and lower states. On the other hand, in real systems relaxation may depend greatly on collisions of the emitting atom with other particles. Thus, there is obviously a need for extension of the theory to the case where collisions play a significant role.

The consideration of collisions usually reduces to the introduction of so-called relaxation terms with appropriately chosen constants into the equation for the density matrix or the probability amplitudes (see [58]). The validity of this procedure, however, is justified only in the case of a small number of models [59, 60], which do not always correspond to the conditions in lasers. In addition, the first investigations of the spectral characteristics of gas lasers in relation to pressure [31, 61] showed that the above-mentioned method of taking collisions into account was inadequate. At the same time, in [31, 61] the conditions were known to be suitable for application of the collision approximation, where it is apparently always possible to operate with relaxation terms. Thus, there is a fairly confused situation, which some authors believe requires a reexamination of the criterion of applicability of the collision model [62].

There is no need, however, for this. The fact is that the existing versions of collision theory have never taken into account the statistical dependence of collisions in which the velocity of translational motion of the atom changes, and collisions which perturb the state of the optical electron. Essentially the same two effects occur in the same collision act, and for this reason they cannot, generally speaking, be regarded as statistically independent. This has been brought up in the literature [63], but has never been taken into account.

It is shown in Chapter IV that the introduction of relaxation terms in the collision approximation is possible only in the case where the statistical dependence of these effects can be neglected. Since this was ignored earlier in linear theory, we develop in § 13-15 a general theory of spectral line broadening with nonlinear effects ignored, but consider collisions in which the internal and external degrees of freedom of the atom are simultaneously perturbed. In § 16, 17 some nonlinear effects are considered and it is shown that the results of [31, 61] find a rational interpretation within the framework of the developed collision theory.

Thus, Part I deals essentially with the old classical problems of radiation theory, which, however, in the case of lasers appear in a different form and have a different solution.

In Chapter II of the dissertation the results of Part I are applied to some problems of laser theory. One of the main questions which crops up here is the correct consideration of the inhomogeneity of the laser medium due to the saturation effect. Despite the extremely small value of the inhomogeneity (change in the dielectric constant of the order of 10^{-9}), it is impossible in the problem in which we are interested to use the standard methods of the electrodynamics of weakly inhomogeneous media (in particular, the geometrical-optics approximation). This is due to the fact that the relative changes in the wave amplitude due to amplification and reflection on inhomogeneities have the same order of magnitude. An approximate method of solution of the electrodynamic problem, where the only simplifying assumption is that of the smallness of the change in the field within a wavelength, is given in § 19. The general formulas derived here are subsequently used to calculate the frequency and output power of gas and solid-state lasers (§ 20, 21).

The last chapter (Chapter VI) is devoted to the problem of stability of monochromatic generation. In this problem the inhomogeneity of the medium due to the saturation effect is of decisive importance. Neglect of inhomogeneity has led some authors to the erroneous conclusion that monochromatic generation is stable in the case of homogeneous broadening [27, 64]. The conducted investigation shows that owing to this inhomogeneity of the medium monochromatic generation is unstable (with a sufficiently high level of excitation of the system) and the critical value of the excitation at which generation on several modes begins depends greatly on the relaxation characteristics of the system.

In the Appendix there is a review of spectroscopic applications of gas lasers. This material is directly related to the main text, but is of a subordinate nature.

The author regards it as a pleasant duty to express his thanks to T. A. Germogenova, T. I. Kuznetsova, G. E. Notkin, G. G. Petrash, I. I. Sobel'man, A. S. Khaikin, and A. A. Feoktistov for collaboration and help in the work.

Part I

INTERACTION OF A QUANTUM SYSTEM AND AN ELECTROMAGNETIC FIELD

CHAPTER I

Induced Emission in a Strong Monochromatic Field

§ 3. Basic Equations

1. Of most interest for laser theory is the case were the frequency ω of the external field differs from the frequency ω_{mn} of the transition between the levels m, n by the order of the width of the corresponding spectral line. For gas systems at moderate pressures and electron densities a typical value of line width in the optical and infrared regions of the spectrum is 10^9-10^{10} sec^{-1}. The transition frequencies are approximately 10^{15} sec^{-1}. Such a relationship between the line widths and transition frequencies allows the use of the so-called two-level scheme. Let the atom* in the absence of an external electromagnetic field have stationary states 1, 2,..., n,..., m,... with energies E_1,..., E_n,..., E_m,... and wave functions ψ_1,..., ψ_n,..., ψ_m,... . If the field spectrum is concentrated close to some natural frequency ω_{mn}, the field in these conditions will lead to transitions only between the states n, m. Hence, the wave function of the atom in the presence of a field can be sought in the form

$$\Psi(t) = a_m(t)\psi_m e^{-i\frac{E_m}{\hbar}t} + a_n(t)\psi_n e^{-i\frac{E_n}{\hbar}t}. \qquad (3.1)$$

All the other states of the atom are manifested either in excitation of the levels m, n, which is reflected in the initial conditions for $a_m(t)$, $a_n(t)$, or in the decay of states m, n due to spontaneous transitions. The Hamiltonian of the system

$$\hat{H} = \hat{H}_0 + \hbar\hat{V}, \quad \hat{V} = \hat{p}\mathbf{E}/\hbar \qquad (3.2)$$

contains the Hamiltonian $\hat{H}_0$ of the atom in the absence of an external field and the term $\hbar\hat{V}$, which describes the interaction of the atom with the field $\mathbf{E}$; $\hat{p}$ denotes the dipole moment operator. Introducing (3.1) and (3.2) into the Schrödinger equation, we obtain a system of equations for a_m, a_n:

$$i\dot{\hat{a}} = \hat{V}\hat{a}, \ \hat{a} = \begin{pmatrix} a_m \\ a_n \end{pmatrix}, \qquad (3.3)$$

*Henceforth for brevity we will use the term "atom," although the results will apply to other quantum systems too.

where

$$\hat{V} = \begin{pmatrix} 0 & V_{mn} \\ V_{nm} & 0 \end{pmatrix}, \quad V_{mn} = V^{*}_{nm} = e^{i\omega_{mn}t}, \quad \boldsymbol{p}_{mn}\boldsymbol{E}/\hbar. \tag{3.4}$$

Equation (3.3) does not include relaxation processes, which determine the finiteness of the lifetime of the states m, n. Henceforth, until Chapter IV, we will assume that the decay of states m, n is due entirely to spontaneous transitions. For most of the questions considered in Chapters I-III extension outside the framework of this assumption will not lead to any qualitatively new effects. In addition, when other causes of relaxation are taken into account the formulas become much more complicated and this makes discussion of the physical aspect of the question difficult. Hence, extension of the theory to the case where the perturbation of the atom due to collisions with other particles is significant is dealt with specially in Chapter IV.

If the relaxation of states m, n is due to spontaneous transitions, then (3.3) will be replaced by the equation

$$i\dot{\hat{a}} = -i\hat{\gamma}\hat{a} + \hat{V}\hat{a}, \quad \hat{\gamma} = \begin{pmatrix} \gamma_m & 0 \\ 0 & \gamma_n \end{pmatrix}, \tag{3.5}$$

where $2\gamma_m$, $2\gamma_n$ are the total probabilities of spontaneous transitions from the levels m, n respectively, i.e., the first Einstein coefficients A_m, A_n.

In lasers excitation is produced by perturbation with a broad spectrum (nonradiative transitions, electron impact, optical excitation by radiation with a broad spectrum, and so on). In these conditions excitation processes can be introduced by means of the following initial conditions for functions $a_m(t)$, $a_n(t)$: at the initial instant t_0 the atom with probability 1 is in state m:

$$\hat{a}(t_0) = \begin{pmatrix} 1 \\ 0 \end{pmatrix} \tag{3.6}$$

or in state n:

$$\hat{a}(t_0) = \begin{pmatrix} 0 \\ 1 \end{pmatrix}. \tag{3.7}$$

If excitation of both states takes place, equation (3.5) has to be solved for the initial conditions (3.6) and (3.7), and the quantities of interest (polarization, work of field, etc.) have to be found for both cases, and then combined with weights proportional to the number of acts of excitation of states m, n in unit time.

We introduce a fundamental matrix $\hat{S}(t)$ of equation (3.5), i.e., a matrix with columns which are linearly independent solutions of (3.5). Using $\hat{S}(t)$ we can write the solution of equation [3.5] in the following way:

$$\hat{a}(t) = \hat{S}(t)\hat{a}(t_0). \tag{3.8}$$

The matrix $\hat{S}(t)$ satisfies the equation

$$\dot{\hat{S}} = -(\hat{\gamma} + i\hat{V})\hat{S} \tag{3.9}$$

with initial conditions

$$\hat{S}(t_0) = \hat{E}, \tag{3.10}$$

where $\hat{E}$ is the unitary matrix.

Since our scheme includes relaxation the matrix S(t) is not unitary. From (3.9) it is easy to find

$$\frac{d}{dt}[\hat{S}^{\dagger}\hat{S}] = -2\hat{S}^{\dagger}\hat{\gamma}\hat{S}, \tag{3.11}$$

from which, in view of the initial conditions (3.10), we obtain

$$\hat{S}^{\dagger}\hat{S} = \hat{E} - 2\int_{t_0}^{t}\hat{S}^{\dagger}\hat{\gamma}\hat{S}\,dt. \tag{3.12}$$

If $t \to \infty$ in (3.12), then $\hat{S} + \hat{S} \to 0$ and

$$2\int_{t_0}^{\infty}\hat{S}^{+}\hat{\gamma}\hat{S}\,dt = \hat{E}.$$

This matrix equation is equivalent to the following:

$$2\gamma_m\int_{t_0}^{\infty}|S_{mm}|^2\,dt + 2\gamma_n\int_{t_0}^{\infty}|S_{mn}|^2\,dt = 1; \tag{3.13}$$

$$2\gamma_m\int_{t_0}^{\infty}|S_{mn}|^2\,dt + 2\gamma_n\int_{t_0}^{\infty}|S_{nn}|^2\,dt = 1; \tag{3.14}$$

$$2\gamma_m\int_{t_0}^{\infty}S_{mm}^{*}\,S_{mn}\,dt + 2\gamma_n\int_{t_0}^{\infty}S_{nm}^{*}\,S_{nn}\,dt = 0. \tag{3.15}$$

Integrals of the type

$$2\gamma_i\int_{t_0}^{\infty}|S_{ij}|^2\,dt \quad (i, j = m, n), \tag{3.16}$$

which come into (3.13) and (3.14), give the probability of relaxation of the atom from level i on excitation of level j (i, j = m, n). Thus, (3.13) and (3.14) simply state that within an infinite time the atom with probability 1 undergoes transitions from levels m, n to other levels. Since we assume for the present that relaxation is due to spontaneous transitions, integrals (3.16) give the corresponding integral probabilities of spontaneous emission.

2. We will be interested in the power of the induced emission (or absorption)

$$P = 2\hbar\omega_{mn}\,\mathrm{Re}\{i a_m^{*} a_n V_{mn}\} \tag{3.17}$$

and the energy emitted by the atom due to induced transitions in its whole lifetime:

$$A = \int_{t_0}^{\infty}P\,dt = 2\hbar\omega_{mn}\,\mathrm{Re}\left\{i\int_{t_0}^{\infty}a_m^{*}(t)\,a_n(t)\,V_{mn}(t)\,dt\right\}. \tag{3.18}$$

Using equation (3.5), we can easily show that

$$A = \hbar\omega_{mn}\left\{|a_m(t_0)|^2 - 2\gamma_m\int_{t_0}^{\infty}|a_m(t)|^2\,dt\right\} = \hbar\omega_{mn}\left\{-|a_n(t_0)|^2 + 2\gamma_n\int_{t_0}^{\infty}|a_n(t)|^2\,dt\right\}. \tag{3.19}$$

It is clear from (3.19) that the value of A may be positive or negative, depending on the initial conditions and sign of ω_{mn}. We will stipulate that the level m is above the level n, so that $\omega_{mn} > 0$. Then $A > 0$ (emission) for initial conditions (3.6) and $A < 0$ (absorption) for initial conditions (3.7). The ratio

$$\frac{A}{\hbar\omega_{mn}} = W_m = 2\gamma_n \int_{t_0}^{\infty} |S_{nm}|^2\, dt = 1 - 2\gamma_m \int_{t_0}^{\infty} |S_{mm}|^2\, dt \qquad (3.20)$$

will be called the emission probability; the ratio

$$-\frac{A}{\hbar\omega_{mn}} = W_n = 2\gamma_m \int_{t_0}^{\infty} |S_{mn}|^2\, dt = 1 - 2\gamma_n \int_{t_0}^{\infty} |S_{nn}|^2\, dt \qquad (3.21)$$

is the absorption probability. It is clear from (3.20), (3.21), and (3.13), (3.14) that W_m (W_n) gives the probability that after excitation of the atom to level m (n) induced emission (absorption) of a photon $\hbar\omega$ will occur.

We denote by Q_m, Q_n the number of acts of excitation of the levels m, n in unit volume in unit time. The average power of the emission of unit volume will then be

$$\hbar\omega\,[Q_m W_m - Q_n W_n]. \qquad (3.22)$$

3. The probability amplitudes of the moving atom satisfy the same equation (3.5), but the matrix element V_{mn} must be regarded as a function of time and the coordinate r of the center of mass of the atom, which, in turn, depends on the time

$$V_{mn}[t, \mathbf{r}(t)] = V_{mn}.$$

In Chapters I-III we will consider the case where collisions of the atom during the lifetime of the excited states can be neglected. Hence, we will assume that

$$\mathbf{r}(t) = \mathbf{r}_0 + \mathbf{v}(t - t_0), \qquad (3.23)$$

where $\mathbf{r}_0$ is the coordinate at the initial instant t_0 when excitation took place, and $\mathbf{v}$ is the velocity of the center of mass of the atom.

The emission (or absorption) power, given by formula (3.17), depends now on $t - t_0$, $\mathbf{r}$, $\mathbf{r}_0$, and $\mathbf{v}$. In accordance with this we can introduce two quantities

$$A(\mathbf{r}_0, \mathbf{v}) = \int_{t_0}^{\infty} P[t - t_0, \mathbf{r}(t), \mathbf{r}_0]\, dt; \qquad (3.24)$$

$$A(\mathbf{r}, \mathbf{v}) = \int_{-\infty}^{t} P[t - t_0, \mathbf{r}, \mathbf{r}_0(t_0)]\, dt_0. \qquad (3.25)$$

The first of them is the total energy emitted (or absorbed) by any single atom after its excitation at $\mathbf{r}_0$, t_0; the second is the mean energy of induced emission (or absorption) at point r, referred to one excited atom with velocity $\mathbf{v}$. The values of (3.24) and (3.25) may be averaged over the velocities of the atoms.

§ 4. Kinetics of Induced Transitions

1. We consider first the case of a stationary atom and a monochromatic field:

$$V_{mn} = G\,[e^{-i(\omega - \omega_{mn})t} + e^{i(\omega + \omega_{mn})t}],$$
$$E = E_0 \cos\omega t, \quad G = \frac{1}{2\hbar} E_0 p_{mn}. \qquad (4.1)$$

It is easy to see that in the expression for V_{mn} we can retain only the term with the difference frequency and discard the rapidly oscillating term with $\omega + \omega_{mn}$. Then the system of equations (3.5) will take the form

$$i(\dot{a}_m+\gamma_m a_m)=Ge^{-i(\omega-\omega_{mn})t}a_n,\quad i(\dot{a}_n+\gamma_n a_n)=Ge^{i(\omega-\omega_{mn})t}a_n. \tag{4.2}$$

It is easy to find the corresponding fundamental matrix $\hat{S}$:

$$\hat{S}=\begin{pmatrix} -\frac{\alpha_2+\gamma_m}{\alpha_1-\alpha_2}e^{\alpha_1 t}+\frac{\alpha_1+\gamma_m}{\alpha_1-\alpha_2}e^{\alpha_2 t} & -\frac{iG}{\alpha_1-\alpha_2}[e^{\alpha_1 t}-e^{\alpha_2 t}] \\ -\frac{iG}{\alpha_1-\alpha_2}[e^{\alpha_1 t}-e^{\alpha_2 t}]e^{i\Omega t} & \left[\frac{\alpha_1+\gamma_m}{\alpha_1-\alpha_2}e^{\alpha_1 t}-\frac{\alpha_2+\gamma_m}{\alpha_1-\alpha_2}e^{\alpha_2 t}\right]e^{i\Omega t} \end{pmatrix}. \tag{4.3}$$

Here α_1, α_2 are the roots of the characteristic equation

$$\alpha_{1,2}=-\frac{i}{2}\Omega-\frac{\Gamma}{2}\pm\sqrt{\left(\frac{i\Omega}{2}+\frac{\gamma}{2}\right)^2-G^2},$$
$$\Omega=\omega-\omega_{mn},\quad \Gamma=\gamma_n+\gamma_m,\quad \gamma=\gamma_n-\gamma_m. \tag{4.4}$$

In formula (4.3) we put $t_0=0$ for simplicity.

Substituting expressions (4.3)-(4.6) in formulas (3.20), (3.21) and performing the integration we find the emission and absorption probabilities

$$W_m=\frac{\Gamma}{\gamma_m}\frac{G^2}{\Omega^2+\Gamma^2+\frac{\Gamma^2}{\gamma_m\gamma_n}G^2},\quad W_n=\frac{\gamma_m}{\gamma_n}W_m. \tag{4.5}$$

Figure 1 shows W_m and W_n as functions of $G^2/\gamma_m\gamma_n$. At low field intensities, $G^2 \ll \gamma_m\gamma_n$, the emission and absorption probabilities are proportional to G^2 and subsequently attain their maximum values

$$W_m\approx\frac{\gamma_n}{\gamma_n+\gamma_m},\quad W_n\approx\frac{\gamma_m}{\gamma_n+\gamma_m}\left(G^2\gg\gamma_m\gamma_n\frac{\Gamma^2+\Omega^2}{\Gamma^2}\right). \tag{4.6}$$

If $\gamma_n \gg \gamma_m$, then practically all the acts of excitation of the state m lead to induced emission of a photon of the external field ($W_m \approx 1$). On excitation of the lower level n the maximum value of the absorption probability is small for $\gamma_n \gg \gamma_m$ and approaches unity when the inequality is reversed.

If G^2 is sufficiently small, then

$$\alpha_1\approx-\gamma_m,\quad \alpha_2\approx-\gamma_n-i\Omega, \tag{4.7}$$

and formula (4.3) becomes the known equation obtained in the first approximation of perturbation theory:

$$\hat{S}=\begin{pmatrix} e^{-\gamma_m t} & -\frac{iG}{\gamma_n-\gamma_m+i\Omega}[e^{-\gamma_m t}-e^{-(\gamma_n+i\Omega)t}] \\ -\frac{iGe^{i\Omega t}}{\gamma_n-\gamma_m+i\Omega}[e^{-\gamma_m t}-e^{-(\gamma_n+i\Omega)t}] & e^{-\gamma_n t} \end{pmatrix}. \tag{4.8}$$

We will first investigate the changes in S_{ij} due to the external field in the special case $\Omega=0$, $\gamma_m=\gamma_n=\Gamma/2$. Then

$$\alpha_{1,2}=-\frac{1}{2}\Gamma\pm iG; \tag{4.9}$$

$$S_{mm}=e^{-\Gamma t/2}\cos Gt,\quad S_{nm}=-ie^{-\Gamma t/2}\sin Gt. \tag{4.10}$$

Thus, the probabilities of finding an atom at levels m, n are oscillating functions of time. These oscillations are obviously due to the fact that the field induces the transition m → n, then the

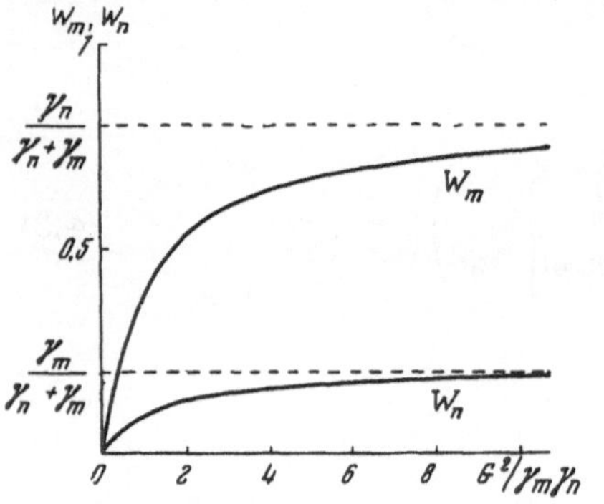

Fig. 1. Probabilities of emission W_m and absorption W_n of photon as functions of $G^2/\gamma_m\gamma_n$.

reverse transition n → m, and so on. Roughly speaking, the atom is at level m for part of the time and at level n for the other part. Since the decay constants of the states m, n are the same ($\gamma_m = \gamma_n = \Gamma/2$) the atom relaxes from both states with equal probability and, hence, the decay of the probabilities $|S_{mm}|^2$ and $|S_{mn}|^2$ is the same as for G = 0.

From the described picture of the transitions we can interpret the saturation effect in the following way. If several oscillations of $|S_{ij}|^2$ take place during the lifetime $1/\Gamma$ (i.e., if $G \gg \Gamma$) the mean populations of the levels are close and the system become less capable of radiating energy.

These arguments apply specifically to the case $\gamma_m = \gamma_n$. This relationship between the decay constants is characteristic for transitions situated in the long-wave region of the spectrum and used in masers. In the optical region fulfillment of the equality $\gamma_m = \gamma_n$ is more the exception than the rule. If $\gamma_m \neq \gamma_n$, the evolution of the system is greatly altered and the intepretation of the saturation effect is different.

To clarify this question we consider the case of greatly differing γ_m and γ_n. For definiteness we will take

$$\gamma_m \ll \gamma_n. \tag{4.11}$$

We will also assume that $\Omega = 0$. If the field is not too great, so that

$$G^2 < \frac{1}{4}\gamma^2 = \frac{1}{4}(\gamma_n - \gamma_m)^2, \tag{4.12}$$

the roots α_1 and α_2 are real and negative

$$-\alpha_{1,2} = \frac{1}{2}\Gamma_{1,2} = \frac{\Gamma}{2} \mp \sqrt{\frac{\gamma^2}{4} - G^2}, \tag{4.13}$$

and S_{ij} do not undergo any oscillations. For instance,

$$\begin{aligned} S_{mm} &= \left\{ \mathrm{ch} \sqrt{\frac{\gamma^2}{4} - G^2}\, t + \frac{\gamma/2}{\sqrt{\frac{\gamma^2}{4} - G^2}} \mathrm{sh} \sqrt{\frac{\gamma^2}{4} - G^2}\, t \right\} e^{-\Gamma t/2}, \\ S_{nm} &= -\frac{iG}{\sqrt{\frac{\gamma^2}{4} - G^2}} \mathrm{sh} \left[\sqrt{\frac{\gamma^2}{4} - G^2}\, t \right] e^{-\Gamma t/2}. \end{aligned} \tag{4.14}$$

If

$$G^2 \ll \gamma^2/4, \tag{4.15}$$

we can expend the radical in the expression $\Gamma_{1,2}$ and retain only the terms G^2/γ only in the indices of the exponents; after simple calculations we obtain

$$\frac{1}{2}\Gamma_1 = \gamma_m + \frac{G^2}{\gamma}, \quad \frac{1}{2}\Gamma_2 = \gamma_n - \frac{G^2}{\gamma}; \tag{4.16}$$

$$S_{mm} = e^{-\Gamma_1 t/2}, \quad S_{nm} = -i\frac{2G}{\gamma}\left[e^{-\Gamma_1 t/2} - e^{-\Gamma_2 t/2}\right]. \tag{4.17}$$

Expressions (4.17) have the same general form as (4.8). The difference is that the probabilities of spontaneous decay γ_m, γ_n are replaced by $\frac{1}{2}\Gamma_1$ and $\frac{1}{2}\Gamma_2$ from (4.16). Hence, when conditions (4.15) are satisfied the induced transitions are manifested in a change in the probabilities of decay of the levels m, n. For the shorter-lived level the decay is reduced, and for the longer-lived level it is increased.

We recall that the degree of nonlinearity is determined by the parameter $G^2/\gamma_m\gamma_n$ (4.5). If relationship (4.11) holds, then saturation may be significant even when condition (4.15) is fulfilled:

$$\frac{G^2}{\gamma_m\gamma_n} = \frac{G^2}{\gamma_n^2}\frac{\gamma_n}{\gamma_m} \approx \frac{G^2}{\gamma^2}\frac{\gamma_n}{\gamma_m} > 1, \tag{4.18}$$

if $\gamma_n \gg \gamma_m$. It also follows from (4.17) and (4.15) that always

$$|S_{nm}|^2 \ll |S_{mm}|^2, \tag{4.19}$$

i.e., the mean populations of the levels m and n are not at all close, as was the case with great saturation for $\gamma_m = \gamma_n$.

These features can be interpreted in the following way. After excitation to level m the atom relaxes by means of spontaneous transitions and, in addition, changes to state n under the action of the field. If γ_n G, then after the transition m → n the atom relaxes from the state n so rapidly that the field does not "manage" to cause the reverse transition n → m. In consequence of this there are no oscillations in $|S_{ij}|^2$ indicative of repeated transitions m → n, n → m. This means that the external field leads to an increase in the rate of depletion of the level m. The increase in the decay constant will be significant if $G^2/\gamma_n \sim \gamma_m$, which is the criterion for the presence of nonlinear effects.

Thus, in the case of greatly differing decay constants saturation is not due to equalization of the mean populations of the levels m, n, but to a reduction in the population of the longer-lived level.

This interpretation explains the maximum values of the emission and absorption probabilities given by formula (4.6). We can assume, for instance, that $W_m \approx \gamma_n/(\gamma_n + \gamma_m) \approx 1$ when $\gamma_n \gg \gamma_m$, owing to the fact that the field greatly reduces the population of the upper level and relaxation proceeds mainly via the level n.

It should be noted that an increase in the rate of decay of the state m owing to induced transitions is not equivalent to broadening of the level. In particular, it is shown in Chapter II that induced transitions in a strong monochromatic field do not lead to broadening, but to splitting, of the spontaneous emission line.

The "aperiodic" transition regime discussed above occurs when $G^2 < (\gamma/2)^2$. In the opposite case, i.e., when the field is sufficiently strong, the roots $\alpha_{1,2}$ become complex

$$\alpha_{1,2} = -\frac{\Gamma}{2} \pm i\sqrt{G^2 - \gamma^2/4} \tag{4.20}$$

and S_{ij} will oscillate:

$$S_{mm} = \left\{\cos\sqrt{G^2 - \gamma^2/4}\,t + \frac{\gamma/2}{\sqrt{G^2 - \gamma^2/4}}\sin\sqrt{G^2 - \gamma^2/4}\,t\right\} e^{-\Gamma t/2},$$
$$S_{nm} = -\frac{iG}{\sqrt{G^2 - \gamma^2/4}}\sin\sqrt{G^2 - \gamma^2/4}\,t e^{-\Gamma t/2}. \tag{4.21}$$

In the limiting case $G^2 \gg \gamma^2/4$ formulas (4.21) are converted to (4.10).

2. We considered above the case of a monochromatic external field. We will show that the main features of the kinetics of transitions, established in section **1**, depend on the relationship between the relaxation constants, and not on the kind of field. For this purpose we will consider two cases in which the solution of system (3.5) can be obtained in quadratures.

Let $\gamma_m = \gamma_n$; we assume, in addition, that

$$\frac{V_{mn}(t)}{V^*_{mn}(t)} = e^{-2i\varphi_0} \tag{4.22}$$

is independent of time. Formula (4.22) means that $V_{mn}(t)$ can be put in the form of a product of a real function of time G(t) and a complex constant

$$V_{mn}(t) = G(t)\, e^{-i\varphi_0}. \tag{4.23}$$

Physically this corresponds to a particular symmetry of the spectrum of the external field relative to the transition frequency. In the indicated conditions we can show that the fundamental matrix $\hat{S}$ of system (3.5) is given by the following formula:

$$\hat{S} = e^{-\frac{\Gamma}{2}t}\begin{pmatrix} \cos f(t) & -ie^{-i\varphi_0}\sin f(t) \\ -ie^{i\varphi_0}\sin f(t) & \cos f(t) \end{pmatrix}, \quad f(t) = \int_{t_0}^{t} G(t')\,dt'. \tag{4.24}$$

Hence, in the case $\gamma_m = \gamma_n$ and fairly general assumptions regarding the kind of external field the probability amplitudes will be oscillating functions of time and for a sufficiently large value of G the mean populations of the levels m and n will be close.

Now let $\gamma_m \ll \gamma_n$. For the case where the inequality

$$\varepsilon^2 = \max\left|\frac{V_{mn}(t)}{\gamma_n - \gamma_m}\right|^2 < \frac{1}{4} \tag{4.25}$$

is satisfied Germogenova [51] devised a method of solving the system of equations (3.5) based on the reduction of (3.5) to a Riccati equation and three linear equations. The Riccati equation is solved by the method of successive approximations and the k-th approximation differs from the exact solution by a value of the order $\varepsilon^{(k+1)}$. In a first approximation the probability amplitudes S_{mm} and S_{mn} are given by the expressions

$$\begin{aligned} S_{mm}(t) &= \exp\left\{-\gamma_m t - \int_0^t\int_0^t V_{mn}(t)\, V^*_{mn}(t'')\, e^{-\gamma_n (t'-t'')}\,dt'\,dt''\right\}, \\ S_{nm}(t) &= -iS_{nm}(t)\int_0^t V^*_{nm}(t')\, e^{-\gamma_n (t-t')}\,dt'. \end{aligned} \tag{4.26}$$

It is easy to see that condition (4.25) is similar to inequality (4.12). In addition, formulas (4.26) are similar to (4.16) and (4.17) – induced transitions lead to an increase in the rate of relaxation of the longer-lived level m. It follows from (4.26) that

$$|S_{nm}(t)|^2 < |S_{mm}(t)|^2\{\max|V_{mn}|^2\}\left|\int_0^t e^{-\gamma_n (t-t')}\,dt'\right| < \varepsilon^2|S_{mm}|^2 \ll |S_{mm}(t)|^2. \tag{4.27}$$

Thus, Germogenova's method can be used to describe the "aperiodic" transition regime. As in the case of a monochromatic field the first approximation (4.26) describes only the transitions m → n and does not contain the reverse transitions. Finally, saturation, which may be great in the case (4.26) too, if $\gamma_m \ll \gamma_n$, is not due to equalization of the populations, but to a reduction of the population of one level due to induced transitions.

If condition (4.25) is not fulfilled the "aperiodic" relaxation regime presumably becomes "oscillatory." This has not been rigorously shown for an arbitrary time dependence $V_{mn}(t)$. However, for very large field amplitudes we can show [65] that

$$\hat{S} = e^{-\frac{\Gamma}{2}(t-t_0)} \begin{pmatrix} e^{-\frac{i}{2}[\varphi(t)-\varphi(t_0)]}\cos f(t) & -ie^{-\frac{i}{2}[\varphi(t)+\varphi(t_0)]}\sin f(t) \\ -ie^{\frac{i}{2}[\varphi(t)+\varphi(t_0)]}\sin f(t) & e^{\frac{i}{2}[\varphi(t)-\varphi(t_0)]}\cos f(t) \end{pmatrix}, \tag{4.28}$$

$$e^{-2i\varphi(t)} = \frac{V_{mn}(t)}{V^*_{mn}(t)}, \quad G(t) = e^{i\varphi(t)}V_{mn}(t), \qquad f(t) = \int_{t_0}^{t} G(t')\,dt'. \tag{4.29}$$

With accuracy to factors with $\exp\left\{\pm\frac{i}{2}\varphi(t) \pm \frac{i}{2}\varphi(t_0)\right\}$, formula (4.28) is the same as (4.24). Thus, sufficiently strong fields lead to oscillation of the probability amplitudes and to equalization of the mean populations of the levels. The decay of the linearly independent solutions of equation (3.5) is determined by the arithmetic mean of the rates of relaxation of the states m, n of the atom in the absence of a field.

3. In sections 1, 2, attention was concentrated on the time course of the transitions induced by the external field. In some problems (in particular, in the investigation of the spectral composition of spontaneous emission) another method of describing these effects is of interest. Substituting in (3.1) expression (4.3) for the probability amplitudes we obtain

$$\Psi(t) = A_1\psi_m e^{-i\left[\frac{E_m}{\hbar} - \alpha_1''\right]t - \alpha_1' t} + A_2\psi_m e^{-i\left[\frac{E_m}{\hbar} - \alpha_2''\right]t - \alpha_2' t} + B_1\psi_n e^{-i\left[\frac{E_n}{\hbar} - (\alpha_1''+\Omega)\right]t - \alpha_1' t} + B_2\psi_n e^{-i\left[\frac{E_n}{\hbar} - (\alpha_2''+\Omega)\right]t - \alpha_2' t}. \tag{4.30}$$

The roots $\alpha_{1,2}$ of the characteristic equation are put in the form $\alpha_{1,2} = -\alpha'_{1,2} + i\alpha''_{1,2}$. In the absence of an external field we have

$$\Psi(t) = A_1\psi_m e^{-\frac{i}{\hbar}E_m t - \gamma_m t} + A_2\psi_n e^{-\frac{i}{\hbar}E_n t - \gamma_n t}. \tag{4.31}$$

Thus, the atom in an external monochromatic field is described by a wave function of the same type as that of a system with quasi-stationary states, the energy of which is $E_m - \alpha''_{1,2}\hbar$, $E_n - (\alpha''_{1,2} + \Omega)\hbar$. The decay constants of the "sublevels" $\alpha'_{1,2}$ depend on the field amplitude $\Omega = \omega - \omega_{mn}$ and γ_m, γ_n.

"Splitting" of the levels of the atom in a monochromatic field can be interpreted from the standpoint of stationary perturbation theory as the removal of degeneration in the atom + field system. To complete the analogy we put $\gamma_m = \gamma_n = 0$ and consider the two states of the atom + field system: the first state corresponds to the atom at the level m and to l photons $\hbar\omega$ in the field; the second state corresponds to the atom at level n and to $l + 1$ photons in the field. The energies of these states are

$$E_1 = E_m + l\hbar\omega, \quad E_2 = E_n + (l+1)\hbar\omega. \tag{4.32}$$

When we take into account the interaction of the atom with the field the corrections to the energies (4.32) are determined from the secular equation

$$\begin{vmatrix} E_m - E_n - \hbar\omega - \Delta E & \hbar V_{mn} \\ \hbar V^{*}_{mn} & -\Delta E \end{vmatrix} = 0,$$

the solution of which is

$$\frac{\Delta E_{1,2}}{\hbar} = -\frac{\omega - \omega_{mn}}{2} \pm \sqrt{\left(\frac{\omega - \omega_{mn}}{2}\right)^2 + |V_{mn}|^2}. \tag{4.33}$$

Formula (4.33) is the same as (4.4) if in the latter we put $\gamma = \gamma_n - \gamma_m = 0$ and $G^2 = |V_{mn}|^2$.

In the case of a nonmonochromatic field the picture of the sublevels of the atom will be more complex. For instance, if in the case of (4.23) function G(t) is periodic, then $\hat{S}e^{+\Gamma t/2}$ can be expended in a Fourier series. The wave function will then have the form

$$\Psi(t) = \psi_m \sum_s A_s e^{-\frac{i}{\hbar}\left[E_m + \hbar\Delta\omega s - \frac{i}{2}\Gamma\right]t} + \psi_n \sum_{s'} B_{s'} e^{-\frac{i}{\hbar}\left[E_n + \hbar\Delta\omega s' - \frac{i}{2}\Gamma\right]t}, \tag{4.34}$$

where $\Delta\omega$ is given by the period T of the function G(t): $\Delta\omega = 2\pi/T$.

§ 5. Induced Emission of Moving Atoms

1. In the conditions typical of the active media of lasers the emitting atom during its lifetime traverses distances which greatly exceed $\lambda/2\pi = 1/k$. This means that the phase of the field will be altered due to the Doppler effect by a value significantly greater than 2π. Hence, the motion of the atoms must play a decisive role.

Consideration of the motion of the atoms greatly complicates the calculations, which must be carried out for each geometric configuration of the field. The exception is a plane traveling monochromatic wave

$$E = E_0 \cos(\omega t - \mathbf{kr}), \tag{5.1}$$

for which the formulas of § 4 are still valid, if the field frequency ω in them is replaced by $\omega - \mathbf{kv}$. In particular, the probability of induced emission, averaged over velocities with a Maxwellian distribution,

$$W_M(\mathbf{v}) = \frac{1}{(\sqrt{\pi}\bar{v})^3} e^{-\mathbf{v}^2/\bar{v}^2}, \quad \bar{v}^2 = \frac{2kT}{m} \tag{5.2}$$

(k is Boltzmann's constant, T is the absolute temperature, m is the mass of the atom) will be equal to

$$\langle W_m \rangle = \frac{\sqrt{\pi}}{\gamma_m k\bar{v}} \frac{G^2}{\sqrt{1 + G^2/\gamma_m\gamma_n}} u\left(\frac{\Omega}{k\bar{v}}, \frac{\Gamma}{k\bar{v}}\sqrt{1 + \frac{G^2}{\gamma_m\gamma_n}}\right). \tag{5.3}$$

Here

$$u(x, y) = \frac{1}{\pi} \int_{-\infty}^{\infty} \frac{y e^{-t^2} dt}{(x - t)^2 + y^2} = \mathrm{Re}\{e^{-(x+iy)^2}[1 - \Phi(x + iy)]\}; \tag{5.4}$$

$\Phi(z)$ is the probability integral.*

*The function u(x, y) is tabulated in [66].

In the limiting case

$$y = \frac{\Gamma}{k\bar{v}} \sqrt{1 + G^2/\gamma_m\gamma_n} \ll 1, \tag{5.5}$$

which occurs most often in gas lasers, $u(x, y) \approx e^{-x^2}$, and it follows from formula (5.3) that

$$\langle W_m \rangle = \frac{\sqrt{\pi}}{\gamma_m k\bar{v}} \frac{G^2}{\sqrt{1 + G^2/\gamma_m\gamma_n}} e^{-\Omega^2/(k\bar{v})^2}. \tag{5.6}$$

Thus, saturation proper does not occur here, but when $G^2 \gg \gamma_m\gamma_n$ the probability of induced emission is proportional to the amplitude G, and not the intensity, of the field. The physical explanation of this is that the number of atoms interacting effectively with the field increases with increase in G. To illustrate this we consider the velocity distribution of the mean population of the lower level after excitation of the upper level:

$$N_{nm}(v) = Q_m(\mathbf{v}) \int_{t_0}^{\infty} |S_{nm}|^2 dt = \frac{Q_m(\mathbf{v})}{2\gamma_n} \frac{\Gamma}{\gamma_m} \frac{G^2}{(\Omega - \mathbf{kv})^2 + \Gamma^2(1 + G^2/\gamma_m\gamma_n)}. \tag{5.7}$$

Assuming that atoms excited to the level m have a Maxwellian velocity distribution, $Q_m(\mathbf{v}) = Q_m W_M(\mathbf{v})$, we obtain

$$N_{nm}(\mathbf{v}) = \frac{Q_m}{2\gamma_n} \frac{\Gamma}{\gamma_m} \frac{G^2 W_M(\mathbf{v})}{(\Omega - \mathbf{kv})^2 + \Gamma^2[1 + G^2/\gamma_m\gamma_n]}. \tag{5.8}$$

The atoms interact most effectively with the field when they have velocities v_p, which satisfy the condition

$$\mathbf{kv}_p = \Omega. \tag{5.9}$$

When (5.5) is satisfied, the width of the distribution $N_{nm}(\mathbf{v})$ is determined by the resonance denominator

$$\mathbf{k}\Delta\mathbf{v} = \Gamma\sqrt{1 + G^2/\gamma_m\gamma_n} = yk\bar{v}. \tag{5.10}$$

Hence, as $G^2/\gamma_m\gamma_n$ increases, the external field causes transitions of atoms with velocities which differ more and more from the "resonance" velocity from (5.9). Accordingly, the velocity distribution of atoms at the upper level will have a dip close to $\mathbf{kv} = \mathbf{kv}_p$:

$$N_{mm}(\mathbf{v}) = \frac{Q_m}{2\gamma_m} W_M(\mathbf{v}) \left[1 - \frac{\Gamma}{\gamma_m} \frac{G^2}{[\mathbf{k}(\mathbf{v} - \mathbf{v}_p)]^2 + \Gamma^2[1 + G^2/\gamma_m\gamma_n]}\right]. \tag{5.11}$$

If $G^2 \gg \gamma_m\gamma_n$ the number of atoms with $\mathbf{kv} = \mathbf{kv}_p$ will be Γ/γ_m times less than when $G = 0$.

The expression for the power of the emission can be put in the same form as in the first approximation of perturbation theory:

$$\hbar\omega Q_m(\mathbf{v}) W_m = 2\pi\hbar\omega G^2 \frac{\Gamma/\pi}{(\Omega - kv)^2 + \Gamma^2} [N_{mm}(\mathbf{v}) - N_{nm}(\mathbf{v})]; \tag{5.12}$$

$$N_{mm}(\mathbf{v}) - N_{nm}(\mathbf{v}) = \frac{Q_m(\mathbf{v})}{2\gamma_m} \frac{1}{1 + \dfrac{G^2}{\gamma_m\gamma_n} \dfrac{\Gamma^2}{(\Omega - \mathbf{kv})^2 + \Gamma^2}}. \tag{5.13}$$

The denominator in (5.13) gives the change in the velocity distribution of the difference of the populations.

2. A traveling monochromatic wave occurs in only a few designs of masers (see [67], for instance). In most cases the steady field is in the form of two opposing traveling monochromatic waves with different (generally speaking) amplitudes:

$$E = E_1 \cos(\omega t - \mathbf{kr} + \delta_1) + E_2 \cos(\omega t - \mathbf{kr} + \delta_2),$$
$$V_{mn}(t) = \frac{1}{2} G_1 e^{-i(\Omega_1 t + \eta_1)} + \frac{1}{2} G_2 e^{-i(\Omega_2 t + \eta_2)}, \tag{5.14}$$

$$\Omega_{1,2} = \omega + \omega_{mn} \pm \mathbf{kv}, \quad \eta_{1,2} = \delta_{1,2} \pm \mathbf{kr}_0. \tag{5.15}$$

In the center-of-mass system of the atom the spectrum of such a field contains two lines with frequencies $\omega \mp \mathbf{kv}$. Even in this relatively simple case equation (3.5) cannot be integrated without additional limitations. Hence, we consider some special cases, which nevertheless help to reveal the nature of the induced emission of the moving atom.

The optical region of the spectrum is characterized by a great difference in the decay constants of the combining levels. Hence, the case (4.25), where formulas (4.26) give approximate expressions for $S_{ij}(t)$ in explicit form, is of interest. We will assume for the sake of definiteness that $\gamma_m \ll \gamma_n$. Substituting (5.15) in (4.26) and discarding terms of the order $G_{1,2}^2/\gamma_n^2$, we find

$$S_{mm} = \exp\left\{-\left[\gamma_m + \frac{1}{4}\left(\frac{G_1^2}{\gamma_n + i\Omega_1} + \frac{G_2^2}{\gamma_n + i\Omega_2}\right)\right]t + \frac{iG_1G_2}{8\mathbf{kv}}\left[\frac{e^{2i\mathbf{kv}t} - 1}{\gamma_n + i\Omega_2} e^{2i\mu} - \frac{e^{-2i\mathbf{kv}t} - 1}{\gamma_n + i\Omega_1} e^{-2i\mu}\right]\right\},$$
$$S_{nm} = -\frac{i}{2} S_{mm}\left\{G_1 \frac{e^{i\Omega_1 t} - e^{-\gamma_n t}}{\gamma_n + i\Omega_1} e^{i\eta_1} + G_2 \frac{e^{i\Omega_2 t} - e^{-\gamma_n t}}{\gamma_n + i\Omega_2} e^{i\eta_2}\right\}, \quad 2\mu = \eta_2 - \eta_1. \tag{5.16}$$

The index of the exponent in the expression for S_{mm} contains two terms. The first term, proportional to the time, corresponds to $\frac{1}{2}\Gamma_1 t$ in formula (4.16) and reflects the effect of the mean value of the field intensity [the factor $\frac{1}{4}$ in (5.16), which is absent in (4.16), is due to a different definition of the parameter G]. The second term, proportional to G_1G_2, is due to inhomogeneity of the field (factor $e^{\pm 2i\mu}$) and leads to effects of an interference nature. If during its lifetime $1/\gamma_m$ the atom traverses a distance $\sim \bar{v}/\gamma_m \ll \lambda/2\pi = 1/k$, i.e., if $kv \ll 1/\gamma_m$, then this term must be taken into account, since the significant factor in these conditions is not the mean, but the local, value of the field intensity at the point where the atom was excited. It is easy to see, however, that in the conditions of interest to us ($\gamma_m \ll \gamma_n \ll k\bar{v}$) the interference term can be discarded in the calculation of the mean (as regards velocity) values. We compare the first and second terms from the viewpoint of dependence on the velocity of the atom. The terms $G_{1,2}^2 \times [\gamma_n + i\Omega_{1,2}]^{-1}$ will have an appreciable value in the velocity interval $|\mathbf{kv} \mp \Omega| \lessgtr \gamma_n$. In the interference terms, however, the effective velocity interval is determined by the factors

$$\{1 - e^{\pm 2i\mathbf{kv}t}\}/\mathbf{kv},$$

which will be greater only when $|\mathbf{kv}| \lessgtr 1/\Delta t$, where Δt is the lifetime of the atom. In our conditions $1/\Delta t \sim \gamma_m + G^2/\gamma_n$; hence, the interference term has an appreciable value in a much smaller velocity interval than the first, and it can be discarded. The physical explanation of this is as follows. If the velocity of the atom is such that during its lifetime $1/[\gamma_m + G^2/\gamma_n]$ it traverses a distance $v/[\gamma_m + G^2/\gamma_n] \sim \lambda/2\pi$ along $\mathbf{k}$, the atom will "see" the mean field and interference effects cannot be significant. Atoms with velocities $|\mathbf{kv}| \sim \gamma_n \gg \gamma_m + G^2/\gamma_n$ interact effectively with the field. Thus, most of the atoms involved in the emission have a fairly large velocity component along the direction $\mathbf{k}$ and are subjected to the mean field intensity. We note that it is essential for the validity of these arguments that the decay of the excited states be monotonic. If the probabilities of finding an atom at levels m, n oscillate in time, there may be resonance between the oscillations of the field intensity and the oscillations of $|S_{ij}|^2$, and interference effects will be considerable. Exact evaluations lead to the same results as the presented qualitative considerations.

In view of what has been said the probability of photon emission is

$$\langle W_m\rangle = 1 - 2\gamma_m \int\limits_0^\infty \langle |S_{mm}|^2\, dt\rangle = \frac{\gamma_n}{4\gamma_m} \int\limits_{-\infty}^{\infty} \left[\frac{G_1^2}{\gamma_n^2+\Omega_1^2} + \frac{G_2^2}{\gamma_n^2+\Omega_2^2}\right] \frac{W_M(v)\,dv}{1+\frac{\gamma_n}{4\gamma_m}\left[\frac{G_1^2}{\gamma_n^2+\Omega_1^2}+\frac{G_2^2}{\gamma_n^2+\Omega_2^2}\right]}. \quad (5.17)$$

Each resonance term in (5.17) gives the contribution of one of the traveling waves to the emission. The nonlinear dependence of $\langle W_m\rangle$ on G_1^2, G_2^2, due to the second term in the denominator of the integrand, is determined additively by the two traveling waves [compare (5.12) and (5.13)].

An equally simple result can be obtained for a field with a spectrum consisting of a set of monochromatic waves with frequencies $\omega_1,\ldots,\omega_j$ and wave vectors $\pm k_1,\ldots,\pm k_j$:

$$\langle W_m\rangle = \frac{\gamma_n}{4\gamma_m}\left\langle \frac{\sum\limits_{i=1}^{j}\left[\frac{G_{1,i}^2}{\gamma_n^2+\Omega_{1,i}^2}+\frac{G_{2,i}^2}{\gamma_n^2+\Omega_{2,i}^2}\right]}{1+\frac{\gamma_n}{4\gamma_m}\sum\limits_{i=1}^{j}\left[\frac{G_{1,i}^2}{\gamma_n^2+\Omega_{1,i}^2}+\frac{G_{2,i}^2}{\gamma_n^2+\Omega_{2,i}^2}\right]}\right\rangle, \quad (5.18)$$

$$\Omega_{1,i} = \omega_i - \omega_{mn} - \mathbf{k}_i\mathbf{v},\quad \Omega_{2,i} = \omega_i - \omega_{mn} + \mathbf{k}_i\mathbf{v}.$$

Formula (5.18) is valid if

$$|\omega_i - \omega_j| \gtrsim \gamma_n, \quad (5.19)$$

i.e., if the distance between the individual lines in the perturbation spectrum is of the same order as the natural width.

Lamb [68] independently obtained results similar to (5.17) and (5.18) by another method. No proof, however, was given in [68] and the limits of applicability of the corresponding expressions were not indicated.

According to formulas (5.17) and (5.18) the factor

$$\frac{W_M(\mathbf{v})}{1+\frac{\gamma_n}{4\gamma_m}\left[\frac{G_1^2}{(\omega-\omega_{mn}-\mathbf{kv})^2+\gamma_n^2}+\frac{G_2^2}{(\omega-\omega_{mn}+\mathbf{kv})^2+\gamma_n^2}\right]} \quad (5.20)$$

in (5.17) can be interpreted as the velocity distribution of the difference of the mean populations of the levels m, n. It is easy to see that this distribution has two minima at velocities

$$\mathbf{kv} = \pm(\omega - \omega_{mn}).$$

When $\omega = \omega_{mn}$ these "dips" merge with one another. With this circumstance is associated the nonmonotonic dependence of $\langle W_m\rangle$ on the frequency, which is clearly seen from the graphs in Fig. 2. At the center of the curves, where $\omega = \omega_{mn}$, there is a minimum with a relative depth which increases with increase in $G^2/\gamma_m\gamma_n$.

For small values of $G^2/\gamma_m\gamma_n$ we can expand the denominator in (5.20) in a series and, retaining the first term, find

$$\langle W_m\rangle = \frac{\sqrt{\pi}}{4\gamma_m k\bar{v}}\, e^{-(\omega-\omega_{mn})^2/(k\bar{v})^2}\left\{(G_1^2+G_2^2) - \frac{G_1^4+G_2^4}{8\gamma_m\gamma_n} - \frac{G_1^2G_2^2}{4\gamma_m\gamma_n}\,\frac{\gamma_n^2}{\gamma_n^2+(\omega-\omega_{mn})^2}\right\}. \quad (5.21)$$

The third term in (5.21) has a much smaller width than $k\bar{v}$ ($\gamma_n \ll k\bar{v}$) and describes the above-mentioned "dip." It is also clear from (5.21) that when $\gamma_n \sim k\bar{v}$ or in the case of a traveling

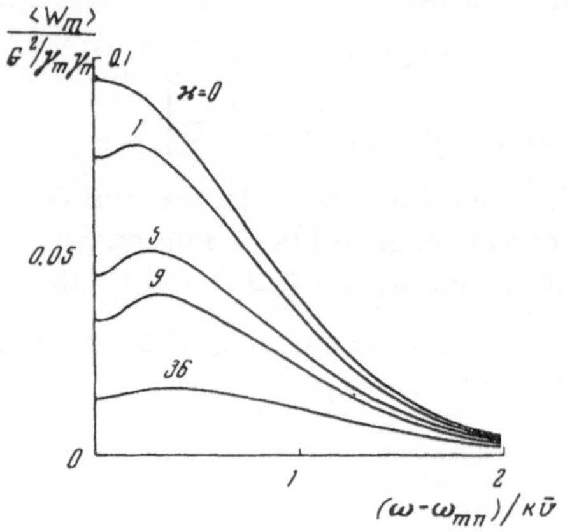

Fig. 2. Dependence of $\langle W_m\rangle$ on frequency ω; $\gamma_n = 0.12\,k\bar{v}$.

wave ($G_1 = 0$ or $G_2 = 0$) there is no dip. The existence of a dip was established in [68] in the approximation (5.21).

For $\omega = \omega_{mn}$ the emission probability is

$$\langle W_m\rangle = \frac{\sqrt{\pi}}{4\gamma_m k\bar{v}} \frac{G_1^2 + G_2^2}{\sqrt{1+\dfrac{G_1^2+G_2^2}{4\gamma_m\gamma_n}}} u(0, z),$$
$$z = \frac{\gamma_n}{k\bar{v}} \sqrt{1 + [G_1^2 + G_2^2]/[4\gamma_m\gamma_n]}, \tag{5.22}$$

where the function $u(0, z)$ is defined in (5.4). The formula differs from (5.3) (when $\omega = \omega_{mn}$) only in the replacement of G^2 by $[G_1^2 + G_2^2]/4$. In particular, when $z \ll 1$ we have $u(0, z) \approx 1$ and

$$\langle W_m\rangle = \frac{\sqrt{\pi}}{4\gamma_m k\bar{v}} \frac{G_1^2 + G_2^2}{\sqrt{1 + [G_1^2 + G_2^2]/4\gamma_m\gamma_n}}. \tag{5.23}$$

In the limiting case $G_1^2 + G_2^2 \gg k\bar{v}$, using the asymptotic expansion of the probability integral, we find

$$\langle W_m\rangle = 1 - 2\frac{\gamma_m}{\gamma_n}\frac{(k\bar{v})^2}{G_1^2 + G_2^2}. \tag{5.24}$$

Thus, for the indicated conditions practically all the atoms, irrespective of their velocity, interact effectively with the radiation, and $\langle W_m\rangle \approx 1$.

3. We now consider another case which can be accurately solved for arbitrary field amplitudes. We have in mind the following conditions:

$$\gamma_m = \gamma_n = \frac{1}{2}\Gamma, \quad V_{mn}[t, \mathbf{r}(t)] = G[t, \mathbf{r}(t)]\, e^{-i\varphi_0}, \tag{5.25}$$

where $G[t, \mathbf{r}(t)]$ is a real function of time and φ_0 is a real constant. In this case the fundamental matrix is given by formula (4.24) and for the emission probability we obtain the following expression:

$$W_m = \Gamma \int_{t_0}^{\infty} \sin^2 f(t)\, e^{-\Gamma(t-t_0)}\, dt, \quad f(t) = \int_{t_0}^{t} G[t', \mathbf{r}(t')]\, dt'. \tag{5.26}$$

Conditions (5.25) are realized, in particular, if the field is a standing monochromatic wave with a frequency the same as the transition frequency:

$$V_{mn}(t, \mathbf{r}) = Ge^{-i\varphi_0} \cos(\mathbf{kr} + \delta) = G(t, \mathbf{r})\, e^{-i\varphi_0};$$
$$\omega = \omega_{mn}, \quad \mathbf{r} = \mathbf{r}_0 + \mathbf{v}(t - t_0). \tag{5.27}$$

Putting (5.27) in (5.26) and averaging over $\mathbf{r}_0$, $\mathbf{v}$, we obtain

$$\langle \overline{W}_m\rangle = \frac{1}{2}\left\{1 - \Gamma\int_0^{\infty}\left\langle J_0\left[\frac{4G}{\mathbf{kv}} \sin\frac{\mathbf{kv}t}{2}\right]\right\rangle e^{-\Gamma t}\, dt\right\}, \tag{5.28}$$

where the stroke above denotes averaging over $\mathbf{r}_0$; J_0 is a Bessel function of the first kind of zero order. Expanding J_0 in a power series and carrying out the integration with respect to t and v term by term, we find

$$\langle \overline{W}_m\rangle = \frac{G^2}{\sqrt{\pi}\Gamma k\bar{v}}\left\{1-\left(\frac{G}{\Gamma}\right)^2+\frac{11}{6}\left(\frac{G}{\Gamma}\right)^4-\frac{151}{36}\left(\frac{G}{\Gamma}\right)^6+\cdots\right\} \quad (\Gamma,\ G \ll k\bar{v}), \tag{5.29}$$

In the region $G > \Gamma$ it is convenient to expand J_0 in a Fourier series

$$J_0\left[\frac{4G}{\mathbf{kv}}\sin\frac{\mathbf{kv}t}{2}\right] = \sum_{-\infty}^{\infty} J_m^2\left(\frac{2G}{\mathbf{kv}}\right)\cos m\mathbf{kv}t.$$

Integrating with respect to t, we obtain

$$\langle \overline{W}_m\rangle = \frac{1}{2}\left\langle 1 - J_0^2 - \sum_{\substack{-\infty \\ m\neq 0}}^{\infty} J_m^2 \frac{\Gamma^2}{\Gamma^2+(m\mathbf{kv})^2}\right\rangle. \tag{5.30}$$

It is easy to show that when $G \gg \Gamma$ the main role in (5.30) is played by $\langle 1 - J_0^2\rangle$, and the series diminishes as $[\Gamma^2/k\bar{v}G^2 \ln G/\Gamma]$. If, in addition, $k\bar{v} \gg G$, then

$$\langle \overline{W}_m\rangle \approx \frac{1}{2}\left\langle 1 - J_0^2\left(\frac{2G}{\mathbf{kv}}\right)\right\rangle = \frac{8}{\pi^{3/2}}\frac{G}{k\bar{v}} \quad (\Gamma \ll G \ll k\bar{v}). \tag{5.31}$$

Finally, when $G \gg k\bar{v} \gg \Gamma$, it follows from (5.30) that

$$\langle \overline{W}_m\rangle \approx \frac{1}{2}\left\{1 - \frac{1}{2\pi^{1/2}}\frac{k\bar{v}}{G}\right\}. \tag{5.32}$$

We compare formulas (5.29), (5.32), with (5.23), (5.24) obtained for the case $\gamma_m \ll \gamma_n$. Expansion of the radical in (5.23) for $G_1 = G_2 = G \ll (\gamma_m\gamma_n)^{1/2}$ gives an expression similar to (5.29):

$$\langle W_m\rangle = \frac{G^2}{\sqrt{\pi}2\gamma_m k\bar{v}}\left\{1-\left(\frac{G}{2\sqrt{\gamma_m\gamma_n}}\right)^2+\frac{3}{2}\left(\frac{G}{2\sqrt{\gamma_m\gamma_n}}\right)^4-\frac{5}{2}\left(\frac{G}{2\sqrt{\gamma_m\gamma_n}}\right)^6+\cdots\right\}. \tag{5.33}$$

The coefficients of the terms with G^4 in (5.33) and (5.29) differ relatively little.

In the case $(\gamma_m\gamma_n)^{1/2} \ll G \ll k\bar{v}$ it follows from (5.23) that

$$\langle W_m\rangle = \sqrt{\frac{\pi}{2}\frac{\gamma_n}{\gamma_m}}\frac{G}{k\bar{v}}. \tag{5.34}$$

This expression differs from (5.31) by the factor $(\gamma_m\gamma_n)^{1/2}$, which is unity when $\gamma_m = \gamma_n$, and by a numerical factor, which in (5.31) is $8\sqrt{2}/\pi^2 = 1.14$ times larger.

Formulas (5.32) and (5.34) differ in two respects. The coefficient $\frac{1}{2}$ in front of the bracket in (5.32) and absent in (5.24) is due to the fact that at the limit $G \to \infty$, $\langle W_m\rangle$ always $\to \gamma \times (\gamma_m + \gamma_n)^{-1}$. The second difference is that the ratio $k\bar{v}/G$ is of the second degree in (5.24) and is of the first degree in (5.32). It should be borne in mind, however, that in both cases these terms are very small, and the observed difference does not greatly affect the value of $\langle W_m\rangle$. The general behavior of $\langle W_m\rangle$ in relation to $2G/k\bar{v}$ is shown in Fig. 3, where curve 1 is plotted from formula (5.30) and curve 2 corresponds to a traveling wave with amplitude $\sqrt{[C_1^2 + G_2^2]/2}$, i.e., is equivalent to expression (5.17).

To determine the role of the motion of the atoms we compare the formulas obtained above with the mean probability of emission of stationary atoms in the field of a standing wave. We will assume in this case that the natural frequencies of the atoms have a Gaussian distribution with a variance equal to $k\bar{v}$. Replacing G in the formula (5.3) by $G\cos\mu$ and averaging over μ,

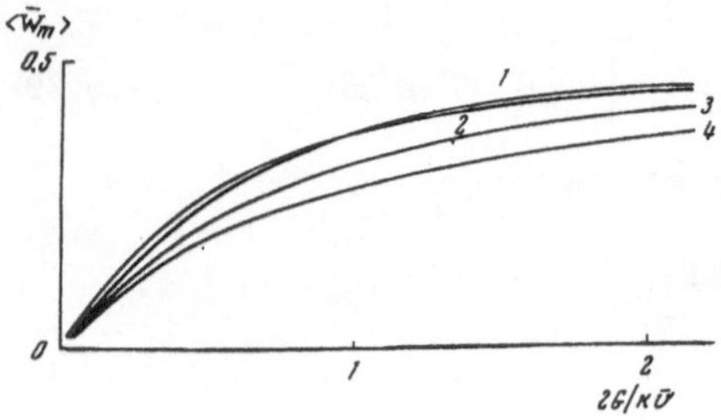

Fig. 3. Plot of $\langle W_m \rangle$ against $2G/k\bar{v}$ ($G, k\bar{v} \gg \Gamma$).

we obtain

$$\langle \overline{W}_m \rangle = \frac{\gamma_n}{\Gamma}\left\{1 - \frac{1}{\sqrt{\pi}k\bar{v}} \cdot \int_{-\infty}^{\infty} e^{-\frac{x^2}{(k\bar{v})^2}} \sqrt{\frac{(\Omega - x)^2 + \Gamma^2}{(\Omega - x)^2 + \Gamma^2(1 + G^2/\gamma_m\gamma_n)}}\, dx\right\}. \qquad (5.35)$$

Curve 4 in Fig. 3 corresponds to this expression. The general nature of the variation of $\langle W_m \rangle$ with $G/k\bar{v}$ is the same as when motion is taken into account, but the graphs differ quantitatively. For instance, in the limiting cases $\Gamma \ll G \ll k\bar{v}$ and $G \gg k\bar{v}$ we can obtain from (5.35)

$$\begin{aligned} \langle \overline{W}_m \rangle &\approx \frac{2}{\sqrt{\pi}} \sqrt{\frac{\gamma_n}{\gamma_m} \frac{G}{k\bar{v}}} \qquad (\Gamma \ll G \ll k\bar{v}), \\ \langle \overline{W}_m \rangle &\approx \frac{\gamma_n}{\Gamma}\left\{1 - \frac{1}{\sqrt{\pi}} \frac{\sqrt{\gamma_m\gamma_n}}{\Gamma} \frac{k\bar{v}}{G}\right\} \qquad (G \gg k\bar{v}). \end{aligned} \qquad (5.36)$$

We return to a comparison of the cases $\gamma_m = \gamma_n$ and $\gamma_m \ll \gamma_n$. We recall that they differ essentially only in the kinetics of the induced transitions. When $\gamma_m \ll \gamma_n \ll k\bar{v}$ the atoms involved in the emission traverse several wavelengths, but relax monotonically in time, so that only the mean field intensity $(G_1^2 + G_2^2)/4$ is significant and its spatial structure is of no importance. In the case of $\gamma_m = \gamma_n$, however, the time dependence of the probabilities $|S_{ij}|^2$ has the form of decaying oscillations and the field geometry has a significant effect. This is illustrated by curve 3 in Fig. 3, which is plotted for a traveling wave with amplitude equal to $G_1 + G_2$. We can conclude from the good agreement of curves 1 and 3 that in saturation conditions the traveling waves forming the standing wave should be "added in amplitude" when $\gamma_m = \gamma_n$, whereas in the case of a weak field and when $\gamma_m \ll \gamma_n$ the "squares of the amplitudes are added."

4. This investigation substantially exhausts the cases where the expression for the fundamental matrix can be obtained in quadratures. It is found that in calculation of induced emission all the models lead to qualitatively the same results with relatively little quantitative difference. Hence, we can infer that if the calculation does not need to be too accurate satisfactory results will be given by any one of the obtained formulas when it is used outside the range of its direct applicability if it can be modified in such a way that it includes all the values of the parameters $\omega - \omega_{mn}$, γ_m, γ_n, G_1 and G_2. In this sense formula (5.17) has the definite advantage of simplicity. For this purpose we can propose the expression

$$\langle \overline{W}_m \rangle = \frac{\Gamma}{4\gamma_m}\left\langle \left[\frac{G_1^2}{\Gamma^2 + \Omega_1^2} + \frac{G_2^2}{\Gamma^2 + \Omega_2^2}\right] \Big/ \left[1 + \frac{\Gamma^2}{4\gamma_m\gamma_n}\left(\frac{G_1^2}{\Gamma^2 + \Omega_1^2} + \frac{G_2^2}{\Gamma^2 + \Omega_2^2}\right)\right]\right\rangle. \qquad (5.37)$$

This formula becomes a rigorous expression in the case of a traveling wave ($G_2 = 0$) and when $\gamma_m + (G_1^2 + G_2^2)/\gamma_n \ll \gamma_n$, and gives results which differ by 10-15% from the actual values when $\gamma_m = \gamma_n$, $G_1 = G_2$, $\omega = \omega_{mn}$. Similar changes can be made in formula (5.18), which will then describe the emission for an arbitrary set of monochromatic lines in the spectrum of the external field.

It should be noted that this type of extrapolation is possible only for the mean probability of induced emission. In several other problems the different models investigated in sections 2 and 3 lead to results which differ not only quantitatively, but qualitatively. This applies, for instance, to the spectral composition of the spontaneous emission (see Chapter II), the inhomogeneity of the medium due to saturation (§ 6), and so on.

§ 6. Spatial Inhomogeneity of Medium Due to Saturation Effect

1. Spatial inhomogeneity of a medium in a strong electromagnetic field can be produced by two physically different causes. In an inhomogeneous electric field there are electrostrictive forces which are proportional to the gradient of the square of the field and lead to compression or rarefaction of the medium. With this is connected, for instance, the so-called induced Mandel'shtam–Brillouin scattering [16] and the self-focusing of an intense beam of rays [15]. Another cause of inhomogeneity is that the field leads to redistribution of the atoms as regards levels and reduces the mean difference of the populations. Since the value of such characteristics as the dielectric constant depends on the difference in the populations it is clear that in a strong inhomogeneous field the medium will be optically inhomogeneous. In this section we consider this effect.

The energy absorbed by unit volume of the medium in unit time in the case of a monochromatic field is given by the expression [3]

$$\frac{\omega\varepsilon''}{8\pi}|E|^2, \tag{6.1}$$

where ε'' is the imaginary part of the dielectric constant.* The same quantity, according to formula (3.22), is connected with the probability of photon emission W_m and with the number of acts of excitation Q_m, Q_n in unit volume in unit time. Hence,

$$\frac{\omega\varepsilon''}{8\pi}|E|^2 = \hbar\omega\,[Q_nW_n - Q_mW_m]. \tag{6.2}$$

Thus, to discuss the optical inhomogeneity of the medium we need only consider the quantity

$$s = \frac{W_m}{2\,|p_{mn}E/2\hbar|^2}. \tag{6.3}$$

2. We consider first of all the case of stationary atoms in the field of a standing monochromatic wave. Using formula (4.5), we find

$$s = \frac{\Gamma}{2\gamma_m}\,\frac{1}{\Omega^2+\Gamma^2\left[1+\dfrac{G^2}{\gamma_m\gamma_n}\cos^2\mu\right]}, \qquad \mu = kr. \tag{6.4}$$

Thus, s is modulated with a period $\pi/k = \lambda/2$.

Of most interest is the mean value of s and the amplitude of the first Fourier harmonic of s:

$$\bar{s} = \frac{1}{\pi}\int_0^\pi s\,d\mu, \quad s_1 = \frac{1}{\pi}\int_0^\pi s\cos 2\mu\,d\mu. \tag{6.5}$$

The quantity s_1, as we will show in Chapter V, gives the coefficient of reflection of a wave propagated in a periodically inhomogeneous medium. We note that

$$\overline{W}_m = (\bar{s}+s_1)\,G^2. \tag{6.6}$$

Performing the simple integration we obtain

$$\overline{W}_m = \frac{\gamma_n}{\Gamma}\left\{1 - \frac{1}{\sqrt{1+\dfrac{G^2}{\gamma_m\gamma_n}\,\dfrac{\Gamma^2}{\Gamma^2+\Omega^2}}}\right\}; \tag{6.7}$$

* The time dependence of the field is taken in the form $e^{-i\omega t}$.

$$\bar{s} = \frac{\Gamma}{2\gamma_m} \frac{1}{[\Gamma^2 + \Omega^2]\sqrt{1 + \frac{G^2}{\gamma_m\gamma_n} \frac{\Gamma^2}{\Gamma^2+\Omega^2}}}; \tag{6.8}$$

$$s = -\bar{s} \frac{\frac{\Gamma^2}{\Gamma^2+\Omega^2} \frac{G^2}{\gamma_m\gamma_n}}{\left[1+\sqrt{1+\frac{G^2}{\gamma_m\gamma_n}\frac{\Gamma^2}{\Gamma^2+\Omega^2}}\right]^2}. \tag{6.9}$$

It is clear from (6.9) that as G increases the value of s_1 increases at first and then decreases. The maximum of s_1 is reached when

$$\frac{G^2}{\gamma_m\gamma_n} \frac{\Gamma^2}{\Gamma^2+\Omega^2} = 2(\sqrt{2}+1). \tag{6.10}$$

The ratio s_1/s for this value of G^2 is

$$\frac{s_1}{s} = -0.393. \tag{6.11}$$

Thus, the degree of modulation of the dielectric constant can attain high values.

3. In gas systems the motion of the atom will reduce the inhomogeneity due to induced transitions. If during its lifetime the atom traverses a distance less than $\lambda/2\pi$ in the direction k the situation will be similar to the case of stationary atoms. In gas lasers we have the reverse relationship. Hence, we can expect that the motion of the atoms will almost completely level out the inhomogeneity of the field. This is actually the case for the conditions analyzed in section 2, § 5:

$$\gamma_m \ll \gamma_n \quad (\gamma_m \gg \gamma_n), \quad G^2 \ll \gamma_n^2. \tag{6.12}$$

The simplification of the problem in that case is due, in particular, to the possibility of neglecting the spatial oscillations of the field [see the discussion of formula (5.16)].

However, as V. P. Chebotaev noted, when γ_m and γ_n have similar values the position is altered somewhat. If $\gamma_m \approx \gamma_n$, $\omega = \omega_{mn}$, the lifetime and velocity range of the atoms at which interaction with the field is effective is determined by the same parameter $\Gamma = \gamma_m + \gamma_n$. Hence, the group of atoms of interest to us traverse during their lifetime a distance

$$v\frac{1}{\Gamma} \approx \frac{\Gamma}{k}\frac{1}{\Gamma} = \frac{\lambda}{2\pi}.$$

Hence, when $\gamma_m \approx \gamma_n$ we can expect the appearance of inhomogeneity in the gas phase too, even if the Doppler width of the line is much greater than the natural width. Nevertheless, as will be shown below, the motion of the atoms significantly reduces the inhomogeneity of the medium.

Of most interest in this connection is the case $\gamma_m = \gamma_n$, $\omega = \omega_{mn}$, considered in section 3. § 5. Using formulas (4.24), (5.27), (6.3), (3.20), and (3.25), we obtain the following expression for s:

$$s = \frac{1}{2G}\int_0^\infty e^{-\Gamma t}\left\langle \frac{\sin\left[\frac{4G}{kv}\sin\frac{kvt}{2}\cos\left(\mu - \frac{kvt}{2}\right)\right]}{\cos\mu}\right\rangle dt. \tag{6.13}$$

Here v is the projection of the velocity in the direction **k**. We consider first of all the case of small G. The expansion of s in powers of G/Γ has the form

$$s = \frac{\sqrt{\pi}}{\Gamma k\bar{v}}\left\{1 - 2\left(\frac{G}{\Gamma}\right)^2 \sum_{m=0}^{\infty}\sum_{l=0}^{m}\sum_{j=0}^{l}(-1)^{m+j}\left(\frac{G}{\Gamma}\right)^{2m}\binom{2m+3}{l+1}\frac{(l+1-j)^{2m+2}}{j!\,(2m+3-j)}\frac{\cos(2m-2l-1)\mu}{\cos\mu}\right\}. \tag{6.14}$$

Multiplying (6.14) by cos 2μ and integrating with respect to μ, we obtain s_1. The first terms of the expansion of s_1 are:

$$s_1 = \frac{\sqrt{\pi}}{\Gamma k\bar{v}}\left\{\frac{1}{12}\left(\frac{G}{\Gamma}\right)^4 - \frac{17}{36}\left(\frac{G}{\Gamma}\right)^6 + \cdots\right\} \qquad (G, \Gamma \ll k\bar{v}). \tag{6.15}$$

Formula (6.15) has two remarkable features. Firstly, the first term of the expansion contains $(G/\Gamma)^4$, whereas in the case of stationary atoms the expansion begins with $(G/\Gamma)^2$ [see (6.9)]. Secondly, when the values of G/Γ are sufficiently small, s_1 is positive. This obviously means that the dielectric constant at the antinodes is greater than at the nodes, contrary to what might be expected from the simple considerations at the beginning of the section. In fact, formula (6.14) at the antinodes ($\cos^2\mu = 1$) and nodes ($\cos\mu = 0$) gives, respectively,

$$\begin{aligned} s &= \frac{\sqrt{\pi}}{\Gamma k\bar{v}}\left\{1 - \left(\frac{G}{\Gamma}\right)^2 + \frac{23}{24}\left(\frac{G}{\Gamma}\right)^4 - \frac{841}{181}\left(\frac{G}{\Gamma}\right)^6 + \cdots\right\} \qquad (\cos^2\mu = 1), \\ s &= \frac{\sqrt{\pi}}{\Gamma k\bar{v}}\left\{1 - \left(\frac{G}{\Gamma}\right)^2 - \frac{19}{24}\left(\frac{G}{\Gamma}\right)^4 - \frac{247}{60}\left(\frac{G}{\Gamma}\right)^6 + \cdots\right\} \qquad (\cos\mu = 0). \end{aligned} \tag{6.16}$$

Thus, inhomogeneity arises in a gas system too, but the motion of the atoms greatly reduces the effect of induced transitions and, in addition, leads to a rather unexpected effect – a relatively greater reduction of the dielectric constant at the nodes of the field in comparison with the antinodes.

For the analysis of the case $\Gamma \ll G \ll k\bar{v}$ it is convenient to use the expansion of $\sin[z\cos(\mu - kvt/2)]$ in (6.13) in a Fourier series. Keeping only the terms which are functions of the velocity v, we obtain

$$\begin{aligned} s &= \frac{1}{G}\sum_{m=0}^{\infty} I_m(-1)^m \frac{\cos(2m+1)\mu}{\cos\mu}, \\ I_m &= \int_0^{\infty}\left\langle I_{2m+1}\left[\frac{4G}{kv}\sin\frac{kvt}{2}\right]\cos(2m+1)\frac{kvt}{2}\right\rangle e^{-\Gamma t}dt. \end{aligned} \tag{6.17}$$

Here I_ν is a Bessel function of the first kind of ν-th order. It follows from formulas

$$\left[\overline{\frac{\cos(2m+1)\mu}{\cos\mu}}\right] = (-1)^m, \quad \left[\overline{\frac{\cos(2m+1)\mu\cos 2\mu}{\cos\mu}}\right] = \begin{cases} 0, & m = 0, \\ (-1)^{m+1}, & m \geqslant 1 \end{cases} \tag{6.18}$$

and (6.17) that

$$\bar{s} = \frac{1}{G}\sum_{m=0}^{\infty} I_m, \qquad s_1 = -\frac{1}{G}\sum_{m=1}^{\infty} I_m. \tag{6.19}$$

The integrals I_m were calculated and were found to be

$$\begin{aligned} I_0 &= \frac{\langle \overline{W}\rangle}{G} = \frac{8}{\pi^{3/2}}\frac{1}{k\bar{v}}, \\ I_1 &= I_0\cdot 0.0330, \qquad I_2 = I_0\cdot 0.0081, \\ I_3 &= I_0\cdot 0.0032, \qquad I_4 = I_0\cdot 0.0014. \end{aligned} \tag{6.20}$$

Substituting (6.20) in (6.19) we obtain

$$\bar{s} = \frac{I_0}{G} 1.046, \quad s_1 = -\frac{I_0}{G} 0.046. \tag{6.21}$$

Thus, the degree of modulation of the dielectric constant is only a few per cent.

To illustrate the fact that the smallness of the inhomogeneity is due to motion of the atoms we calculate s for the case where the atoms are stationary, and their natural frequencies have a Gaussian distribution with a variance equal to the Doppler width $k\bar{v}$. It follows from formulas (6.8), (6.9) that

$$\bar{s} = \frac{1}{\sqrt{\pi}\gamma_m k\bar{v}} \frac{1}{\sqrt{1 + G^2/\gamma_m\gamma_n}} K(p), \quad p^2 = \frac{G^2/\gamma_m\gamma_n}{1 + G^2/\gamma_m\gamma_n},$$

$$s_1 = \frac{1}{\sqrt{\pi}\gamma_m k\bar{v}} \frac{1}{\sqrt{1 + G^2/\gamma_m\gamma_n}} \left\{K(p) - 2\frac{K(p) - E(p)}{p^2}\right\}. \tag{6.22}$$

Here K(p) and E(p) are complete elliptic integrals of the first and second kind, respectively. Using their expansion in a series of powers of $p' = (1 - p^2)^{\frac{1}{2}}$, we obtain the following approximate formulas:

$$\bar{s} = \frac{1}{\sqrt{\pi}\gamma_m k\bar{v}} \frac{\sqrt{\gamma_m\gamma_n}}{G} \ln \frac{4G}{\sqrt{\gamma_m\gamma_n}}$$

$$s_1 = -\frac{1}{\sqrt{\pi}\gamma_m k\bar{v}} \frac{\sqrt{\gamma_m\gamma_n}}{G} \ln \frac{4G}{e^2\sqrt{\gamma_m\gamma_n}} \qquad \left(\frac{G^2}{\gamma_m\gamma_n} \gg 1\right). \tag{6.23}$$

It is clear from expressions (6.23) that $\bar{s}$ and s_1 differ very slightly. This clearly shows that the opposite result, contained in (6.21), is due to the motion of the atoms.

CHAPTER II

Spontaneous Emission with Induced Transitions Taken into Account

§ 7. General Theory

1. The theory of spontaneous emission is usually derived for an isolated atom (see [56, 69], for instance). In this case the probability of finding an atom in excited states decreases exponentially with time and, consequently, the spontaneous emission line has a dispersion form. It was shown in Chapter I that the kinetics of the decay of excited states of an atom in a strong electromagnetic field is substantially different from the case for an isolated atom. Hence, the spectral composition of the spontaneous emission will also undergo changes. Thus, we must revise the theory of spontaneous emission and include induced transitions in it. This question is considered in this chapter.

2. We will start from the system of equations for the probability amplitudes

$$\dot{\hat{a}}' = -\hat{\gamma}\hat{a}' - i(\hat{V} + \hat{V}_\mu)\hat{a}',$$

$$\hat{V}_\mu = \begin{pmatrix} 0 & V_\mu \\ V_\mu^* & 0 \end{pmatrix}, \quad V_\mu = G_\mu e^{-i[(\omega_\mu - \omega_{mn})t - \mathbf{k}_\mu \mathbf{r}]}. \tag{7.1}$$

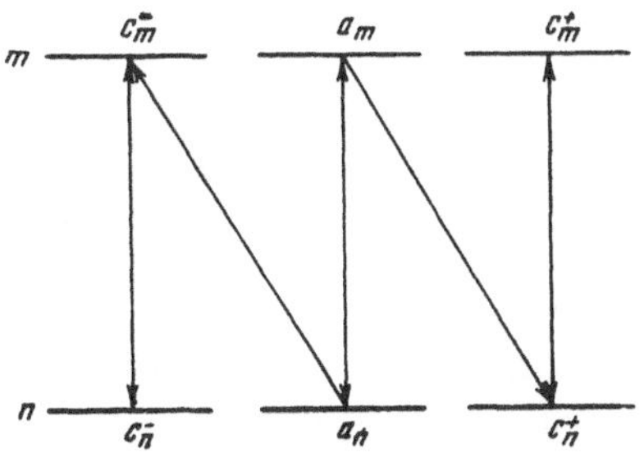

Fig. 4. Scheme of transitions between states m, n of an atom under the action of strong (vertical arrows) and weak fields.

As distinct from system (3.5), a small perturbation, corresponding to a traveling monochromatic wave with wave vector $\mathbf{k}_\mu = (\omega_\mu/c)\mathbf{n}_\mu$, is introduced. The solution of system (7.1) will be sought by the method of successive approximations with V_μ regarded as a small quantity. The solution of (7.1) with $V_\mu = 0$ is the zero approximation. The equation for the first approximation is obtained from (7.1) if the function $\hat{a}'$ in the term $V_\mu \hat{a}'$ is replaced by the zero approximation. Then the solution of system (7.1) in a first approximation in V' will contain three terms:

$$\hat{a} + \hat{c}^+ + \hat{c}^-, \qquad \hat{c}^{\pm} = \begin{pmatrix} c_m^{\pm} \\ c_n^{\pm} \end{pmatrix}, \tag{7.2}$$

where $\hat{a}$ denotes the zero approximation for $\hat{a}'$; $\hat{c}^+$ and $\hat{c}^-$ denote corrections to the first approximation proportional to V_μ^* and V_μ, respectively. The physical sense of the function $c_{mn}^{\pm}$ is illustrated by the scheme depicted in Fig. 4. Under the action of the strong field the atom undergoes transitions from the state m to the state n and back (vertical arrows in Fig. 4). In addition, the atom can pass from the state m to the state n after emitting a photon of frequency ω_μ of the "weak" field V_μ. The contribution of such transitions to the amplitude of the states m, n is given by the functions c_m^+, c_n^+. Transitions accompanied by absorption of photons $\hbar\omega_\mu$ are also possible. Their contribution is given by functions c_m^-, c_n^-.

The probabilities of emission and absorption of photons $\hbar\omega_\mu$ are, respectively,

$$\begin{aligned} W_\mu^e &= 2\gamma_m \int_{t_0}^{\infty} |c_m^+|^2 dt + 2\gamma_n \int_{t_0}^{\infty} |c_n^+|^2 \, dt, \\ W_\mu^a &= 2\gamma_m \int_{t_n}^{\infty} |c_m^-|^2 dt + 2\gamma_n \int_{t_0}^{\infty} |c_n^-|^2 dt. \end{aligned} \tag{7.3}$$

The actual emission (or absorption) which is observed experimentally is given by the difference of these quantities.

The spontaneous emission probability, which is what concerns us in this chapter, can be calculated from W_μ^e. In this case $|V_\mu|^2$ is the spectral density of the perturbation for the interaction of the atom with zero oscillations of the field

$$|V_\mu|^2 = 4\pi\hbar\omega \frac{p_{mn}^2}{\hbar^2} F(\mathbf{k}_\mu) \frac{\Delta \mathbf{k}_\mu}{(2\pi)^3} = \frac{2\gamma_{mn}}{8\pi^2} F(\mathbf{k}_\mu) \Delta\omega \, \Delta O, \tag{7.4}$$

where $2\gamma_{mn} = A_{mn}$ is the Einstein coefficient for spontaneous emission in the transition m → n, ΔO is an element of solid angle. The function $F(\mathbf{k}_\mu)$ is connected with the number of field oscillations Δn of unit volume in the interval Δk_μ:

$$\Delta n = F(\mathbf{k}_\mu) \frac{\Delta \mathbf{k}_\mu}{(2\pi)^3}. \tag{7.5}$$

For free space $F(\mathbf{k}_\mu) \equiv 1$. In all the following formulas we will omit this factor.

3. We express W_μ^e directly by the perturbation V_μ. We note in advance that spontaneous emission from the upper level m can occur after either excitation of the level m or the level n, since a strong field can induce the transition n → m, after which there is a finite probability of spontaneous decay of the level m. Hence, we consider two types of initial conditions: $a_m(t_0) = 1$ and $a_n(t_0) = 1$. We introduce the matrix

$$\hat{c} = \begin{pmatrix} c^+_{mm} & c^+_{mn} \\ c^+_{nm} & c^+_{nn} \end{pmatrix}, \tag{7.6}$$

where the second subscript indicates to which of the two levels the atom was excited. With the aid of the matrix $\hat{c}$ the emission probability can be written in the form

$$W^e_\mu = 2\int_{t_0}^{\infty} \mathrm{Sp}\,\{\hat{\rho}\hat{c}^\dagger\hat{\gamma}\hat{c}\}\,dt, \tag{7.7}$$

where $\hat{c}^\dagger$ denotes the conjugate matrix of $\hat{c}$, and the elements of the diagonal $\hat{\rho}$ give the relative number of acts of excitation to the levels m and n:

$$\hat{\rho} = \begin{pmatrix} \rho_m & 0 \\ 0 & \rho_n \end{pmatrix} = \frac{1}{Q_m + Q_n}\begin{pmatrix} Q_m & 0 \\ 0 & Q_n \end{pmatrix}. \tag{7.8}$$

Matrix $\hat{c}$ satisfies the equation

$$\dot{\hat{c}} = -(\hat{\gamma} + i\hat{V})\,\hat{c} - i\hat{V}^\dagger_\mu\hat{S}, \tag{7.9}$$

where, as distinct from (7.1),

$$\hat{V}_\mu = \begin{pmatrix} 0 & V_\mu \\ 0 & 0 \end{pmatrix}, \tag{7.10}$$

and $\hat{S}$ is the fundamental matrix of the equation (7.1) when $V_\mu = 0$, and satisfies the initial conditions

$$\hat{S}(t_0) = \hat{E}. \tag{7.11}$$

$\hat{E}$ is the unitary matrix. The initial conditions for $\hat{c}$ are zero conditions.

The solution of equation (7.9) which satisfies zero initial conditions is given by the formula (see [70], for instance)

$$\hat{c}(t) = -i\hat{S}(t)\int_{t_0}^{t} \hat{S}^{-1}(t')\,\hat{V}^\dagger_\mu(t')\,\hat{S}(t')\,dt'. \tag{7.12}$$

Substituting (7.12) in (7.7) and using the relationship

$$[\hat{A}\hat{B}]^\dagger = B^\dagger A^\dagger, \tag{7.13}$$

we can write formula (7.7) in the following way:

$$W^e_\mu = 2\mathrm{Sp}\int_{t_0}^{\infty} \hat{\rho}\hat{U}^\dagger\hat{\Gamma}\,\hat{U}\,dt, \tag{7.14}$$

where we have introduced the symbols

$$\hat{\Gamma} = S^\dagger\hat{\gamma}\hat{S}, \quad \hat{U} = \int_{t_0}^{t} \hat{S}^{-1}\hat{V}^\dagger_\mu\hat{S}\,dt'. \tag{7.15}$$

According to (3.11), we have the relationship

$$\hat{\Gamma} = -\frac{1}{2}\frac{d}{dt}(\hat{S}^\dagger \hat{S}); \tag{7.16}$$

in view of the condition

$$\hat{S}(\infty) = 0 \tag{7.17}$$

we have

$$\int_t^\infty \hat{\Gamma}\, dt' = \frac{1}{2}\hat{S}^\dagger \hat{S}. \tag{7.18}$$

Relationship (7.18) clearly enables us to perform one integration with respect to time in (7.14). Substituting (7.15) in (7.14) and changing the order of integration so that integration with respect to t is first, and using (7.18), we can find

$$W_\mu^e = 2\,\mathrm{Re}\left\{\mathrm{Sp}\int_{t_0}^\infty\int_{t'}^\infty dt'\,dt''\hat{\rho}\hat{S}^\dagger(t'')\,\hat{V}_\mu(t'')\,\hat{S}(t'')\,\hat{S}^{-1}(t')\,\hat{V}_\mu^\dagger(t')\,\hat{S}(t')\right\}. \tag{7.19}$$

We put $\hat{V}_\mu$ in the following form:

$$\hat{V}_\mu = G_\mu e^{-i\Omega_\mu t}\hat{\beta},\ \hat{\beta} = \begin{pmatrix}0 & 1\\ 0 & 0\end{pmatrix},\quad |G_\mu|^2 = \frac{\gamma_{mn}}{4\pi^2}\Delta\omega\Delta O. \tag{7.20}$$

Then the spectral density of spontaneous emission $w_\mu = W_\mu^e/\Delta\omega\Delta O$ can be written as:

$$w_\mu = \frac{\gamma_{mn}}{2\pi^2}\,\mathrm{Re}\left\{\mathrm{Sp}\,\hat{\rho}\int_{t_0}^\infty dt'\int_{t'}^\infty dt''\,\hat{S}^\dagger(t'')\,\hat{\beta}\hat{S}(t'')\,\hat{S}^{-1}(t')\,\hat{\beta}^\dagger\hat{S}(t')\,e^{-i\Omega_\mu(t''-t')}\right\}. \tag{7.21}$$

This is the final expression for the spectral composition of the spontaneous emission.

We calculate the correlation function corresponding to (7.21):

$$\Phi(\tau) = \int_{-\infty}^\infty w_\mu(\Omega_\mu)\,e^{-i\Omega_\mu\tau}\,d\Omega_\mu. \tag{7.22}$$

Integration with respect to Ω_μ will give $2\pi\delta[\tau \pm (t'' - t')]$; performing the integration with respect to t', we obtain

$$\Phi(\tau) = \frac{\gamma_{mn}}{2\pi}\,\mathrm{Sp}\left\{\hat{\rho}\int_{t_0}^\infty \hat{S}^\dagger(t)\,\hat{\beta}\hat{S}(t)\,\hat{S}^{-1}(t+\tau)\,\hat{\beta}^\dagger\hat{S}(t+\tau)\,dt\right\}. \tag{7.23}$$

The integral spontaneous emission probability is obtained, according to (7.22), from (7.23) with $\tau = 0$:

$$\Phi(0) = \int_{-\infty}^\infty w_\mu\,d\Omega_\mu = \frac{\gamma_{mn}}{2\pi}\,\hat{\mathrm{S}}\mathrm{p}\left\{\hat{\rho}\int_{t_1}^\infty \hat{S}^\dagger(t)\,\hat{\beta}\hat{\beta}^\dagger\hat{S}(t)\,dt\right\}. \tag{7.24}$$

Multiplying the matrices contained in the integrand in (7.24) we find

$$\Phi(0) = \frac{\gamma_{mn}}{2\pi}\left\{\rho_m\int_{t_0}^\infty |S_{mm}|^2\,dt + \rho_n\int_{t_0}^\infty |S_{mn}|^2\,dt\right\},$$

i.e., the integral spontaneous emission probability is expressed by the population N_m of the level m, as was to be expected from the formulas of § 3.

4. We return to formula (7.21) for the spectral density of spontaneous emission and consider a particular case of interest for several applications. It was shown in § 3 that in many cases of practical importance the fundamental matrix is expressed in the form of a linear combination of exponential functions or, in other words, the wave function is the superposition of the wave functions describing the quasi-stationary states. In this case the integration in formula (7.21) can be carried out and the structure of the line can be determined.

Let the matrix $\hat{S}(t)$ be given by the formula

$$\hat{S}(t) = \begin{pmatrix} \sum_s A_{ms} e^{-\alpha_s t} & \sum_r A_{nr} e^{-\alpha_r t} \\ \sum_{s'} B_{ms'} e^{-\beta_{s'} t} & \sum_{r'} B_{nr'} e^{-\beta_{r'} t} \end{pmatrix}. \tag{7.25}$$

Substitution of (7.25) in (7.21) leads to the following result:

$$\frac{2\pi^2}{\gamma_{mn}} w_\mu = \rho_m \left[\sum_{sr'} \frac{C_{sr'}}{i(\Omega_\mu - \alpha_s'' + \beta_{r'}'') + \alpha_s' + \beta_{r'}'} + \sum_{ss'} \frac{C_{ss'}'}{i(\Omega_\mu - \alpha_s'' + \beta_{s'}'') + \alpha_s' + \beta_{s'}'} \right] + \rho_n \left[\sum_{rr'} \frac{D_{rr'}}{i(\Omega_\mu - \alpha_r'' + \beta_{r'}'') + \alpha_r' + \beta_{r'}'} + \sum_{rs'} \frac{D_{rs'}'}{i(\Omega_\mu - \alpha_r'' + \beta_{s'}'') + \alpha_r' + \beta_{s'}'} \right], \tag{7.26}$$

$$\alpha = \alpha' + i\alpha'', \quad \beta = \beta' + i\beta''.$$

The coefficients $C_{sr'}$, $C'_{ss'}$, $D'_{rr'}$, $D'_{rs'}$, which are independent of Ω_μ, can be expressed in the form of series, but we will not do this, since we are interested at present only in the relationship between w_μ and the frequency. It is clear from (7.26) that the spontaneous emission line consists of components of dispersion form; the position of the maxima of the individual components depend on the difference in the energies of the quasi-stationary sublevels into which the levels m, n of the isolated atom are split; the widths of the components are equal to the sum of the real parts of α and β. Thus, as regards frequencies and widths of the components the position is the same as it would be if there was a transition from each sublevel of the state m to each sublevel of the state n. The intensities of the components are determined by the amplitudes A_{ms}, $A_{ns'}$, B_{mr}, $B_{nr'}$ and we cannot draw any general conclusions regarding them.

5. Expression (7.21) relates to spontaneous emission by the same transition where the strong external field acts. In a similar way we can consider other transitions in which only one of the two levels m, n, perturbed by the external field, joins. Formula (7.21) is still valid in these cases, if the following substitutions are made:

$$\gamma_{mn} \to \gamma_{mj}, \quad \Omega_\mu \to \omega - \omega_{mj},$$

$$\hat{\beta} \to \begin{pmatrix} 0 & 0 & 1 \\ 0 & 0 & 0 \\ 0 & 0 & 0 \end{pmatrix}, \quad \hat{\rho} \to \begin{pmatrix} \rho_m & 0 & 0 \\ 0 & \rho_n & 0 \\ 0 & 0 & 0 \end{pmatrix}, \quad \hat{S} \to \begin{pmatrix} S_{mm} & S_{mn} & 0 \\ S_{nm} & S_{nn} & 0 \\ 0 & 0 & e^{-\gamma_j (t-t_0)} \end{pmatrix} \tag{7.27}$$

for transition m → j and

$$\gamma_{mn} \to \gamma_{nk}, \quad \Omega_\mu \to \omega \to \omega_{nk},$$

$$\hat{\beta} \to \begin{pmatrix} 0 & 0 & 0 \\ 0 & 0 & 1 \\ 0 & 0 & 0 \end{pmatrix}, \quad \hat{\rho} \to \begin{pmatrix} \rho_m & 0 & 0 \\ 0 & \rho_n & 0 \\ 0 & 0 & 0 \end{pmatrix}, \quad \hat{S} \to \begin{pmatrix} S_{mm} & S_{mn} & 0 \\ S_{nm} & S_{nn} & 0 \\ 0 & 0 & e^{-\gamma_k (t-t_0)} \end{pmatrix} \tag{7.27'}$$

for transition $n \to k$. Here γ_j and γ_k are the rates of decay of the levels j, k. In the case of (7.25) the line will have a structure similar to (7.26) but $\beta' + i\beta''$ will be replaced by γ_j ($m \to j$) or $\alpha' + i\alpha''$ by γ_k ($n \to k$).

Similar changes can be made for the spontaneous transitions in which the levels m, n are low.

6. The region of application of the theory expounded in this section is prescribed by the assumption of the insignificance of the term $i\hat{V}_\mu \hat{c}$, which is discarded in equation (7.9). To illustrate the physical sense of this assumption we note that the amplitudes c_m^+, c_n^+ describe the states of the "atom + field" system in which the atom is at levels m, n, and a photon of frequency ω_μ appears in the field (right side of Fig. 4). The mean population of these states is approximately $(\gamma_m + \gamma_n)/\gamma_{mn}$ times less than the populations of states described by amplitudes a_m, a_n (middle of Fig. 4). It is obvious that state c_m^+ may give rise to new states in which there will be two photons with frequencies ω_μ, $\omega_{\mu'}$, three photons with frequencies ω_μ, $\omega_{\mu'}$, $\omega_{\mu''}$, and so on, in the field. Their mean populations will diminish in the ratio $\gamma_{mn}/(\gamma_m + \gamma_n)$. In this section none of these states are considered. Hence, all the results are applicable if

$$\gamma_{mn} \ll (\gamma_m + \gamma_n), \tag{7.28}$$

and the method of calculation can be regarded as a first approximation in an expansion in terms of the parameter $\gamma_{mn}/(\gamma_m + \gamma_n)$. In view of the incoherence of the spontaneous emission it is probable that the exact expression will differ from the obtained expression only in the integral spontaneous emission probability and the shape of the line will not be significantly altered.

§ 8. Spontaneous Emission of Stationary Atoms in an External Field

We will begin to particularize the general formulas of § 7 with the simplest case of stationary atoms. Using formulas (4.3) and (7.12) we can calculate the functions c_m^+, c_n^+. They are

$$\begin{gathered} c_{mm}^+ = -\frac{GG_\mu}{\alpha_1 - \alpha_2}\left[A_{1m}(\varphi_{11} - \varphi_{12}) + A_{2m}(\varphi_{21} - \varphi_{22})\right], \\ c_{nm}^+ = -\frac{iG_\mu}{\alpha_1 - \alpha_2}\{A_{1m}[(\alpha_1 + \gamma_m)\varphi_{11} - (\alpha_2 + \gamma_m)\varphi_{12}] + A_{2m}[(\alpha_1 + \gamma_m)\varphi_{21} - (\alpha_2 + \gamma_m)\varphi_{22}]\}. \end{gathered} \tag{8.1}$$

Here φ_{ij} denotes functions

$$\varphi_{ij} = \frac{1}{i\Omega_\mu + \alpha_i - \alpha_j}\{e^{(i\Omega_\mu + \alpha_i)t} - e^{\alpha_j t}\}, \quad i, j = 1, 2; \tag{8.2}$$

constants A_{1m}, A_{2m} are

$$A_{1m} = -\frac{\alpha_2 + \gamma_m}{\alpha_1 - \alpha_2}, \quad A_{2m} = \frac{\alpha_1 + \gamma_m}{\alpha_1 - \alpha_2}. \tag{8.3}$$

Functions c_{mn}^+, c_{nn}^+, which relate to the case of excitation of the lower level n, are obtained from (8.1) by the replacement of A_{1m}, A_{2m} by the constants A_{1n}, A_{2n}:

$$A_{1n} = -A_{2n} = -\frac{iG}{\alpha_1 - \alpha_2}\,\frac{(\alpha_1 + \gamma_m)(\alpha_2 + \gamma_m)}{iG(\alpha_1 - \alpha_2)}. \tag{8.4}$$

For arbitrary values of the constants γ_m, γ_n the expression obtained for w_μ is very awkward and difficult to visualize. For our purpose it will be sufficient to consider some special cases

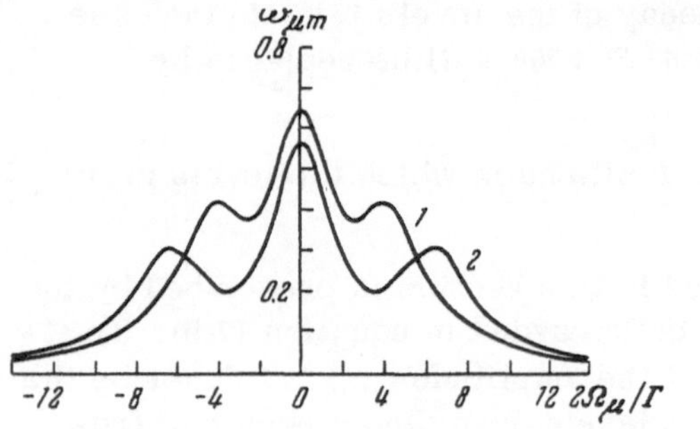

Fig. 5. Frequency dependence of spectral density of spontaneous emission probability (in units of $2\gamma_{mn}/4\pi^2\Gamma^2$) for $\gamma_m = \gamma_n = \Gamma/2$; $\omega = \omega_{mn}$. 1) $G = \Gamma$; 2) $G = {}^3\!/_2\Gamma$

which illustrate the general relationships dealt with in § 7. If

$$\gamma_m = \gamma_n = \frac{1}{2}\Gamma, \quad \omega = \omega_{mn}, \tag{8.5}$$

the spectral density of spontaneous emission is given by the expression (when the atom is excited to the upper level m)

$$w_{\mu m} = \frac{\gamma_{mn}}{2\pi^2}\frac{1}{4}\left\{\left[1 + \frac{\Gamma^2}{\Gamma^2 + 4G^2}\right]\cdot\right.$$
$$\cdot\left[\frac{1}{\Omega_\mu^2 + \Gamma^2} + \frac{1/2}{(\Omega_\mu - 2G)^2 + \Gamma^2} + \frac{1/2}{(\Omega_\mu + 2G)^2 + \Gamma^2}\right] -$$
$$\left. - \frac{G}{\Gamma^2 + 4G^2}\left[\frac{\Omega_\mu - 2G}{(\Omega_\mu - 2G)^2 + \Gamma^2} + \frac{\Omega_\mu + 2G}{(\Omega_\mu + 2G)^2 + \Gamma^2}\right]\right\}, \tag{8.6}$$
$$\Omega_\mu = \omega_\mu - \omega_{mn}.$$

It is clear from formula (8.6) that the spontaneous emission line contains three components with maxima situated at frequencies [the three terms of the first line in (8.6)]

$$\omega_{mn} - 2G, \quad \omega_{mn}, \quad \omega_{mn} + 2G.$$

The antisymmetric terms of the second line (their sum is symmetric relative to $\Omega_\mu = 0$), as can easily be shown, comprise not more than 1/5 of the terms of the first line. The intensity of each side component is a half of the intensity of the unshifted component; the absolute intensity, however, is halved owing to the factor $1 + \Gamma^2/(\Gamma^2 + 4G^2)$ when G changes from 0 to ∞. The results of numerical calculations from formula (8.6) are given in Fig. 5. These changes in the line shape can easily be related to the special features of the kinetics of induced transitions. When conditions (8.5) are fulfilled the probabilities of finding an atom at levels m, n oscillate with frequency 2G and decay with a decay constant Γ [see (4.9)]. The frequency of the oscillations determines the splitting and Γ determines the width of the line components.

The interpretation of (8.6) is of interest also from the viewpoint of quasi-stationary states of the atom in an electromagnetic field. As noted in § 4, when conditions (8.5) are fulfilled, each of the levels of the atom is split into two sublevels and the distance between them is 2G. We can assume that the line represented by formula (8.6) is the result of transitions between these sublevels (Fig. 6). The unshifted component, which corresponds to two transitions, is naturally twice as intense as the side components.

Another relatively simple case corresponds to the conditions

$$\gamma_m \ll \gamma_n, \quad G^2 \ll \gamma_n^2, \tag{8.7}$$

where relaxation is aperiodic and the effect of the strong field reduces to a change in the decay constants (see § 4). Here we can use expression (4.16) for the characteristic roots $\alpha_{1,2}$. Substituting (4.16) and (8.1) in (7.3) and performing the integration, we obtain (for excitation of the upper level)

$$w_{\mu m} = \frac{1}{4\pi^2}\frac{\gamma_{mn}}{\gamma_m}\left\{\frac{1}{1 + \varkappa\dfrac{\gamma_n^2}{\gamma_n^2 + \Omega^2}}\frac{\gamma_n}{\Omega_\mu^2 + \gamma_n^2} - \varkappa\frac{\gamma_n}{\Omega^2 + \gamma_n^2}\frac{(2\gamma_m)^2}{\eta^2 + (\Omega_\mu - \Omega)^2}\left[\frac{\gamma_n^2}{\Omega^2 + \gamma_n^2}\left(1 - \frac{\Omega(\Omega_\mu - \Omega)}{\eta\gamma_n}\right) - \frac{1}{2}\right]\right\}; \tag{8.8}$$

$$\varkappa = \frac{G^2}{\gamma_m\gamma_n}, \quad \Omega = \omega - \omega_{mn}, \quad \Omega_\mu = \omega_\mu - \omega_{mn}, \quad \eta = 2\gamma_m\left[1 + \varkappa\frac{\gamma_n^2}{\Omega^2 + \gamma_n^2}\right]. \tag{8.9}$$

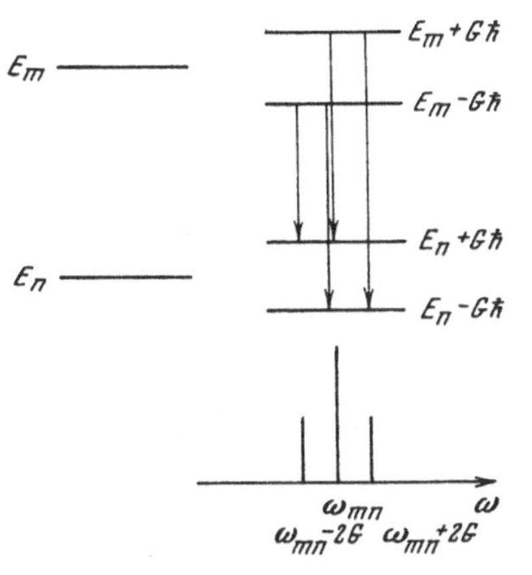

Fig. 6. Scheme of spontaneous transitions between the sublevels of states m, n.

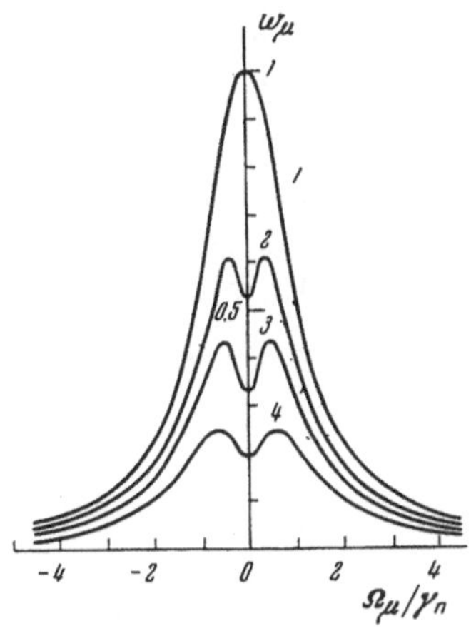

Fig. 7. Frequency dependence of spectral density w_μ (in units $\gamma_{mn}/4\pi^2\gamma_m\Gamma$; $\omega = \omega_{mn}$, $\gamma_m = 0.1\ \gamma_n$. 1) $G = 0$; 2) $G^2/\gamma^2 = 0.04$; 3) $G^2/\gamma_n^2 = 0.08$; 4) $G^2 = (\gamma_n - \gamma_m)^2/4$.

The first term in braces in (8.8) describes the usual line shape with width γ_n [in view of (8.7) $\gamma_m + \gamma_n$ is replaced by γ_n] and the value of this term is reduced by a factor

$$1\Big/\left[1+\varkappa\frac{\gamma_n^2}{\Omega^2+\gamma_n^2}\right]$$

owing to the decrease in the population of the upper level due to induced transitions. The second term in the braces reflects the change in the time course of relaxation in the presence of a strong field. Its width as a function of Ω_μ is determined by η, which, in view of (8.7), is much less than the width of the line as a whole. Hence, near the frequency $\Omega_\mu = \Omega$ there appears a relatively distinct structure, asymmetric when $\Omega \neq 0$. If $\Omega = 0$, then

$$w_{\mu m}=\frac{1}{4\pi^2}\frac{\gamma_{mn}}{\gamma_m}\left\{\frac{1}{1+\varkappa}\frac{\gamma_n}{\Omega_\mu^2+\gamma_n^2}-\frac{\varkappa}{2\gamma_n}\frac{(2\gamma_m)^2}{\Omega_\mu^2+\eta^2}\right\}, \tag{8.10}$$

i.e., in the center of the line of width γ_n there is a narrow dip, the width of which is η, and the relative depth of which depends on $\varkappa = G^2/\gamma_m\gamma_n$ and reaches ½ when $\varkappa \gg 1$. The results of numerical calculations for the case (8.7) are given in Fig. 7. A comparison of Figs. 7 and 5 shows that the shape of the spontaneous emission line is sensitive to the decay curve of the excited states.

We considered above two cases for which the expressions for $w_\mu(\Omega_\mu)$ are relatively simple. The calculations can be completed without imposing any restrictions on γ_m, γ_n, G, and ω. However, the expressions obtained are awkward and difficult to visualize. It is clear from expression (7.26) that in the general case w_μ is expressed by a combination of four resonance factors

$$\begin{gathered}\frac{1}{i(\Omega_\mu-\Omega)-2\alpha_1'},\quad \frac{1}{i(\Omega_\mu-\Omega)-2\alpha_2'},\\ \frac{1}{i(\Omega_\mu-\Omega-\alpha_1''+\alpha_2'')-(\alpha_1'+\alpha_2')},\quad \frac{1}{i(\Omega_\mu-\Omega+\alpha_1''-\alpha_2'')-(\alpha_1'+\alpha_2')}.\end{gathered} \tag{8.11}$$

The appearance of these factors can be attributed to transitions between the sublevels which are formed in interaction of the atom and the field and which, according to (4.30), are:

$$\begin{gathered}E_{1m}=E_m-\alpha_1''\hbar,\quad E_{2m}=E_m-\alpha_2''\hbar,\\ E_{1n}=E_n-(\alpha_1''+\Omega)\hbar,\quad E_{2n}=E_n-(\alpha_2''+\Omega)\hbar.\end{gathered} \tag{8.12}$$

The real parts of α_1, α_2 determine the decay of the sublevels and the width of the factors (8.11). In the case of resonance, i.e., $\Omega = 0$, two factors in (8.11) are the same and the line is symmetric. When $\Omega \neq 0$ all four factors of (8.11) are different and the line is symmetric. The relative role of a particular factor depends significantly on the relationships between γ_m, γ_n, G, and Ω.

§ 9. Spontaneous Emission of Moving Atoms

1. When motion is taken into account the coordinate **r** of the atom, which is contained in the expression (7.1) for the perturbation V_μ, must be regarded as a function of time. In the absence of collisions

$$V_\mu = G_\mu e^{-i(\Omega_\mu - \mathbf{k}_\mu \mathbf{v})t}, \quad \mathbf{r} = \mathbf{r}_0 + \mathbf{v}(t - t_0). \tag{9.1}$$

Below we will be interested in the spontaneous emission in an arbitrary direction $\mathbf{k}_\mu$. We will consider only those cases where the strong field consists of a traveling or standing monochromatic wave. Hence, on averaging over the velocities it is convenient to use the projection **v** on the wave vector **k** of the strong field and a projection perpendicular to **k**. We will assume that at the instant of excitation there is a Maxwellian velocity distribution:

$$W_M(v, u) = \frac{1}{\pi \bar{v}^2} e^{-(v^2+u^2)/\bar{v}^2}, \tag{9.2}$$

where v and u are the indicated projections of **v**.

It is clear from the general considerations expounded in § 7 and 5 that in the spectrum of the spontaneous emission of moving atoms we can expect the appearance of two factors connected with the external field – a change in the law of relaxation of the excited states and a change in the velocity distribution of the atoms. Below we will analyze some special cases in which these factors lead to interesting effects.

2. We consider first the case where the strong field is a traveling wave. We assume that inequalities (8.7) are fulfilled. Then the spectral density of spontaneous emission will obviously be given by formula (8.8) if Ω and Ω_μ in it are replaced, respectively, by

$$\Omega - \mathbf{kv} = \Omega - kv, \quad \Omega_\mu - \mathbf{k}_\mu \mathbf{v} = \Omega_\mu - k_\mu (v \cos\theta + u \sin\theta), \tag{9.3}$$

where θ is the angle between **k** and $\mathbf{k}_\mu$. Hence,

$$w_{\mu m} = \frac{1}{4\pi^2} \frac{\gamma_{mn}}{\gamma_m \gamma_n} \frac{1}{1 + \varkappa \dfrac{\gamma_n^2}{(\Omega - \mathbf{kv})^2 + \gamma_n^2}} \frac{\gamma_n^2}{(\Omega_\mu - \mathbf{k}_\mu v)^2 + \gamma_n^2} -$$

$$- \frac{\varkappa \gamma_n^2}{(\Omega - \mathbf{kv})^2 + \gamma_n^2} \frac{(2\gamma_m)^2}{(\Omega_\mu - \Omega + \mathbf{kv} - \mathbf{k}_\mu \mathbf{v})^2 + \eta^2} \left[\frac{\gamma_n^2}{(\Omega - \mathbf{kv})^2 + \gamma_n^2} \left(1 - \frac{(\Omega - \mathbf{kv})(\Omega_\mu - \Omega + \mathbf{kv} - \mathbf{k}_\mu \mathbf{v})}{\gamma_n \eta} \right) - \frac{1}{2} \right] \Bigg\}, \tag{9.4}$$

$$\eta = 2\gamma_m \left[1 + \varkappa \frac{\gamma_n^2}{(\Omega - \mathbf{kv})^2 + \gamma_n^2} \right].$$

We will average expression (9.4) over the velocities on the assumption that $\gamma_n \ll k\bar{v}$ and, firstly, for the cases $\theta = \pi/2$ and $\theta = 0$. It is easy to show that when $\theta = \pi/2$ the ratio of the second line in (9.4) to the first after averaging is a value of the order G^2/γ_n^2, i.e., we can consider only the first term in the braces in (9.4):

$$\langle w_{\mu m} \rangle = \frac{1}{4\pi^2} \frac{\gamma_{mn}}{\gamma_m} \left\langle \frac{1}{1 + \varkappa \dfrac{\gamma_n^2}{(\Omega - kv)^2 + \gamma_n^2}} \right\rangle \left\langle \frac{\gamma_n}{(\Omega_\mu - k_\mu v)^2 + \gamma_n^2} \right\rangle. \tag{9.5}$$

Thus, in the considered approximation the spontaneous emission line has the usual form [second factor in (9.5)] and the external field alters only the integral intensity of the line. The latter is connected with the induced emission probability $\langle W_m \rangle$

$$\int_{-\infty}^{\infty} \langle w_{\mu m}(\Omega_\mu)\rangle\, d\Omega_\mu = \frac{1}{4\pi}\frac{\gamma_{mn}}{\gamma_m}[1-\langle W_m\rangle], \tag{9.6}$$

where $\langle W_m \rangle$ is given by expression (5.3).

A more interesting situation arises when $\theta = 0$. Putting $k_\mu = (\omega_\mu/\omega)k \approx k$, we obtain from (9.4)

$$\langle w_{\mu m}\rangle = \frac{1}{4\pi^2}\frac{\gamma_{mn}}{\gamma_m}\int_{-\infty}^{\infty}\frac{1}{\sqrt{\pi}\bar{v}}\frac{e^{-v^2/\bar{v}^2}\,dv}{1+\varkappa\dfrac{\gamma_n^2}{(\Omega-kv)^2+\gamma_n^2}}\left\{\frac{\gamma_n}{(\Omega_\mu-kv)^2+\gamma_n^2}-\right.$$

$$\left.-\frac{\varkappa\gamma_n}{(\Omega-kv)^2+\gamma_n^2}\frac{2\gamma_m\eta}{(\Omega_\mu-\Omega)^2+\eta^2}\left[\frac{\gamma_n^2}{(\Omega-kv)^2+\gamma_n^2}\left(1-\frac{(\Omega-kv)(\Omega_\mu-\Omega)}{\eta\gamma_n}\right)-\frac{1}{2}\right]\right\}. \tag{9.7}$$

The expression in the braces in (9.7) can be regarded as the spontaneous emission of atoms with velocity v, and the factor

$$\frac{e^{-v^2/\bar{v}^2}}{\sqrt{\pi}\bar{v}}\Bigg/\left[1+\varkappa\frac{\gamma_n^2}{(\Omega-kv)^2+\gamma_n^2}\right]$$

as the velocity distribution of the atoms altered by the induced transitions [see formula (5.13)].

We consider more fully the case where we have the inequality

$$\gamma_n\sqrt{1+\varkappa} \ll k\bar{v}. \tag{9.8}$$

When this is true, the factor $e^{-v^2/\bar{v}^2}$ changes much more slowly than the rest of the integrand in (9.7), and $\langle w_{\mu m}\rangle$ can be put in the form

$$\langle w_{\mu m}\rangle = \frac{1}{4\pi}\frac{\gamma_{mn}}{\gamma_m}\frac{1}{\sqrt{\pi}k\bar{v}}e^{-\Omega_\mu^2/(k\bar{v})^2}\left\{1-\frac{\varkappa}{1+\varkappa+\sqrt{1+\varkappa}}\frac{\Gamma_1^2}{\Gamma_1^2+(\Omega_\mu-\Omega)^2}+\right.$$

$$\left.+\frac{\varkappa}{2}\operatorname{Re}\left[\frac{1}{1+i\xi}\frac{\sqrt{1+\varkappa}\sqrt{1+\varkappa/(1+i\xi)}-1}{\sqrt{1+\varkappa}\sqrt{1+\varkappa/(1+i\xi)}\,[\sqrt{1+\varkappa}+\sqrt{1+\varkappa/(1+i\xi)}]}\right]\right\} \tag{9.9}$$

$$\Gamma_1 = \gamma_n[1+\sqrt{1+\varkappa}], \quad \xi = \frac{\Omega_\mu-\Omega}{2\gamma_m}.$$

Figure 8 shows graphs calculated from formula (9.9). Their characteristic features are as follows. Near $\Omega_\mu = \Omega$ there is a relatively broad dip in the Doppler line (width of dip Γ_1), and in the dip there is a narrow peak with width of the order $2\gamma_m$. The dip of width Γ_1 is described by the second term in the braces in (9.9). It is due to the change in the distribution of the atoms as regards projection of the velocity on the wave vector $\boldsymbol{k}$. The sharp "peak" in the center is connected with the change in the velocity distribution and with the law of decay of the excited states of the atom. The shape of the "peak" is expressed in a rather complex way, but its main parameters can easily be determined from (9.9). The value of the corresponding term in (9.9) when $\xi = 0$ is

$$\frac{\varkappa^2}{4(1+\varkappa)^{3/2}}. \tag{9.10}$$

For $\varkappa \ll 1$ the height of the "peak," proportional to $\varkappa^2$, is much less than the depth of the dip, which is $\varkappa/2$. When $\varkappa \gg 1$, however, expression (9.10) increases as $\sqrt{\varkappa}/4$, and when $\varkappa \gtrsim 16$ the height of the "peak" is greater than the depth of the dip.

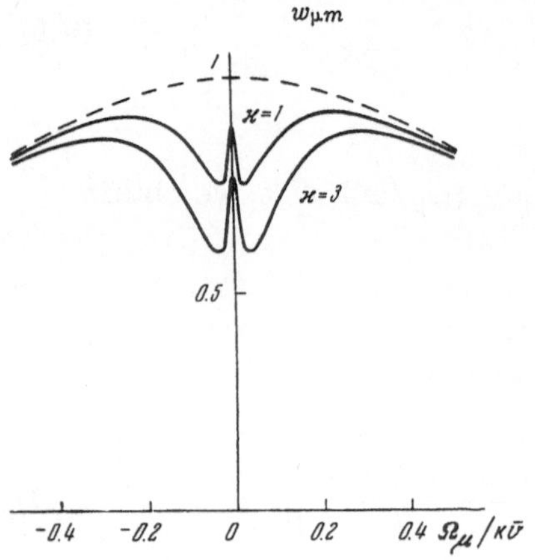

Fig. 8. Spectral density $\langle w_{\mu m}\rangle$ as a function of frequency (in units of $\gamma_{mn}/4\pi^{3/2}\gamma_m k\bar{v}$). $\omega = \omega_{mn}$, $\gamma_m = 0.2\gamma_n$, $\gamma_m = 1/30\, k\bar{v}$.

The contribution of the "sharp peak" to the integral emission probability can be determined directly from formula (9.4) by integrating the corresponding term with respect to frequency and averaging over the velocity:

$$\frac{1}{4\pi}\frac{\gamma_{mn}}{\gamma_m}\frac{1}{\sqrt{\pi}k\bar{v}}e^{-\Omega_\mu^2/(k\bar{v})^2}\frac{\varkappa^2}{4\sqrt{1+\varkappa}\,[1+\sqrt{1+\varkappa}]^2}\gamma_m.$$

We note that when $\varkappa \gg 1$ this quantity increases as $\sqrt{\varkappa} \propto G$. The ratio of the areas of peak and dip is

$$\frac{\gamma_m}{4\gamma_n}\frac{\varkappa}{[1+\sqrt{1+\varkappa}]^2} \ll 1. \tag{9.11}$$

A fairly complete representation of the width of the "peak" is given by the ratio of its integral intensity to the intensity at the maximum:

$$\pi\gamma_m \frac{4}{\left[1+\frac{1}{\sqrt{1+\varkappa}}\right]^2}. \tag{9.12}$$

It is clear from (9.12) that the width of the sharp peak is of the order of γ_m and increases by a factor of four when $\varkappa$ changes from zero to infinity.

3. Similar results are given by a calculation of the spontaneous emission in the case of a standing wave, if conditions (8.7) are fulfilled. Here we can set out from formulas (4.26) for the solution of the system of equations (3.5). In this case we must put $V_{mn}(t)$ in (4.26) in the form of a sum of matrix elements which determine the contribution of the "strong" and "weak" fields. Elementary, but cumbersome, calculations lead to the following expression for $\langle w_{\mu m}\rangle$:

$$\langle w_{\mu m}\rangle = \frac{1}{4\pi^2}\frac{\gamma_{mn}}{\gamma_m}\Big\langle\frac{2\gamma_m}{\eta}\frac{\gamma_n}{(\Omega_\mu-\mathbf{k}_\mu\mathbf{v})^2+\gamma_n^2} - \frac{\varkappa}{4}\frac{\gamma_n}{(\Omega-\mathbf{kv})^2+\gamma_n^2}\times$$

$$\times\frac{(2\gamma_m)^2}{(\Omega_\mu-\Omega-\mathbf{k}_\mu\mathbf{v}+\mathbf{kv})^2+\eta^2}\left[\frac{\gamma_n^2}{(\Omega-\mathbf{kv})^2+\gamma_n^2}\left(1-\frac{(\Omega-\mathbf{kv})(\Omega_\mu-\Omega-\mathbf{k}_\mu\mathbf{v}+\mathbf{kv})}{\eta\gamma_n}\right)-\frac{1}{2}\right]\Big\rangle; \tag{9.13}$$

$$\eta = 2\gamma_m\left\{1+\frac{\varkappa}{4}\left[\frac{\gamma_n^2}{(\Omega-\mathbf{kv})^2+\gamma_n^2}+\frac{\gamma_n^2}{(\Omega+\mathbf{kv})^2+\gamma_n^2}\right]\right\}.$$

By comparing formulas (9.13) and (9.4) we can easily derive the following conclusion. If $\Omega \gg \gamma_n$, then in the expression for η in (9.13) we can keep only one of the two terms. Then (9.13) corresponds with (9.4) with $\varkappa$ replaced by $\varkappa/4$. If $\Omega = 0$, however, $\varkappa$ must be replaced by $\varkappa/2$ and, in addition, the second term in expression (9.13) for $\langle w_{\mu m}\rangle$ must be halved in comparison with the corresponding term in (9.4). Thus, all the qualitative conclusions which can be derived from an analysis of (9.4), are still valid in the case of a standing wave too. Only the height of the narrow "peak" close to frequency $\Omega_\mu = \Omega$ is altered.

4. We consider now the spontaneous emission of atoms moving in a strong standing wave in the case where

$$\gamma_m = \gamma_n = \frac{1}{2}\Gamma, \quad \omega = \omega_{mn}. \tag{9.14}$$

In this case the fundamental matrix $\hat{S}$ of equation (3.5) is given by formula (4.24) and substitution of the corresponding values in (7.19) leads to the following expression for the spontaneous emission probability

$$w_\mu = \frac{\gamma_{mn}}{8\pi^2}\int_0^\infty dt\int_t^\infty dt' e^{-\Gamma t}\{1+\cos 2[f(t)-f(t')] \pm \cos 2f(t) \pm \cos 2f(t')\}\cos(\Omega_\mu - \mathbf{k}_\mu\mathbf{v})(t'-t); \qquad (9.15)$$

$$f(t) = \frac{G}{\mathbf{kv}}[\sin(\mathbf{kv}t+\mu) - \sin\mu]; \quad \mu = \mathbf{kr}+\delta. \qquad (9.16)$$

The plus sign in (9.15) corresponds to excitation of the upper level and the minus sign to excitation of the lower level. After averaging over the coordinates (i.e., over μ) formula (9.15) takes the form

$$w_\mu = \frac{\gamma_{mn}}{8\pi^2}\int_0^\infty dt\int_t^\infty dt' e^{-\Gamma t}\left\{1 + J_0\left[\frac{4G}{\mathbf{kv}}\sin\frac{\mathbf{kv}(t-t')}{2}\right] \pm \right.$$
$$\left. \pm J_0\left[\frac{4G}{\mathbf{kv}}\sin\frac{\mathbf{kv}t}{2}\right] \pm J_0\left[\frac{4G}{\mathbf{kv}}\sin\frac{\mathbf{kv}t'}{2}\right]\right\}\cos(\Omega_\mu - \mathbf{k}_\mu\mathbf{v})(t-t'). \qquad (9.17)$$

For further analysis it is convenient to expand the Bessel functions in Fourier series. After this, integration with respect to t, t' is carried out in explicit form:

$$\langle w_\mu\rangle = \frac{1}{4\pi}\frac{2\gamma_{mn}}{\Gamma}\frac{1}{4}\left\langle \frac{\Gamma/\pi}{(\Omega_\mu - \mathbf{k}_\mu\mathbf{v})^2+\Gamma^2} + \sum_{-\infty}^{\infty} J_m^2\left(\frac{2G}{\mathbf{k}v}\right)\frac{\Gamma/\pi}{(\Omega_\mu - \mathbf{k}_\mu\mathbf{v} - m\mathbf{k}\mathbf{v})^2+\Gamma^2} \pm \right.$$
$$\left. \pm \sum_{-\infty}^{\infty} J_m^2\left(\frac{2G}{\mathbf{kv}}\right)\frac{\Gamma/\pi}{(m\mathbf{kv})^2+\Gamma^2}\left[\frac{\Gamma^2}{(\Omega_\mu - \mathbf{k}_\mu\mathbf{v})^2+\Gamma^2} + \frac{m\mathbf{kv}(\Omega_\mu - \mathbf{k}_\mu\mathbf{v} - m\mathbf{kv})+\Gamma^2}{(\Omega_\mu - \mathbf{k}_\mu\mathbf{v} - m\mathbf{kv})^2+\Gamma^2}\right]\right\rangle. \qquad (9.18)$$

It is clear from (9.18) that the spontaneous emission line consists of components of dispersion form with width Γ. The individual components are equidistant at the distance of the Doppler shift **kv**. In accordance with § 7, this line structure can be interpreted as a consequence of transitions between systems of the sublevels of the atom formed by its motion in a strong standing-wave field. In fact, in the considered case the wave function of the atom can be put in the form [see (4.34)]

$$\Psi = \psi_m \sum_s A_s e^{-i\left[\frac{E_m}{\hbar} - s\mathbf{kv}\right]t - \frac{\Gamma}{2}t} + \psi_n \sum_{s'} B_{s'} e^{-i\left[\frac{E_n}{\hbar} - s'\mathbf{kv}\right]t - \frac{\Gamma}{2}t}.$$

Hence, the atom is described by a wave function of the same type as the system with quasi-stationary states m, s; n, s'. The energies of these states, $E_m - s\hbar\mathbf{kv}$, $E_n - s'\hbar\mathbf{kv}$ form two systems of equidistant sublevels with splitting **kv**. The amplitudes A_s, $B_{s'}$ depend on the field, velocity, and initial conditions. It is easy to show that $|A_s|$, $|B_{s'}|$ differ significantly from zero for $|s|, |s'| < G/|\mathbf{kv}|$, and $|A_s|$ and $|B_{s'}|$ attain their maximum values close to the boundaries of this region (Fig. 9). Each term in (9.18) corresponds to a transition between the sublevels $E_m - s\hbar\mathbf{kv}$, $E_n - s'\hbar\mathbf{kv}$; $m = s' - s$. The width of the components is determined by the decay of the states m, s; n, s'. The series of interference terms in the last line of formula (9.18) is obviously due to the fact that the states m, s; n, s' are not independent.

When $G \to 0$, all $J_m \to 0$, except the Bessel functions of zero order, and from (9.18) we obtain the usual expression for the spectral density of spontaneous emission

$$\langle w_{\mu m}\rangle = \frac{1}{4\pi}\frac{2\gamma_{mn}}{\Gamma}\left\langle\frac{\Gamma/\pi}{(\Omega_\mu \mathbf{k}_\mu\mathbf{v})^2+\Gamma^2}\right\rangle, \quad \langle w_{\mu n}\rangle = 0.$$

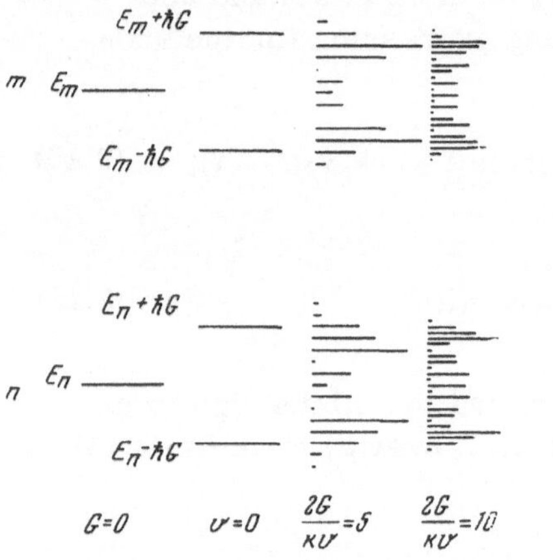

Fig. 9. Scheme of sublevels s, s' of atom moving in the field of a plane monochromatic standing wave.

With increase in the field function $J_0^2(2G/kv)$ decreases, and functions $J_m^2(2G/\mathbf{kv})$, $m \neq 0$, become different from zero. We consider first of all how the integral spontaneous emission probability varies with G. Integrating (9.18) with respect to Ω_μ and extracting the terms with m = 0, we can find

$$\int_{-\infty}^{\infty} \langle w_\mu \rangle \, d\Omega_\mu = \frac{1}{4\pi} \frac{2\gamma_{mn}}{\Gamma} \frac{1}{4} \Big\langle 1 + J_0^2 + \sum_{m\neq 0} J_m^2 \pm 2J_0^2 \pm 2 \sum_{m\neq 0} J_m^2 \frac{\Gamma^2}{(mkv)^2+\Gamma^2} \Big\rangle. \qquad (9.19)$$

In view of the identity

$$J_0^2 + \sum_{m\neq 0} J_m^2 = 1,$$

and also formula (5.30) for the induced emission probability $\langle \overline{W}_m \rangle$, we can write (9.19) in the form

$$\int_{-\infty}^{\infty} \langle w_{\mu m} \rangle \, d\Omega_\mu = \frac{1}{4\pi} \frac{2\gamma_{mn}}{\Gamma} \langle 1 - \overline{W}_m \rangle, \quad \int_{-\infty}^{\infty} \langle w_{\mu n} \rangle \, d\Omega_\mu = \frac{1}{4\pi} \frac{2\gamma_{mn}}{\Gamma} \langle \overline{W}_m \rangle. \qquad (9.20)$$

The physical sense of these formulas is obvious. On excitation of the upper level m the integral spontaneous emission probability decreases by $\langle \overline{W}_m \rangle$; in the case of excitation of the lower level n spontaneous emission is due entirely to absorption of photons of the external field (absorption probability $W_n = W_m$). In the case of equally probable excitation of both levels the integral intensity of spontaneous emission does not depend on G at all:

$$\int_{-\infty}^{\infty} \langle w_{\mu m} + w_{\mu n} \rangle \, d\Omega_\mu = \frac{1}{4\pi} \frac{2\gamma_{mn}}{\Gamma}.$$

This case is of most interest from the viewpoint of the question in which we are interested – how the law of decay of the excited states affects the spectral composition of spontaneous emission. It was noted at the start of this section that $\langle w_\mu \rangle$ will vary because the velocity distributions of the mean populations also vary under the influence of the external field. In the case of excitation of the levels m, n with equal velocities ($Q_m = Q_n$) the velocity distribution of the atoms is independent of the field. For instance,

$$N_m(\mathbf{v}) = N_{mm}(\mathbf{v}) + N_{mn}(\mathbf{v}) = \Big\{ Q_m \int_{t_0}^{\infty} |S_{mm}|^2 dt + Q_n \int_{t_0}^{\infty} |S_{mn}|^2 dt \Big\} W_M(\mathbf{v}).$$

Using expressions (4.24) with $Q_m = Q_n$ we find

$$N_m(\mathbf{v}) = \frac{Q_m}{\Gamma} W_M(\mathbf{v}). \qquad (9.21)$$

There will be an exactly similar expression for $N_n(v)$. When $Q_m = Q_n$ it follows from (9.18) that

$$\langle w_\mu \rangle = \frac{1}{4\pi} \frac{2\gamma_{mn}}{\Gamma} \frac{1}{2} \Big\langle \frac{\frac{\Gamma}{\pi}\left[1 + J_0^2\left(\frac{2G}{\mathbf{kv}}\right)\right]}{(\Omega_\mu - \mathbf{k}_\mu \mathbf{v})^2 + \Gamma^2} + \sum_{m\neq 0} \frac{\frac{\Gamma}{\pi} J_m^2\left(\frac{2G}{\mathbf{kv}}\right)}{(\Omega_\mu - \mathbf{k}_\mu \mathbf{v} - m\mathbf{kv})^2 + \Gamma^2} \Big\rangle. \qquad (9.22)$$

We consider first the expression (9.22) for the direction $\mathbf{k}_\mu$, perpendicular to $\mathbf{k}$. Averaging over $u = \mathbf{k}_\mu \mathbf{v}/k$ gives

$$\langle w_\mu \rangle = \frac{1}{4\pi} \frac{2\gamma_{mn}}{\Gamma} \frac{1}{\sqrt{\pi} k\bar{v}} \frac{1}{2} \left\langle [1 + J_0^2]\, e^{-\frac{\Omega_\mu^2}{(k\bar{v})^2}} + \sum_{m \neq 0} J_m^2 e^{-\frac{(\Omega_\mu - mkv)^2}{(k\bar{v})^2}} \right\rangle . \qquad (9.23)$$

The first term in (9.23) gives a line of Gaussian form with width $k\bar{v}$ and maximum at frequency $\Omega_\mu = 0$. This part has the same form as the spontaneous emission line of the isolated atom. The only difference is that in this case the line intensity depends on G and is halved when G changes from zero to $G \gg k\bar{v}$.

The terms of the series in (9.23) depend on the frequency, generally speaking, in a different way. We write the m-th term of the series in the following form:

$$\left\langle J_m^2 e^{-\left(\frac{\Omega_\mu - mkv}{k\bar{v}}\right)^2} \right\rangle = \frac{1}{\sqrt{\pi} k\bar{v}} \int_{-\infty}^{\infty} J_m^2 \left(\frac{2G}{z}\right) e^{-[z^2 + (\Omega_\mu - mz)^2]/(k\bar{v})^2} dz =$$

$$= \frac{1}{\sqrt{\pi} k\bar{v}} e^{-\frac{\Omega_\mu^2}{[(1+m^2)k\bar{v}]^2}} \int_{-\infty}^{\infty} J_m^2 \left(\frac{2G}{z}\right) e^{-\left[z - \frac{m}{1+m^2}\Omega_\mu\right]^2 (1+m^2)/(k\bar{v})^2} dz. \qquad (9.24)$$

The integral in (9.24) has its greatest value when the maxima of the exponent and the Bessel function coincide. It is known that $J_m^2(x)$ attains its greatest value when the argument x is approximately equal to the order, i.e., when $z = 2G/m$. Hence, this coincidence will occur for $|\Omega_\mu| \approx [1 + (1/m^2)]2G$. Since this condition is practically independent of m for $|m| \geq 1$, the integrals in all the terms of the series in (9.23) will have their greatest values in the frequency region $\Omega_\mu \approx \pm 2G$.

Thus, the spectral distribution of the intensity of spontaneous emission, given by expression (9.23), is made up of three lines with maximum at $\omega_\mu \approx \omega_{mn} - 2G$, $\omega_\mu = \omega_{mn}$, and $\omega_\mu \approx \omega_{mn} + 2G$. For the ratio of the integral intensities of the shifted and unshifted parts of the profile it is easy to obtain: $[1 - \langle J_0^2 \rangle]/[1 + \langle J_0^2 \rangle]$. In a weak field $\langle J_0^2 \rangle \to 1$, i.e., all the energy is concentrated in the unshifted part of the line. In the limiting case $G \gg k\bar{v}$ we have $\langle J_0^2 \rangle \to 0$ and the intensity ratio is close to unity.

We consider now the case where $\mathbf{k}_\mu$ and $\mathbf{k}$ are parallel. Here formula (9.22) has the form

$$\langle w_\mu \rangle = \frac{1}{4\pi} \frac{2\gamma_{mn}}{\Gamma} \frac{1}{2} \left\langle \frac{\Gamma/\pi}{\Gamma^2 + (\Omega_\mu - kv)^2} \left[1 + J_0^2 \left(\frac{2G}{kv}\right)^2\right] + \sum_{m \neq 0} \frac{\Gamma/\pi}{\Gamma^2 + [\Omega_\mu - (1+m)kv]^2} J_m^2 \left(\frac{2G}{kv}\right) \right\rangle . \qquad (9.25)$$

The most remarkable feature of this distribution is the presence of a narrow line with radiation width Γ and a maximum at $\omega_\mu = \omega_{mn}$. This line corresponds to the term $m = -1$

$$\frac{1}{4\pi} \frac{2\gamma_{mn}}{\Gamma} \frac{1}{2} \left\langle J_1^2 \left(\frac{2G}{kv}\right) \right\rangle \frac{\Gamma/\pi}{\Omega_\mu^2 + \Gamma^2} \qquad (9.26)$$

and owes its origin to compensation of the Doppler shifts $\mathbf{kv}$ and $\mathbf{k}_\mu \mathbf{v}$ for transitions $s - s' = 1$ between the sublevels of the atom (see Fig. 9).

The quantity $\frac{1}{2}\langle J_1^2 \rangle$, which gives the integral intensity of the narrow line, becomes zero when $G \to 0$; it is easy to calculate its values in limiting cases:

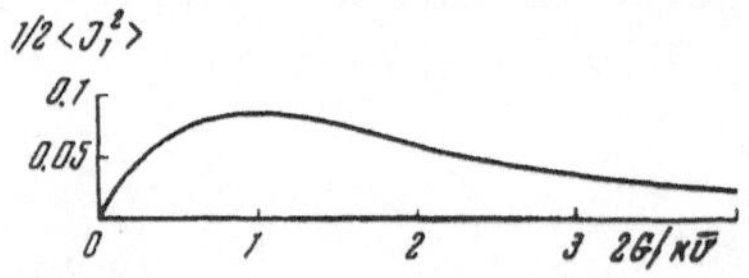

Fig. 10. Amplitude of "narrow line" as a function of $2G^2/k\bar{v}$.

$$\frac{1}{2}\langle J_1^2\rangle \approx \frac{4}{3\pi^{3/2}}\frac{2G}{k\bar{v}},\quad \frac{2G}{k\bar{v}}\ll 1,$$
$$\frac{1}{2}\langle J_1^2\rangle \approx \frac{1}{\pi^{3/2}}\frac{k\bar{v}}{2G},\quad \frac{2G}{k\bar{v}}\gg 1. \tag{9.27}$$

Figure 10 shows the general behavior of $\frac{1}{2}\langle J_1^2\rangle$ as a function of $2G/k\bar{v}$. The maximum value of $\frac{1}{2}\langle J_1^2\rangle$, attained in the region of $2G \sim k\bar{v}$, is 0.085, i.e., the narrow line may contain 8.5% of the integral intensity of the whole line. The ratio of the "peak" intensities will be of the order $0.085\,k\bar{v}/\Gamma$, i.e., may be very large, since $k\bar{v} \gg \Gamma$.

We write formula (9.25) in the following way:

$$\langle w_\mu\rangle = \frac{1}{4\pi}\frac{2\gamma_{mn}}{\Gamma}\frac{1}{2}\left\{\frac{1}{\sqrt{\pi}k\bar{v}}e^{-\Omega_\mu^2/(k\bar{v})^2} + \frac{\Gamma/\pi}{\Gamma^2+\Omega_\mu^2}\langle J_1^2\rangle + \sum_{\substack{-\infty\\ m\neq -1}}^{\infty}\left\langle\frac{\Gamma/\pi}{[\Omega_\mu-(1+m)kv]^2+\Gamma^2}J_m^2\left(\frac{2G}{kv}\right)\right\rangle\right\}. \tag{9.28}$$

The first line here contains the terms for which the dependence on Ω_μ is clear. We calculate the series in the second line for a case of practical interest

$$\Gamma \ll G \ll k\bar{v}. \tag{9.29}$$

At the point $\Omega_\mu = 0$ we have the following estimate:

$$\frac{\Gamma}{\pi}\sum_{m\neq -1}\left\langle\frac{J_m^2}{\Gamma^2+(m+1)^2(kv)^2}\right\rangle \approx \frac{1}{\sqrt{\pi}k\bar{v}}\frac{\Gamma}{G} \ll \frac{1}{\sqrt{\pi}k\bar{v}},$$

which can easily be found by the same method as in the case of the series in formula (5.30). Thus, close to $\Omega_\mu = 0$ the series in (9.28) can be neglected. When $\Omega_\mu \gtrsim G$ we can use the fact that close to $kv = \Omega_\mu/(m+1)$ functions $J_m^2(2G/kv)$ and $\exp\{-v^2/\bar{v}^2\}$ change much more slowly than the resonance factors. Hence, we can put

$$\sum_{\substack{-\infty\\ m\neq -1}}^{\infty}\frac{\Gamma}{\pi^{3/2}\bar{v}}\int_{-\infty}^{\infty}\frac{J_m^2\left(\frac{2G}{kv}\right)e^{-v^2/\bar{v}^2}dv}{\Gamma^2+[\Omega_\mu-(m+1)kv]^2} \approx \frac{1}{\sqrt{\pi}k\bar{v}}\sum_{m\neq -1}\frac{1}{|m+1|}e^{-\left[\frac{\Omega_\mu}{(m+1)k\bar{v}}\right]^2}J_m^2\left[\frac{2G}{\Omega_\mu}|m+1|\right]. \tag{9.30}$$

If $\Omega_\mu > 2G$, then all the J_m^2 are small, except $J_0^2 \approx 1$. Hence, at large frequencies the line will have the usual Gaussian profile. The results of numerical calculations from formulas (9.30) and (9.28) are shown in Fig. 11. Superimposed on the Doppler line close to $\Omega_\mu = 0$ (broken curve) there is a dip with a width of the order G and a narrow line (9.26). We recall that in the considered case the integral emission probability is independent of the field. Hence, the areas of the dip and the narrow line are practically equal and the "peak" intensity of the narrow line is greater than the depth of the dip by the ratio of the widths G/Γ.

Outwardly the profile of the line in Fig. 11 resembles the profile in Fig. 8 obtained in the case $\gamma_m \ll \gamma_n$. The causes leading to the distribution (9.23) and (9.9), however, are very different. In the case of (9.9) the dip is due to a change in the velocity distribution of the atoms from the Maxwellian form. In the case of (9.28) the velocity distribution (as was shown above) is independent of G [see (9.21)]. The quantitative difference between Figs. 8 and 11 is that in the case of (9.28) the area of the sharp line is approximately equal to the area of the dip, whereas in the case $\gamma_m \ll \gamma_n$ the ratio of these areas, according to (9.11), is much less than unity.

We considered above two directions of observation of the spontaneous emission – parallel and perpendicular to **k**. We will show how the sharp line (9.26) changes in the intermediate cases. If the angle between **k** and $\mathbf{k}_\mu$ is denoted by θ, the term $m = -1$ in (9.25) has the form

$$\frac{\Gamma}{\pi^2\bar{v}^2}\int\limits_{-\infty}^{\infty}\int\limits_{-\infty}^{\infty}\frac{J_1^2\left(\frac{2G}{kv}\right)e^{-(v^2+u^2)/\bar{v}^2}\,dv\,du}{[\Omega_\mu - ku\sin\theta + kv(1-\cos\theta)]^2+\Gamma^2}. \qquad (9.31)$$

If $k\bar{v}\sin\theta < \Gamma$ the resonance factor as a function of u has a greater width than $e^{-u^2/\bar{v}^2}$ and the terms which depend on θ can be discarded. Thus, in the range of angles

$$\theta < \Gamma/k\bar{v}$$

the line shape will be practically the same as when $\theta = 0$. If, $k\bar{v}\sin\theta > \Gamma$, the ratio of the widths of the resonance factor and the exponent is the reverse and the term of interest to us is

$$\frac{1}{\pi\bar{v}^2 k\sin\theta}\int\limits_{-\infty}^{\infty} J_1^2\left(\frac{2G}{kv}\right)\exp\left\{-\left[v^2+\frac{\{\Omega_\mu+kv(1-\cos\theta)\}^2}{(k\sin\theta)^2}\right]\Big/\bar{v}^2\right\}dv. \qquad (9.32)$$

If angle θ is not too large, the term $kv(1-\cos\theta) \approx kv\theta^2/2$ can be discarded, and (9.32) becomes the formula

$$\frac{1}{\sqrt{\pi}k\bar{v}\sin\theta}e^{-\Omega_\mu^2/[k\bar{v}\sin\theta]^2}\langle J_1^2\rangle. \qquad (9.33)$$

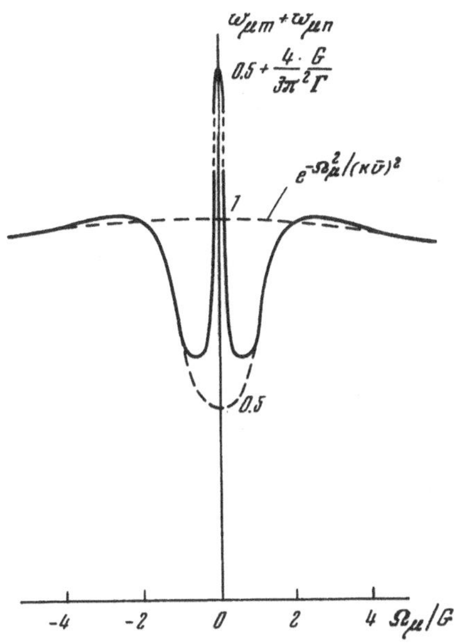

Fig. 11. Spectral density of spontaneous emission probability (in units of $\gamma_{mn}/2\pi^{3/2}\Gamma k\bar{v}$), as a function of frequency. $\gamma_m = \gamma_n = \Gamma/2$; $\omega = \omega_{mn}$.

Hence, in this case the shape of the narrow line (9.26) is altered, but its width is much less than $k\bar{v}$. If the external field is large enough, $2G > k\bar{v}$, then when $\tan(\theta/2) > k\bar{v}/2G$ the line is split into two components, the maxima of which are shifted into the region $\Omega_\mu = \pm 2G$. This can easily be shown by the same arguments as were used in the analysis of formula (9.24).

5. In conclusion we consider experimental ways of verifying the developed theory. We note first of all that so far there has not been a single experimental investigation of the spectral composition of spontaneous emission in the presence of intense induced transitions. There are several reasons for this. One of the most important is probably the relative difficulty of the experiment. The fact is that the most interesting effects occur in fairly intense fields, when $G > \gamma_m, \gamma_n$. On the other hand, in a short laser, where monochromatic generation occurs, the power of emission of most present-day lasers is low. Another complicating circumstance is the need to observe spontaneous emission in directions close to the direction of generation (the difference in the angles is not more than $\Gamma/k\bar{v} \sim 0.1$-0.01) and in this case, of course, it is difficult to get rid of scattered light. Finally, the actual measurements with a resolution of the order of $\omega_{mn}/\Gamma \sim 10^7$-$10^8$ present a considerable problem and necessitate the use of very precise and stable equipment.

Despite these difficulties, the investigation of spontaneous emission is of great interest since it provides a way of determining such quantities as the natural line widths, the lifetimes of highly excited states, the amplitudes of the fields inside the laser, and so on. This range of questions is discussed more fully in the Appendix.

CHAPTER III

Induced Emission and Absorption of a "Weak" Field in the Presence of a "Strong" Field

§10. Formulation of Problem

To describe the interaction of a quantum system with a weak field we can proceed from the system of equations (7.1), where the weak field is described by the term $\hat{V}_\mu Q$. A weak field, in contrast to a strong one, is a field which does not affect the mean populations of the levels m, n. This means that the system of equations (7.1) has to be solved in a first approximation in V_μ. However, as distinct from Chapter II, we are now interested in the difference in the values of W_μ^e and W_μ^a given by formula (7.3). This difference, as will be shown below, can alter the sign within the line width. Hence, it is more convenient to operate with an emission (or absorption) coefficient and not with the probabilities W_μ^e, W_μ^a. In the symbols adopted here the emission coefficient is written as

$$\alpha_\mu = \frac{\lambda^2}{4\pi}\gamma_{mn}\frac{Q_m(W^e_{\mu m} - W^a_{\mu m}) + Q_n(W^e_{\mu n} - W^a_{\mu n})}{|V_\mu|^2}. \tag{10.1}$$

The coefficient α_μ has the dimension of inverse length and gives the relative change in intensity of the traveling monochromatic wave in unit length. In the case of amplification α_μ is positive, and in the case of absorption it is negative. If moving atoms are considered, α_μ must be averaged over the velocities.

Calculations similar to those carried out in § 7 for W_μ^e lead to the following expression for α_μ:

$$\alpha_\mu = \frac{\lambda^2}{4\pi}\gamma_{mn}(Q_m + Q_n)\,\mathrm{Sp}\Big\{\hat{\rho}\int\limits_{t_0}^{\infty}\int\limits_{t'}^{\infty}\hat{S}^\dagger(t'')\hat{\beta}\hat{S}(t'')\hat{S}^{-1}(t')\hat{\beta}^\dagger\hat{S}(t')e^{-i\Omega_\mu(t''-t')}dt'dt'' -$$
$$- \hat{\rho}\int\limits_{t_0}^{\infty}\int\limits_{t'}^{\infty}\hat{S}^\dagger(t'')\hat{\beta}^\dagger\hat{S}(t'')\hat{S}^{-1}(t')\hat{\beta}\hat{S}(t')e^{i\Omega_\mu(t''-t')}dt'dt'' + \text{к. с.}\Big\}. \tag{10.2}$$

Integrating (10.2) with respect to Ω_μ, we easily obtain a formula for the integral emission coefficient

$$\alpha_{\mu\infty} = \int\limits_{-\infty}^{\infty}\alpha_\mu d\Omega_\mu = \frac{\lambda^2}{4}2\gamma_{mn}(N_m - N_n). \tag{10.3}$$

Thus, we still have the usual expression for $\alpha_{\mu\infty}$, the only difference being that the populations N_m, N_n depend on the strong field. Hence, when saturation occurs we can introduce integral Einstein coefficients for induced emission and absorption, and the relationship between them and the first integral Einstein coefficient is as before [compare formulas (10.3) and (7.24)]. In linear theory, where all fields are so weak that there is no saturation, we can introduce the spectral densities of the first and second Einstein coefficients, and the relationships for them are the same as those for the integral quantities [57]. When nonlinear effects are taken into account the position is considerably different. As already mentioned, the coefficient α_μ can change sign when the frequency changes within the order of the line width (see, for instance, Fig. 12). In other words, α_μ as a function of the frequency is by no means proportional to the density of spontaneous emission and the concept of spectral Einstein coefficients ceases to make sense.

Formula (10.2) relates to the case where the weak field is monochromatic. In specific physical conditions this is usually not the case – the spectrum of the weak field is continuous or consists of several monochromatic components. It was shown in [52, 68, 71] that in the presence of a strong field weak fields cause polarization of the atom not only at its "own" frequency ω_μ, but also at the combination frequency $2\omega - \omega_\mu$, where ω is the frequency of the strong field. Hence, in the analysis of the amplification or attenuation of the Fourier component of the field with frequency ω_μ the field component with frequency $2\omega - \omega_\mu$ must be taken into account. The result, however, depends on the phase relationships between the components. If the phase difference between them is random and takes any value with equal probability, the interference effect is insignificant and the result of the interaction of the quantum system with any Fourier component of the weak field does not depend on the other components. Here we will consider only such cases.

§ 11. Forced Emission and Absorption of Weak Field by Stationary Atoms in a Strong Monochromatic Field

1. In the case indicated in the title α_μ can be calculated completely for any relationships of the parameters. The expression obtained for α_μ, however, is very cumbersome. Hence, here we will confine ourselves to an analysis of the special case where the frequency of the strong field is the same as the transition frequency. In this case it follows from (10.2) and (4.3) that

$$\alpha_\mu = \frac{\lambda^2}{4\pi} 2\gamma_{mn}\left(\frac{Q_m}{2\gamma_m} - \frac{Q_n}{2\gamma_n}\right)\frac{1}{1 + G^2/\gamma_m\gamma_n}\left\{\frac{\Gamma}{\Omega_\mu^2 + \Gamma^2} - \frac{2G^2}{\Gamma}\operatorname{Re}\left[\left(1 + \frac{\Gamma}{\Gamma + i\Omega_\mu}\right)\frac{1}{(\Gamma + i\Omega_\mu)^2 - (\gamma_n - \gamma_m)^2 + 4G^2}\right]\right\}. \quad (11.1)$$

We compare (11.1) with the expression for α_μ when $G = 0$:

$$\alpha_\mu = \frac{\lambda^2}{4\pi} 2\gamma_{mn}\left(\frac{Q_m}{2\gamma_m} - \frac{Q_n}{2\gamma_n}\right)\frac{\Gamma}{\Gamma^2 + \Omega_\mu^2}. \quad (11.2)$$

The factor $[1 + G^2/\gamma_m\gamma_n]^{-1}$ in (11.1) is obviously due to the change in the mean difference of the populations of the levels m, n owing to transitions induced by the strong field. If the effect of the strong field consisted only in alteration of the population difference, the braces in (11.1) would contain only the first term. The second term in the braces reflects the variation not only of the populations, but also of the shape of the curve of α_μ as a function of frequency. When $\omega_\mu \to \omega_{mn}$ we have

$$\alpha_\mu = \frac{\lambda^2}{4\pi}\gamma_{mn}\left(\frac{Q_m}{\gamma_m} - \frac{Q_n}{\gamma_n}\right)\frac{1}{\Gamma}\frac{1}{[1 + G^2/\gamma_m\gamma_n]^2}\,(\Omega_\mu \to 0). \quad (11.3)$$

At the same time, the gain α for a strong field, like α_μ, is equal to

$$\alpha = \frac{\lambda^2}{4\pi}\gamma_{mn}\left(\frac{Q_m}{\gamma_m} - \frac{Q_n}{\gamma_n}\right)\frac{1}{\Gamma}\frac{1}{1 + G^2/\gamma_m\gamma_n}. \quad (11.4)$$

Thus, when $\omega_\mu \to \omega = \omega_{mn}$, the coefficient α_μ is not equal to α, and α_μ is always less than α by a factor $1 + G^2/\gamma_m\gamma_n$. We note that the same discontinuity of the gain occurs when $\omega \neq \omega_{mn}$, but the factor $1 + G^2/\gamma_m\gamma_n$ in this case is replaced by

$$1 + \frac{G^2}{\gamma_m\gamma_n}\frac{\Gamma^2}{\Gamma^2 + \Omega^2}.$$

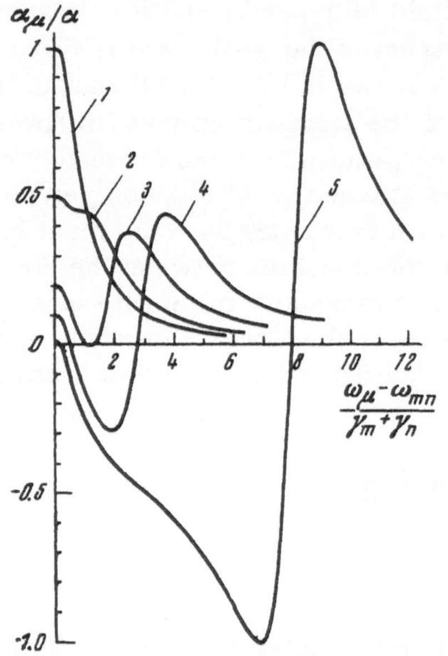

Fig. 12. Graph of gain of "weak" field $\omega = \omega_{mn}$, $\gamma_m = \gamma_n = \Gamma/2$. 1) G = 0; 2) G = $\Gamma/2$; 3) G = Γ; 4) G = $3\Gamma/2$; 5) G = 4Γ.

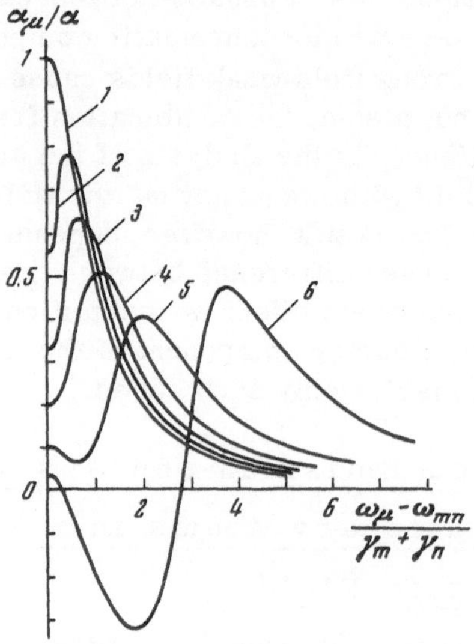

Fig. 13. Graph of gain of "weak" field; $\gamma_m = 0.1\gamma_n$, $\omega = \omega_{mn}$. 1) G = 0; 2) $G^2 = 0.8\gamma_m\gamma_n$; 3) $G^2 = 2\gamma_m\gamma_n$; 4) $G^2 = 4\gamma_m\gamma_n$; 5) $G^2 = 9\gamma_m\gamma_n$; 6) $G^2 = 29\gamma_m\gamma_n$.

The behavior of α_μ when the difference between ω_μ and ω_{mn} increases depends on the relationship between γ_m and γ_n and on the value of the parameter $G^2/\gamma_m\gamma_n$. Figures 12 and 13 show the results of calculations of the ratio α_μ/α. We consider first the case $\gamma_m = \gamma_n = \Gamma/2$ (Fig. 12). At relatively low values of G^2, $G^2 \lessapprox \gamma_m\gamma_n$, the coefficient α_μ is a monotonic function of the frequency (curve 2, Fig. 12). With further increase in G side maxima appear at frequencies $\Omega_\mu \approx \pm(2G+\Gamma)$ and between them and the frequency ω_{mn} there is a region of negative values of α_μ, i.e., at these frequencies the weak field is absorbed, and not emitted, by the medium (curves 3, 4, 5, Fig. 12). It is significant that the magnitude of the side maximum of the ratio α_μ/α increases with increase in G. When $\Omega_\mu = \pm(2G+\Gamma)$ and $G^2 \gg \Gamma^2$ it is easy to obtain from (11.1)

$$\frac{\alpha_\mu}{\alpha} \approx \frac{G}{4\Gamma}. \tag{11.5}$$

Hence, when $G > 4\Gamma$ the amplification of the weak field at frequencies $\Omega_\mu = \pm(2G+\Gamma)$ becomes greater than the amplification of the strong field (curve 5, Fig. 12).

We turn now to the case $\gamma_m \neq \gamma_n$ and, for the sake of definiteness, we let $\gamma_m < \gamma_n$. Interesting features arise in the cases $\gamma_m \ll \gamma_n$ and $G^2 < (\gamma_n - \gamma_m)^2/4$. In these conditions increasing difference from ω_{mn} first leads to an increase in α_μ (curves 2, 3, Fig. 13) and then to a decrease in α_μ. This behavior persists to some extent even when $G^2 > (\gamma_n - \gamma_m)^2/4$ (curve 4, Fig. 13). In all these cases α_μ at all frequencies is less than α. When G increases significantly, i.e., when $G^2 \gg (\gamma_n - \gamma_m)^2/4$, the function $\alpha_\mu(\Omega_\mu)$ will have approximately the same form as when $\gamma_m = \gamma_n$. This is easily seen from formula (11.1) and from a comparison of curve 6 in Fig. 13 and curve 4 in Fig. 12.

The curves in Figs. 12, 13 give a good illustration of the difference between $\alpha_\mu(\Omega_\mu)$ and the spectral densities of spontaneous emission – there is nothing in common between the graphs in Figs. 12, 13 and Figs. 8, 11. Yet it is clear from (11.1) that the contribution of the second term in the braces in (11.1) to the integral emission coefficient is zero. In fact, all the poles of the second term lie in the upper half-plane of the complex frequency Ω_μ. Hence, the integral of the second term with respect to Ω_μ is zero. Hence, the value of $\alpha_{\mu\infty}$ depends only on the first term, the integral of which is π. Thus,

$$\alpha_{\mu\infty} = \frac{\lambda^2}{4\pi}\gamma_{mn}'\left(\frac{Q_m}{\gamma_m} - \frac{Q_n}{\gamma_n}\right)\frac{1}{1 + G^2/\gamma_m\gamma_n} \tag{11.6}$$

in complete correspondence with the general formula (10.3).

In [72-74] the change in the sign of α_μ was experimentally confirmed. It is true that the experimental conditions differed a little from those assumed above, and in [72-74] the weak field consisted of several monochromatic components strictly related in phase. As was noted at the end of § 10, this is extremely important. Hence, the results of [72-74] cannot be used for a thorough verification of the theory. Nevertheless, the most important conclusion, viz., the change in the sign of α_μ, is reliably confirmed.

§ 12. Induced Emission and Absorption of Weak Field by Moving Atoms

As in the analysis of spontaneous emission, the perturbation V_μ when the motion of the atoms is taken into account must be put in the form (9.1). We will also average over the velocities and coordinates of the atoms in the same way as was done in Chapter II.

In this section we will consider the case of a great difference in the decay constants of the levels m, n and relatively small values of G:

$$\gamma_m + \frac{G^2}{\gamma_n} \ll \gamma_n. \tag{12.1}$$

The strong field is assumed to be a plane monochromatic wave. In the case of a strong field in the form of a traveling wave substitution of the fundamental matrix $\hat{S}$ from (4.3) in (10.2), using condition (12.1), leads to the following formula:

$$\langle\alpha_\mu\rangle = \frac{\lambda^2}{4\pi}\gamma_{mn}\left(\frac{Q_m}{\gamma_m} - \frac{Q_n}{\gamma_n}\right)\left\langle\frac{2\gamma_m}{\eta}\frac{\gamma_n}{(\Omega_\mu - \mathbf{k}_\mu\mathbf{v})^2 + \gamma_n^2} - \right.$$

$$\left. - \frac{\varkappa\gamma_n^3}{[(\Omega - \mathbf{kv})^2 + \gamma_n^2]^2}\frac{(2\gamma_m)^2}{(\Omega_\mu - \Omega - \mathbf{k}_\mu\mathbf{v} + \mathbf{kv})^2 + \eta^2}\left[1 - \frac{(\Omega - \mathbf{kv})(\Omega_\mu - \Omega - \mathbf{k}_\mu\mathbf{v} + \mathbf{kv})}{\eta\gamma_n}\right]\right\rangle; \tag{12.2}$$

where

$$\eta = 2\gamma_m\left[1 + \varkappa\frac{\gamma_n^2}{(\Omega - \mathbf{kv})^2 + \gamma_n^2}\right], \quad \varkappa = \frac{G^2}{\gamma_m\gamma_n}. \tag{12.3}$$

If $\mathbf{k}_\mu$ is perpendicular to $\mathbf{k}$, the second term in (12.2) is of the order of G^2/γ_n^2 and can be neglected. Hence,

$$\langle\alpha_\mu\rangle = \frac{\lambda^2}{4\pi}\gamma_{mn}\left(\frac{Q_m}{\gamma_m} - \frac{Q_n}{\gamma_n}\right)\left\langle\frac{1}{1 + \varkappa\dfrac{\gamma_n^2}{\gamma_n^2 + (\Omega - kv)^2}}\right\rangle\left\langle\frac{\gamma_n}{\gamma_n^2 + (\Omega_\mu - kv)^2}\right\rangle. \tag{12.4}$$

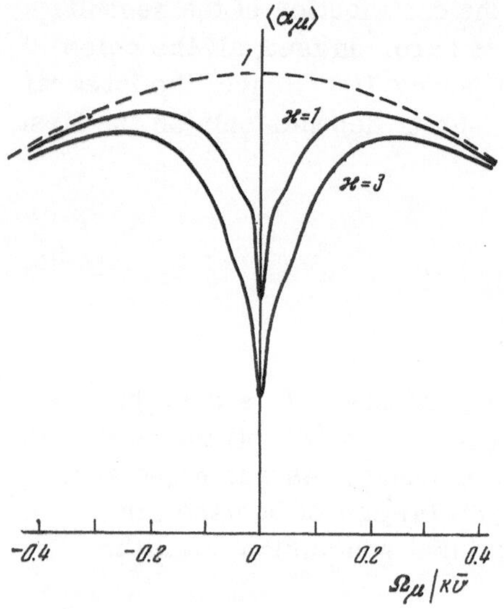

Fig. 14. Graph of gain of weak field $\langle \alpha_\mu \rangle$ in units of

$$\frac{\lambda^2}{4} \frac{\gamma_{mn}}{\sqrt{\pi} k\bar{v}} \left(\frac{Q_m}{\gamma_n} - \frac{Q_n}{\gamma_n} \right)$$

The broken curve corresponds to $\exp\{-\Omega_\mu^2/(k\bar{v})^2\}$, $\gamma_m = 0.06\gamma_n$, $\gamma_n = \frac{1}{30} k\bar{v}$.

Thus, as in the case of spontaneous emission [compare (9.5)], the dependence of $\langle \alpha_\mu \rangle$ on Ω_μ in the approximation used is the same as when $\varkappa = 0$. The external field alters only the absolute value of $\langle \alpha_\mu \rangle$.

The most interesting case for practical application is when $\mathbf{k}_\mu$ is parallel to $\mathbf{k}$. Putting $\mathbf{k}_\mu = \mathbf{k}$, we obtain from (12.2)

$$\langle \alpha_\mu \rangle = \frac{\lambda^2}{4\pi} \frac{\gamma_{mn}}{k\bar{v}} \left(\frac{Q_m}{\gamma_m} - \frac{Q_n}{\gamma_n} \right) e^{-\frac{\Omega_\mu^2}{(k\bar{v})^2}} \left\{ 1 - \frac{\varkappa}{1+\varkappa+\sqrt{1+\varkappa}} \frac{\Gamma_1^2}{(\Omega_\mu - \Omega)^2 + \Gamma_1^2} - \right.$$
$$\left. - \mathrm{Re} \left[\frac{\varkappa}{(1+i\xi)\sqrt{1+\varkappa}\sqrt{1+\varkappa/(1+i\xi)}\,[\sqrt{1+\varkappa}+\sqrt{1+\varkappa/(1+i\xi)}]} \right] \right\},$$

$$\Gamma_1 = \gamma_n [1 + \sqrt{1+\varkappa}], \quad \xi = \frac{\Omega_\mu - \Omega}{2\gamma_m}. \qquad (12.5)$$

Here we use the same symbols as in formula (9.9). We also assume that $k\bar{v} \gg \Gamma_1$. The terms in the first line in (12.5) are identical with the corresponding terms in (9.9). They describe a dip of dispersion form with width Γ_1 and depth $\varkappa/[1 + \varkappa + (1 + \varkappa)^{\frac{1}{2}}]$ superimposed on the Doppler line. The center of this dip coincides with the frequency of the strong field. The term in the second line of formula (12.5) is similar in general form to the corresponding term in (9.9), but differs from it in sign. Hence, if there is a narrow peak in the dip of the spontaneous emission line (see Fig. 8) there will be an even narrower dip in the center of the dip in the profile of $\langle \alpha_\mu \rangle$ (Fig. 14). Its main parameters are: The depth is

$$\frac{\varkappa}{2[1+\varkappa]^{3/2}}, \qquad (12.6)$$

the area

$$2\pi\gamma_m \frac{1+\varkappa}{1+\sqrt{1+\varkappa}+\varkappa}; \qquad (12.7)$$

the effective width, equal to the ratio of (12.7) and (12.6), is

$$4\pi\gamma_m \frac{1+\varkappa}{1+\sqrt{1+\varkappa}}. \qquad (12.8)$$

The relative depth of the narrow dip is relatively small (maximum value ~19%), but when $\omega_\mu \approx \omega$ the gain of the weak field is still less than for the strong field:

$$\langle \alpha_\mu \rangle < \langle \alpha \rangle.$$

In the case of a strong field in the form of a standing wave, calculation leads to very similar results. The value of $\langle \alpha_\mu \rangle$ will be given by formula (12.2), if $\varkappa$ in it is replaced by $\varkappa/4$, and η by

$$2\gamma_m \left\{ 1 + \frac{\varkappa}{4} \left[\frac{\gamma_n^2}{\gamma_n^2 + (\Omega - kv)^2} + \frac{\gamma_n^2}{\gamma_n^2 + (\Omega + kv)^2} \right] \right\}.$$

The consequences of such a replacement were investigated in § 9 for the case of spontaneous emission. Hence, we will not dwell on this here.

CHAPTER IV

Effect of Atomic Collisions

§ 13. Classical Theory of Doppler Broadening of Spectral Lines with Collisions Taken into Account. General Theory

1. In the theory of broadening of spectral lines it is convenient to proceed from the correlation function $\Phi(t)$, which is connected with the intensity distribution $I(\Omega)$ in the spectral line by the following relationships (see, for instance, [57], § 36):

$$I(\Omega)=\frac{1}{\pi}\operatorname{Re}\int_0^\infty \Phi(t)e^{i\Omega t}dt,\quad \Phi(t)=\int_{-\infty}^{\infty} I(\Omega)e^{-i\Omega t}d\Omega. \tag{13.1}$$

In the collision approximation, within which all the subsequent analysis is conducted, the correlation function has the form

$$\Phi(t)=\langle e^{-i\mathbf{kr}(t)}e^{i\varphi(t)}\rangle. \tag{13.2}$$

The quantity $-i\mathbf{kr}$ in the index of the exponent [$\mathbf{k}$ is the wave vector, $\mathbf{r}(t)$ is the displacement of the atom in time t] is due to the Doppler effect, i.e., it determines the change in phase of the emission of the oscillator due to the motion as a whole; $\varphi(t)$ describes the shift in phase of the atomic oscillator occurring in time t due to disturbance of the intraatomic motion by collisions. The phase $\varphi(t)$ may be complex, $\varphi = \eta + i\beta$, and the imaginary part β describes the quenching due to collisions. The averaging in (13.2) is over the ensemble of oscillators.

The extensive literature on the broadening of spectral lines includes thorough treatment of cases of purely Doppler broadening and broadening due solely to interactions:

$$\Phi_c(t)=\langle e^{i\varphi(t)}\rangle;\quad I_c(\Omega)=\frac{1}{\pi}\operatorname{Re}\int_0^\infty \Phi_c(t)e^{i\Omega t}dt; \tag{13.3}$$

$$\Phi_D(t)=\langle e^{-i\mathbf{kr}(t)}\rangle;\quad I_D(\Omega)=\frac{1}{\pi}\operatorname{Re}\int_0^\infty \Phi_D(t)e^{i\Omega t}dt. \tag{13.4}$$

In what follows we will need the following results of calculations of $\Phi(t)$ and $I(\Omega)$ in the collision approximation (see, for instance, [57], Chapter X):

$$\Phi_c(t)=e^{-(\Gamma+i\Delta)t};\quad \Gamma+i\Delta=\int P(g)\left[1-e^{i\psi(g)}\right]dg,$$
$$I_c(\Omega)=\frac{\Gamma/\pi}{(\Omega-\Delta)^2+\Gamma^2}. \tag{13.5}$$

Here Γ and Δ are the width and shift of the line, g denotes the set of collision parameters, $P(g)dg$ denotes the number of collisions of kind g in unit time, and $\psi(g)$ is the phase shift due to collisions with parameters g. In formula (13.5) and throughout § 13-15 the frequencies will be measured from the natural frequency of the emitting system.

The case of purely Doppler broadening has been investigated only for the Brownian-motion model [57, 75, 76]. If the velocity distribution of the atoms is Maxwellian the correlation function has the form

$$\Phi_D(t) = \exp\left\{-\frac{(k\bar{v})^2}{2\nu_d^2}[\nu_d t - 1 + e^{-\nu_d t}]\right\}, \tag{13.6}$$

where ν_d is the effective collision frequency, which is connected with the diffusion coefficient D by the relationship $\nu_d = \bar{v}^2/2D$. In the limiting case $\nu \to 0$ it follows from (13.6) that

$$\Phi_D(t) = \exp\left\{-\frac{1}{4}(k\bar{v}t)^2\right\},\ I_D(\Omega) = \frac{1}{\sqrt{\pi}k\bar{v}}\exp\left\{-\frac{\Omega^2}{(k\bar{v})^2}\right\}. \tag{13.7}$$

When $\nu_d \neq 0$ the line becomes narrower and in the limiting case $\nu_d \gg k\bar{v}$ in the frequency region $|\Omega| \ll \nu_d$ formula (13.6) gives

$$\Phi_D(t) = \exp\{-\gamma_d t\},\ I_D(\Omega) = \frac{\gamma_d/\pi}{\gamma_d^2 + \Omega^2},\ \gamma_d = \frac{(k\bar{v})^2}{2\nu_d}. \tag{13.8}$$

The physical reason for the change in the line shape from (13.7) to (13.8) is quite clear. In accordance with (13.4) the width $\Phi_D(t)$ as a function of t is determined by the characteristic time t_k required for displacement of the oscillator through a distance of the order of $1/k = \lambda/2\pi$. Any factor limiting or retaining the motion of the atom will lead to broadening of $\Phi_D(t)$ and, hence, to narrowing of the profile $I_D(\Omega)$. A well-known example of such narrowing is the Mossbauer effect, due to localization of the emitting atom in a region which is small in comparison with $\lambda/2\pi$. Collisions of gas atoms obviously lead to an increase in the characteristic time t_k. Hence, collisions will be manifested in a reduction in the width of the spectrum in comparison with $k\bar{v}$. This effect will be particularly significant if the mean free path l is much less than $\lambda/2\pi$. In this limiting case the time of displacement of the atom through a distance $\lambda/2\pi$ is given by the diffusion law $t_k = \left(\frac{\lambda}{2\pi}\right)^2 \frac{1}{2D}$, and the line width will be

$$\frac{1}{t_k} = 2D\left(\frac{2\pi}{\lambda}\right)^2 = \frac{\bar{v}^2}{\nu_d}k^2 = 2\gamma_d \tag{13.9}$$

in correspondence with (13.8).

When the motion of the atoms and broadening due to interaction are both taken into account the statistical independence of $e^{-i\mathbf{kr}}$ and $e^{i\varphi}$ is usually assumed. Then

$$\Phi(t) = \langle e^{-i\mathbf{kr}}e^{i\varphi}\rangle = \langle e^{-i\mathbf{kr}}\rangle\langle e^{i\varphi}\rangle = \Phi_D(t)\,\Phi_c(t), \tag{13.10}$$

$$I(\Omega) = \int_{-\infty}^{\infty} I_D(\Omega')\,I_c(\Omega - \Omega')\,d\Omega'. \tag{13.11}$$

However, there are no grounds for assuming the statistical independent of $e^{-i\mathbf{kr}}$ and $e^{i\varphi}$ in the general case [63, 77]. In fact, in the same act of collision there may be a shift of levels and a change in the velocity of the atom. Moreover, the shift of the levels and the change in velocity may be related to one another [57]. Hence, (13.10) and (13.11) correspond only to some limiting case.

The role played by the statistical dependence of the two cofactors in (13.2) depends on the relationship between two parameters – the time t_φ, during which the phase φ acquires an increment of the order of 2π, and the time t_v, during which the velocity of the atom changes significantly. If $t_\varphi \ll t_v$, the effect of collisions on the Doppler effect can generally be neglected. In this case the resulting profile will be given by the convolution (13.11) of the profiles (13.7) and (13.5). This case is exemplified by collisions of atoms with electrons: Owing to their small mass electrons have hardly any effect on the velocity of the atom even in collisions where the phase is considerably altered. If $t_\varphi \sim t_v$ or $t_\varphi \gg t_v$, the assumption of statistical independence of $e^{-i\mathbf{kr}}$ and $e^{i\varphi}$ lacks any basis.

The actual fact of statistical dependence of Doppler broadening and broadening due to interaction was reported in [77]. Yet this has not been taken into account in any investigation. In this section we develop a theory of spectral line broadening in which this effect is given due consideration.

2. We introduce a distribution function $f(\mathbf{r}, \mathbf{v}, \varphi, t, \mathbf{v}_0)$. This function specifies the relative number of atoms which at $t = 0$ were at the point $\mathbf{r} = 0$, had velocity $\mathbf{v}_0$, phase $\varphi = 0$ and during time t moved through a distance $\mathbf{r}$ and acquired velocity $\mathbf{v}$ and phase φ. Using this function we can write expression (13.2) in the form

$$\Phi(t) = \int d\mathbf{v}_0 d\mathbf{v} \int e^{-i(\mathbf{kr}-\varphi)} f(\mathbf{r}, \mathbf{v}, \varphi, t, \mathbf{v}_0)\, d\mathbf{r}\, d\varphi. \tag{13.12}$$

This distribution function f satisfies the kinetic equation

$$\frac{\partial f}{\partial t} + \mathbf{v}\nabla f = S, \tag{13.13}$$

where S is the collision integral. If $\mathbf{r}, \varphi$ as arguments of f are to have the same sense as $\mathbf{r}(t)$ and $\varphi(t)$ in (13.2), the solution of equation (13.13) must satisfy the initial conditions:

$$f(\mathbf{r}, \mathbf{v}, \varphi, 0, \mathbf{v}_0) = W(\mathbf{v}_0)\,\delta(\mathbf{v} - \mathbf{v}_0)\,\delta(\mathbf{r})\,\delta(\varphi); \tag{13.14}$$

If the spatially inhomogeneous problem is considered, boundary conditions for f must be specified. We confine ourselves entirely to the homogeneous problem. It is obvious that it will correspond to reality if the dimensions of the real region are much greater than the characteristic lengths (wavelength, mean free path), which is the case in most optical problems.

We will assume that during the collision time the velocity $\mathbf{v}$ of the atom and the phase φ change, but the coordinate does not. This corresponds with collision theory, in which the duration of the collision is neglected. Then the collision integral can be written in the following way:

$$S = -f(\mathbf{r}, \mathbf{v}, \varphi, t, \mathbf{v}_0) \int A(\mathbf{v}, \mathbf{v}', \varphi' - \varphi)\, d\mathbf{v}'\, d\varphi' + \int A(\mathbf{v}', \mathbf{v}, \varphi - \varphi') f(\mathbf{r}, \mathbf{v}', \varphi', t, \mathbf{v}_0)\, d\mathbf{v}'\, d\varphi'. \tag{13.15}$$

Here $A(\mathbf{v}, \mathbf{v}', \varphi' - \varphi)$ is the probability (in unit time) of collisions accompanied by a change in velocity $\mathbf{v} \to \mathbf{v}'$ and phase $\varphi \to \varphi'$. Function A depends only on the difference $\varphi - \varphi'$, which also corresponds with the collision approximation.

It is clear from formulas (13.12)-(13.15) that averaging in (13.12) over $\mathbf{v}_0$ can be carried out in general form prior to the solution of equation (13.13).* We introduce a new distribution function

$$f(\mathbf{r}, \mathbf{v}, \varphi, t) = \int f(\mathbf{r}, \mathbf{v}, \varphi, t, \mathbf{v}_0)\, d\mathbf{v}_0. \tag{13.16}$$

Averaging over $\mathbf{v}_0$ does not alter the form of the kinetic equation and the collision integral. Only the initial conditions are altered. Thus, we finally obtain

$$\Phi(t) = \int d\mathbf{v} \int e^{-i(\mathbf{kr}-\varphi)} f(\mathbf{r}, \mathbf{v}, \varphi, t)\, d\mathbf{r}\, d\varphi; \tag{13.17}$$

---

*Prior averaging over $\mathbf{v}_0$ is characteristic for the considered linear problem. Looking forward, we note that in the analysis of nonlinear effects averaging over $\mathbf{v}_0$ should not be performed at this stage of the calculations (see § 17).

$$I(\Omega)=\frac{1}{\pi}\operatorname{Re}\int d\mathbf{v}\int e^{i(\Omega t-\mathbf{kr}+\varphi)}f(\mathbf{r},\mathbf{v},\varphi,t)\,d\mathbf{r}\,d\varphi\,dt; \tag{13.18}$$

$$\frac{\partial f}{\partial t}+\mathbf{v}\nabla f=-\int\{A(\mathbf{v},\mathbf{v}',\varphi'-\varphi)f(\mathbf{r},\mathbf{v},\varphi,t)-A(\mathbf{v}',\mathbf{v},\varphi-\varphi')f(\mathbf{r},\mathbf{v}',\varphi',t)\}\,d\mathbf{v}'\,d\varphi'; \tag{13.19}$$

$$f(\mathbf{r},\mathbf{v},\varphi,0)=W(\mathbf{v})\,\delta(\mathbf{r})\,\delta(\varphi). \tag{13.20}$$

It is clear from (13.17) and (13.18) that the correlation function and the spontaneous emission spectrum are given by the Fourier components of the distribution function averaged over the velocities. In the case of $\Phi(t)$ a Fourier transformation of $\mathbf{r}$, φ is made and in the case of $I(\Omega)$ a Fourier transformation of $\mathbf{r}$, φ, t is made.

Formulas (13.17)-(13.20) leave out another cause of line broadening – radiation damping. It can be taken into consideration by the addition of a term $-\gamma f$ to the collision integral, where γ is an appropriately chosen damping constant. This leads to a factor $e^{-\gamma t}$ in the expression for $\Phi(t)$. Hence, radiation damping can always be taken into account in the last stage of the calculations.

3. All the main results mentioned in section **1** follow from (13.17)-(13.20). If the atom is at rest and undergoes collisions which alter the phase, then

$$W(\mathbf{v})=\delta(\mathbf{v}),\quad A(\mathbf{v},\mathbf{v}',\varphi'-\varphi)=\delta(\mathbf{v}-\mathbf{v}')\,B(\varphi'-\varphi).$$

The Fourier transformation of the kinetic equation in φ gives

$$\frac{d\Phi}{dt}=-(\Gamma+i\Delta)\,\Phi,\quad \Gamma+i\Delta=\int B(\psi)\,[1-e^{i\psi}]\,d\psi. \tag{13.21}$$

The solution of this equation with initial conditions (13.20) leads to expression (13.5).

If the atom is in motion and does not undergo any collisions, then

$$A(\mathbf{v},\mathbf{v}',\varphi'-\varphi)=\delta(\mathbf{v}-\mathbf{v}')\,\delta(\varphi'-\varphi),\quad S\equiv 0,$$

and the solution of equation (13.19) with initial conditions (13.20) is

$$f(\mathbf{r},\mathbf{v},\varphi,t)=W(\mathbf{v})\,\delta(\mathbf{r}-\mathbf{v}t)\,\delta(\varphi). \tag{13.22}$$

When the atoms have a Maxwellian velocity distribution we obtain expression (13.7).

If in the collisions the phase φ is not altered, then

$$A(\mathbf{v},\mathbf{v}',\varphi'-\varphi)=A(\mathbf{v},\mathbf{v}')\,\delta(\varphi'-\varphi)$$

and $f(\mathbf{r},\mathbf{v},\varphi,t)$ at any instant is proportional to $\delta(\varphi)$. It is clear from (13.17) and (13.18) that in this case we simply drop φ as an argument of the distribution function. Thus, equations (13.17)-(13.20) take the form

$$\begin{gathered}\Phi(t)=\int d\mathbf{v}\int e^{-i\mathbf{kr}}f(\mathbf{r},\mathbf{v},t)\,d\mathbf{r},\\ I(\Omega)=\frac{1}{\pi}\operatorname{Re}\int d\mathbf{v}\int e^{i(\Omega t-\mathbf{kr})}f(\mathbf{r},\mathbf{v},t)\,d\mathbf{r}\,dt,\\ \frac{\partial f}{\partial t}+\mathbf{v}\nabla f=-f(\mathbf{r},\mathbf{v},t)\int A(\mathbf{v},\mathbf{v}')\,d\mathbf{v}'+\int A(\mathbf{v}',\mathbf{v})f(\mathbf{r},\mathbf{v}',t)\,d\mathbf{v}',\\ f(\mathbf{r},\mathbf{v},0)=W(\mathbf{v})\,\delta(\mathbf{r}).\end{gathered} \tag{13.23}$$

In the Brownian-motion model of the emitting atom we use the collision integral in Chandrasekhar's form [78]

$$\frac{\partial f}{\partial t}+\mathbf{v}\nabla f=\nu_d\left\{\operatorname{div}_{\mathbf{v}}(\mathbf{v}f)+\frac{\bar{v}^2}{2}\Delta_{\mathbf{v}}f\right\}. \tag{13.24}$$

The parameter ν_d is the effective collision frequency. This frequency determines the time $t_v = 1/\nu_d$ during which the atom "forgets" its initial velocity [see formula (131) in [78]],

$$\langle \mathbf{v}(t)\rangle = \mathbf{v}_0 e^{-\nu_d t}.$$

It is convenient to solve equation (13.24) by a Fourier transformation with respect to **r** and **v**. For the function

$$F(\mathbf{k}, \varkappa, t) = \int e^{-i(\mathbf{kr}+\varkappa\mathbf{v})} f(\mathbf{r}, \mathbf{v}, t)\, d\mathbf{r}d\mathbf{v} \tag{13.25}$$

it follows from (13.24) and (13.20) that

$$\frac{\partial F}{\partial t} + (\nu_d\varkappa - \mathbf{k})\frac{\partial F}{\partial \varkappa} = -\frac{\nu_d\overline{v^2}}{2}\varkappa^2 F,$$

$$F(\mathbf{k}, \varkappa, 0) + \int W(\mathbf{v})\, e^{-i\varkappa\mathbf{v}} d\mathbf{v}. \tag{13.26}$$

In the case of a Maxwellian velocity distribution we easily obtain from (13.26)

$$F(\mathbf{k}, \varkappa, t) = \exp\left\{-\frac{1}{2}\left[G\varkappa^2 + 2H\varkappa\mathbf{k} + P\mathbf{k}^2\right]\right\}, \tag{13.27}$$

where

$$G = \frac{\overline{v^2}}{2}, \quad H = \frac{\overline{v^2}}{2\nu_d}(1 - e^{-\nu_d t}), \quad P = \frac{\overline{v^2}}{\nu_d^2}[\nu_d t - 1 + e^{-\nu_d t}]. \tag{13.28}$$

If we put $\varkappa = 0$ in (13.27), then, according to (13.25) and (13.17), we obtain the correlation function

$$\Phi(t) = F(\mathbf{k}, 0, t) = \exp\left\{-\frac{\overline{(kv)^2}}{2\nu_d^2}[\nu_d t - 1 + e^{-\nu_d t}]\right\} \tag{13.29}$$

in accordance with (13.6).

Thus, equations (13.17)-(13.20) enable us to consider different cases of spectral line broadening from a single viewpoint and the previously known results can be obtained relatively simply. Below we will consider broadening due to interaction and the Doppler effect with due regard to their statistical dependence (§ 15). In § 14 the case of purely Doppler broadening is analyzed more thoroughly than has been done in the literature.

§ 14. Doppler Broadening

1. We consider first the Brownian-motion model of the emitting atom. The correlation function is given by formula (13.29) and the corresponding intensity distribution $I(\Omega)$ is expressed by a degenerate hypergeometric function [79]:

$$I(\Omega) = \operatorname{Re} J(\Omega); \tag{14.1}$$

$$J(\Omega) = \frac{1}{\pi}\int_0^\infty \exp\left\{i\Omega t - \frac{\overline{(kv)^2}}{2\nu_d}[\nu_d t - 1 + e^{-\nu_d t}]\right\}dt = \frac{1}{\pi}\frac{1}{\gamma_d - i\Omega}\Phi\left(1.1 + \frac{\gamma_d - i\Omega}{\nu_d};\ \frac{\gamma_d}{\nu_d}\right), \quad \gamma_d = \frac{\overline{(kv)^2}}{2\nu_d}, \tag{14.2}$$

where

$$\Phi(\alpha, \gamma; z) = 1 + \frac{\alpha}{\gamma}\frac{z}{1!} + \frac{\alpha(\alpha+1)}{\gamma(\gamma+1)}\frac{z}{2!} + \dots. \tag{14.3}$$

We trace the changes in $I(\Omega)$ which occur at low collision frequencies, $\nu_d \ll k\bar{v}$. For the center of the line $\Omega = 0$ we find

$$I(0) \approx \frac{1}{\sqrt{\pi}k\bar{v}}\left\{1 + \frac{2}{3\sqrt{\pi}}\frac{\nu_d}{k\bar{v}}\right\}. \tag{14.4}$$

For the far wing of the line the asymptotic expansion for $I(\Omega)$ has the form

$$I(\Omega) \approx \frac{1}{\sqrt{\pi}k\bar{v}}\frac{\nu_d(k\bar{v})^3}{2\sqrt{\pi}\Omega^4} \quad (\Omega \gg k\bar{v} \gg \nu_d). \tag{14.5}$$

A comparison of (14.4) and (14.5) with (13.7) shows that the introduction of collisions leads to an increase in the intensity of the line at the center and on the wings, which decay in accordance with a power law Ω^{-4}.

As the collision frequency increases, the intensity of the line at the center increases monotonically (Fig. 15, curve 1). In the intermediate case $\nu_d \sim k\bar{v}$ and for $\nu_d \gg k\bar{v}$ it is convenient to use the following expansions for $I(\Omega)$:

$$I(\Omega) = \frac{1}{\pi}\operatorname{Re}\left\{\frac{1}{\gamma_d - i\Omega}\left[1 + \frac{\gamma_d}{\nu_d + \gamma_d - i\Omega} + \frac{\gamma_d^2}{(\nu_d + \gamma_d - i\Omega)(2\nu_d - \gamma_d - i\Omega)} + \dots\right]\right\} =$$

$$= \frac{e^{\gamma_d/\nu_d}}{\pi}\left\{\frac{\gamma_d}{\Omega^2 + \gamma_d^2} - \frac{\nu_d + \gamma_d}{\Omega^2 + (\gamma_d + \nu_d)^2}\frac{\gamma_d}{\nu_d} + \frac{2\nu_d + \gamma_d}{\Omega^2 + (2\nu_d + \gamma_d)^2}\frac{1}{2}\left(\frac{\gamma_d}{\nu_d}\right)^2 - \dots\right\}. \tag{14.6}$$

Figure 16 shows graphs of $I(\Omega)$ for some values of $\nu_d/k\bar{v}$. This figure clearly shows the narrowing of the central part of the line and the appearance of wings of greater intensity than purely Gaussian. The broken curve in Fig. 16 represents the dispersion line with the same width as curve 3. A comparison with curve 3 shows that the actual line has less intense wings than the dispersion curve, which indicates a more rapid decrease of intensity with frequency (Ω^{-4}, and not Ω^{-2}).

In the limiting case $\nu_d \gg k\bar{v}$ we can restrict ourselves to the first two terms in the expansion (14.6). Since in this case $\gamma_d = \frac{k\bar{v}}{2}\frac{k\bar{v}}{\nu_d} \ll k\bar{v} \ll \nu_d$, the expression for $I(\Omega)$ takes the form

$$I(\Omega) = \frac{1}{\pi}\frac{\gamma_d\nu_d^2}{[\Omega^2 + \gamma_d^2][\Omega^2 + \nu_d^2]}. \tag{14.7}$$

Near the center of the line, $\Omega = \nu_d$, we can neglect Ω^2 in comparison with ν_d^2, and (14.7) gives

$$I(\Omega) = \frac{1}{\pi}\frac{\gamma_d}{\Omega^2 + \gamma_d^2}, \tag{14.8}$$

i.e., the line has a simple dispersion shape with width

$$\gamma_d = \frac{(k\bar{v})^2}{2\nu_d}. \tag{14.9}$$

Since the mean free path $l = \bar{v}/\nu_d$, then

$$2\gamma_d = k\bar{v}k\frac{\bar{v}}{\nu_d} = k\bar{v}\frac{2\pi l}{\lambda} \ll k\bar{v}, \tag{14.10}$$

i.e., the central part of the line is $2\pi l/\lambda$ times narrower than the Doppler width $k\bar{v}$, which is perfectly consistent with § 13.

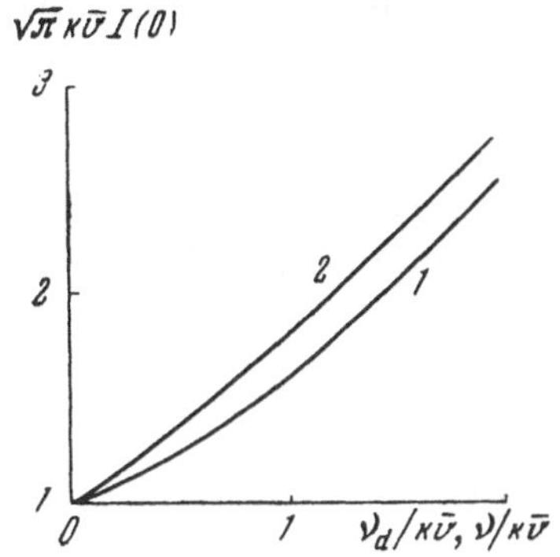

Fig. 15. Intensity in center of line as a function of frequency of elastic collisions. Curve 1 corresponds to the Brownian-motion model; curve 2 is for the hard-collision model.

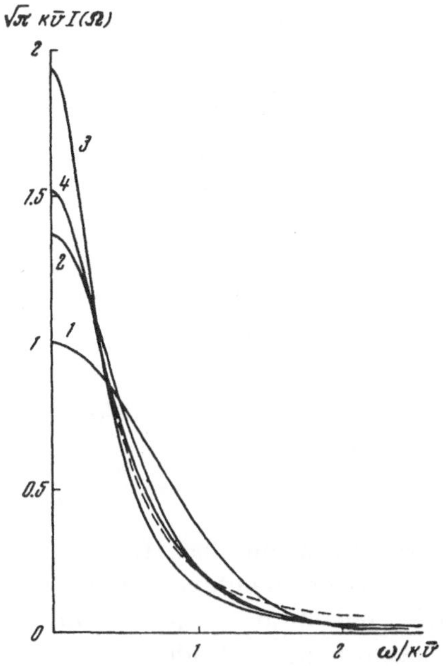

Fig. 16. Profile of Doppler-broadened line. 1) $\nu_d = 0$; 2) $\nu_d = 0.7 k\bar{v}$; 3) $\nu_d = \sqrt{2}k\bar{v}$; 4) hard-collision model, $\nu = 0.7\ k\bar{v}$; the broken curve is the dispersion line with a width equal to that of line 3.

At large frequencies, $\Omega \gg \nu_d \gg k\bar{v}$, expression (14.7) becomes the asymptotic formula (14.5), which, consequently, is true for any ν_d.

It should be noted that the relative intensity of the wings is comparatively low. As can be seen from the estimate of the region of applicability of formula (14.5) the ratio of the intensity of the wing at $\Omega = \nu_d$ to the line intensity at $\Omega = 0$ is $(k\bar{v}/\Omega)^3 \ll 1$. Nevertheless, in a whole series of problems (particularly astrophysical problems) the presence of a wing proportional to Ω^{-4} must be taken into account, since the absolute intensity may be fairly high.

The above-discovered distribution of intensity in the wings of the line differs from that obtained in [75] (see also [57]):

$$I(\Omega) \propto e^{-\Omega^2/(k\bar{v})^2}, \quad \Omega > \nu_d. \tag{14.11}$$

The arguments put forward in [75, 57] in support of (14.11) reduce to the following. Since the integrand in expression (14.2) for $I(\Omega)$ contains an oscillating factor $e^{i\Omega t}$, the value of $I(\Omega)$ at large Ω will be determined by the behavior of the correlation function at low t. Expanding the index of the exponent in formula (14.2) in powers of t and confining ourselves to the first nonvanishing term ($\propto t^2$) we arrive at (14.11). A similar argument is used in [80] for the analysis of the limiting expressions for the spectrum $I(\Omega)$.

At the same time, it was shown above that on the wing of the line $I(\Omega) \propto \Omega^{-4}$, irrespective of the relationships between ν_d and $k\bar{v}$, the last result can actually be derived from general theorems of Fourier analysis. In fact, in the case of steady-state random processes the correlation function is a function of $|t|$. On the other hand, expansion of (14.2) in a series of powers of t when $t \to 0$ will have the form

$$\Phi(t) = 1 - \frac{(k\bar{v})^2}{2}\left[\frac{t^2}{2!} - \frac{|t|^3}{3!} + \ldots\right]. \tag{14.12}$$

It is clear from (14.12) that at the point $t = 0$ the third derivative of function $\Phi(t)$ has a discontinuity. According to the well-known theorem of the order of decrease of the Fourier expansion coefficients (see [81], for instance), the Fourier component of the function $\Phi(t)$, i.e., $I(\Omega)$, will not decrease more rapidly than Ω^{-4}. Thus, the variation of the intensity at large frequencies depends (when $t \to 0$) on the behavior of the derivatives of the correlation function, and not on this function itself.

In this paper we consider the broadening of the emission lines. It is of interest to note that the collisions also cause the appearance of a wing with an exponential decrease of intensity in the Doppler profile of the line of Rayleigh scattering in a gas. This was shown in [82]. However, in this case at large frequencies $I(\Omega) \propto \Omega^{-6}$. The difference in the index of the power is due to the specific nature of Rayleigh scattering. Since the total momentum of the colliding particles is conserved and both particles are implicated in the scattering, then it is second, and not the first, derivative of the scattering field of these particles which has a discontinuity. Hence, the spectrum of the field decreases as Ω^{-3}, and the intensity distribution as Ω^{-6}.

2. In the above-discussed Brownian-motion model it is assumed that a significant change in velocity is produced by a large number of relatively soft collisions [78]. It is not clear in advance how applicable this approximation is to the case of hard collisions, where the velocity is appreciably altered in one collision. Hence, we return to the general equation (13.23). We write the first term of the collision integral in (13.23) in the form $-\nu f$, where ν is the effective collision frequency

$$\nu(\mathbf{v}) = \int A(\mathbf{v}, \mathbf{v}')\, d\mathbf{v}'. \tag{14.13}$$

For the space–time Fourier component $F(\mathbf{k}, \mathbf{v}, \Omega)$ of the distribution function we can obtain from (13.23) the following Fredholm integral equation of the second kind:

$$F(\mathbf{k}, \mathbf{v}, \Omega) = \frac{1}{\nu - i(\Omega - \mathbf{kv})} \int A(\mathbf{v}', \mathbf{v})\, F(\mathbf{k}, \mathbf{v}', \Omega)\, d\mathbf{v}' + \frac{W(\mathbf{v})}{\nu - i(\Omega - \mathbf{kv})}. \tag{14.14}$$

The second term on the right side of equation (14.14) takes the initial conditions into account.

To solve equation (14.14) we need to know the specific form of the kernal $A(\mathbf{v}', \mathbf{v})$. We use a relatively simple hard-collision model which enables us to find the spectrum $I(\Omega)$ in closed form. The model is based on the assumption that $A(\mathbf{v}', \mathbf{v})$ is independent of $\mathbf{v}'$, i.e., $A(\mathbf{v}', \mathbf{v}) = A(\mathbf{v})$. In other words, it is assumed that the velocity $\mathbf{v}$ of the particle after collision is independent of its velocity $\mathbf{v}'$ before collision. In contrast to the Brownian-motion model (13.24), this model reflects the main qualitative features of scattering of light particles by heavy ones [83-85].

The specific form of the kernel $A(\mathbf{v})$ is determined uniquely by the condition $S \equiv 0$ for statistical equilibrium. It is easy to show that $A(\mathbf{v})$ will be given by the formula

$$A(\mathbf{v}) = \nu W_M(\mathbf{v}), \tag{14.15}$$

where ν is independent of the velocity. Such a form of collision integral means that after the first collision the probability of different values of the velocity of the atom is determined by the equilibrium distribution and that the arbitrary initial distribution becomes an equilibrium distribution in a time of order $1/\nu$. Hence, the quantity ν has the same physical sense as ν_d – the quantities $1/\nu$ and $1/\nu_d$ determine the time in which the atom completely "forgets" its initial velocity.

With such a kernel $A(\mathbf{v})$ the equations for the distribution function $f(\mathbf{r}, \mathbf{v}, t)$ and its Fourier component $F(\mathbf{k}, \mathbf{v}, \Omega)$ take the form

$$\frac{\partial f}{\partial t} + \mathbf{v}\nabla f = -\nu\left[f - W_M(\mathbf{v}) \int f(\mathbf{r}, \mathbf{v}', t)\, d\mathbf{v}'\right], \tag{14.16}$$

$$F(\mathbf{k}, \mathbf{v}, \Omega) = \frac{\nu W_M(\mathbf{v})}{\nu - i(\Omega - \mathbf{kv})} \int F(\mathbf{k}, \mathbf{v}', \Omega)\, d\mathbf{v}' + \frac{W(\mathbf{v})}{\nu - i(\Omega - \mathbf{kv})}. \tag{14.17}$$

Integrating the right and left sides of equation (14.17) with respect to **v**, we obtain

$$I(\Omega)=\frac{1}{\pi}\operatorname{Re}\left\{\int\frac{W(\mathbf{v})\,d\mathbf{v}}{\nu-i(\Omega-\mathbf{kv})}\Big/\left[1-\nu\int\frac{W_M(\mathbf{v})\,d\mathbf{v}}{\nu-i(\Omega-\mathbf{kv})}\right]\right\}. \tag{14.18}$$

Formula (14.18) can be interpreted in the following way. The numerator in (14.18) corresponds to the situation in which the emission of the oscillator is divided into a series of incoherent trains of length $1/\nu$ and within each train the frequency emitted is shifted by the amount of the Doppler shift **kv**. However, the real trains into which the emission is divided due to the change in the velocity of the atom in collisions cannot be regarded as incoherent. As already mentioned (§ 13), coherence is disturbed in times t_k, corresponding to displacement of the atom through a distance $\lambda/2\pi$. Hence, all the emission in time t_k, irrespective of how many trains it is divided into, can interfere. The role of this interference is reflected by the denominator in (14.18). Interference between adjacent trains can be neglected only in the case where the train length $l \gg \lambda/2\pi$. But this means that $\nu \ll k\bar{v}$. When this condition is satisfied the second term in the denominator is much less than unity, and the numerator becomes the usual Doppler intensity distribution corresponding to the initial velocity distribution of the atoms. If ν is not small in comparison with $k\bar{v}$, the interference of the different trains is significant.

We assume that W(v) is an equilibrium distribution function. It follows from (14.18) that

$$I(\Omega)=\frac{1}{\sqrt{\pi}k\bar{v}}\operatorname{Re}\left\{\frac{w\left(\frac{\Omega}{k\bar{v}},\frac{\nu}{k\bar{v}}\right)}{1-\sqrt{\pi}\frac{\nu}{k\bar{v}}w\left(\frac{\Omega}{k\bar{v}},\frac{\nu}{k\bar{v}}\right)}\right\}, \tag{14.19}$$

where the function

$$w(x,y)=\frac{i}{\pi}\int_{-\infty}^{\infty}\frac{e^{-t^2}\,dt}{x+iy-t}=\frac{1}{\sqrt{\pi}}\int_0^{\infty}e^{-\frac{z^2}{4}+i(x+iy)z}\,dz \tag{14.20}$$

can be expressed by a probability integral of complex argument [66]:

$$w(x,y)=e^{-(x+iy)^2}\left\{1-\frac{2}{\sqrt{\pi}}\int_0^{i(x+iy)}e^{-t^2}\,dt\right\}. \tag{14.21}$$

It is easy to see from (14.19) and (14.20) that in the limiting case $\nu \to 0$ we obtain the Gaussian intensity distribution (13.7). When $\nu \neq 0$, but $\nu \ll k\bar{v}$, we can easily obtain an expansion of $I(\Omega)$ in powers of $\nu/k\bar{v}$, $\Omega/k\bar{v}$. In the center of the line, $\Omega = 0$, we have from (14.19) and (14.20)

$$I(\Omega)\approx\frac{1}{\sqrt{\pi}k\bar{v}}\left\{1+\frac{\pi-2}{\sqrt{\pi}}\frac{\nu}{k\bar{v}}\right\}. \tag{14.22}$$

This expression differs from (14.4) only in the numerical factor in front of $\nu/k\bar{v}$, which in formula (14.22) is approximately twice as large. Hence, when $\nu = \nu_d$ hard collisions are approximately twice as effective, in the sense of line narrowing, as soft collisions. We can show that when $\nu/k\bar{v}$ increases, the intensity at the point $\Omega = 0$ increases monotonically (curve 2, Fig. 15).

The behavior of the intensity on the wings of the line can be determined by using the asymptotic expansion of the probability integral for large values of the modulus of the argument:

$$I(\Omega)\approx\frac{1}{\sqrt{\pi}k\bar{v}}\frac{1}{2\sqrt{\pi}}\frac{\nu(k\bar{v})^3}{\left[\Omega^2+\frac{(k\bar{v})^4}{4\nu^2}\right][\Omega^2+\nu^2]}\qquad(\nu^2+\Omega^2\gg(k\bar{v})^2). \tag{14.23}$$

Thus, both models discussed in this section lead at high collision frequencies (ν, $\nu_d \gg k\bar{v}$) to the same line shape in the whole region of frequencies. At not too high values of ν, ν_d the wings of the line are also the same. The central part of the line is a little different for the same values of the parameters ν and ν_d. The general nature of the differences can be seen from Fig. 16, where curves 2 and 4 are calculated for $\nu = \nu_d = 0.7k\bar{v}$.

To discover the reason for the closeness of the results in the hard- and soft-collision models we compare the first and second moments for the displacement **r** and velocity **v**. Formula (13.27) shows that in the Brownian-motion model **r** and **v** are normally distributed random quantities (see, for instance, [80], § 58) with second moments

$$\langle \mathbf{r}^2 \rangle = 3P = 3\frac{\bar{v}^2}{\nu_d^2}[\nu_d t - 1 + e^{-\nu_d t}],$$
$$\langle \mathbf{rv} \rangle = 3H = 3\frac{\bar{v}^2}{2\nu_d}[1 - e^{-\nu_d t}], \tag{14.24}$$
$$\langle \mathbf{v}^2 \rangle = 3G = 3\frac{\bar{v}^2}{2};$$

as regards the first moments, then

$$\langle \mathbf{r} \rangle = 0, \quad \langle \mathbf{v} \rangle = 0. \tag{14.25}$$

In the hard-collision model **r** and **v** are, generally speaking, not distributed normally. However, it is easy to show that in this case the first and second moments are also given by expressions (14.24), (14.25) if ν_d in them is replaced by ν. Thus, the two considered models correspond to different distributions for **r**, **v**, but give the same mean values, variances, and corrrelation moment of these quantities. Since in the case of a large number of collisions any distribution tends to normal, then it is quite understandable that when $\nu \gg k\bar{v}$, i.e., $l \ll \lambda/2\pi$, the two models lead to identical results. Conversely, the greatest difference can be expected when $\nu \ll k\bar{v}$, i.e., when collisions are relatively rare, and this is the case.

The relation between the hard- and soft-collision models can be investigated more fully if we assign a kernel $A(\mathbf{v}, \mathbf{v}')$ which contains these two models as limiting cases. Following [83], we put

$$A(\mathbf{v}, \mathbf{v}') = a(\mathbf{v}' - \gamma\mathbf{v}). \tag{14.26}$$

In view of the fact that the collision integral becomes zero in the case of the equilibrium distribution we derive an explicit expression for $a(\mathbf{v}' - \gamma\mathbf{v})$:

$$a(\mathbf{v}' - \gamma\mathbf{v}) = \frac{\mu}{[\pi(1-\gamma^2)\bar{v}^2]^{3/2}} \exp\left\{-\frac{(\mathbf{v}' - \gamma\mathbf{v})^2}{(1-\gamma^2)\bar{v}^2}\right\}, \tag{14.27}$$
$$\mu = \int a(\mathbf{v}' - \gamma\mathbf{v})\,d\mathbf{v}'.$$

The physical sense of the parameter γ can easily be determined if we calculate the mean velocity $\{\mathbf{v}'\}$ after the collision:

$$\{\mathbf{v}'\} = \frac{1}{\mu}\int \mathbf{v}' a(\mathbf{v}' - \gamma\mathbf{v})\,d\mathbf{v}' = \gamma\mathbf{v}.$$

Hence, γ is the ratio of the mean velocity after collision to the velocity of the particle before collision. Thus, the main assumption of model (14.26) is that this ratio is independent of **v**. The value of the constant γ must be chosen in accordance with the specific nature of the collisions. If the masses of the two colliding particles are close, then $\gamma \approx 0$. For collisions of heavy particles with light ones γ is close to unity. As is shown in [83], when $1 - \gamma \ll 1$ the collision integral with kernel (14.27) corresponds to the Brownian-motion model with accuracy to terms of order $(1-\gamma)^2$.

For model (14.27) it is easy to calculate the second moments for **r** and **v**: they are

$$\langle \mathbf{r}^2\rangle = 3\frac{\bar{v}^2}{\mu(1-\gamma)}\left[t - \frac{1-e^{-\mu(1-\gamma)t}}{\mu(1-\gamma)}\right],$$
$$\langle \mathbf{rv}\rangle = 3\frac{\bar{v}^2}{2\mu(1-\gamma)}\left[1-e^{-\mu(1-\gamma)t}\right], \tag{14.28}$$
$$\langle \mathbf{v}^2\rangle = 3\frac{\bar{v}^2}{2}.$$

Hence, the variance and correlation moment in this more general model have the same form as before [compare (14.24)]. The only difference is that as the effective collision frequency, $1/t_v$ (14.28) contains the quantity $\mu(1-\gamma)$. This quantity, then, is equivalent to the parameters ν, ν_d in the hard-collision model and the Brownian-motion, i.e., the soft-collision $(1-\gamma \ll 1)$, model.

3. We now consider the case where hard and soft collisions occur simultaneously. Since we consider only paired collisions the collision integral S can be written in the form

$$S = \nu_d\left[\operatorname{div}_{\mathbf{v}}(\mathbf{v}f) + \frac{\bar{v}^2}{2}\Delta_{\mathbf{v}} f\right] - \nu\left[f - W_M(\mathbf{v})\int f(\mathbf{r}, \mathbf{v}', t)\, d\mathbf{v}'\right]. \tag{14.29}$$

The solution of the kinetic equation with collision integral (14.29) can be found by the same methods as were used above. The final result for function $I(\Omega)$ has the form

$$I(\Omega) = \frac{1}{\pi}\frac{\dfrac{1}{\nu+\gamma_d-i\Omega}\Phi\left[1.1+\dfrac{\nu+\gamma_d-i\Omega}{\nu_d};\ \dfrac{\gamma_d}{\nu_d}\right]}{1-\dfrac{\nu}{\nu+\gamma_d-i\Omega}\Phi\left[1.1+\dfrac{\nu+\gamma_d-i\Omega}{\nu_d};\ \dfrac{\gamma_d}{\nu_d}\right]}. \tag{14.30}$$

This expression contains as special limiting cases (14.2) and (14.18), obtained from (14.30) when $\nu \to 0$ and $\nu_d \to 0$, respectively.*

If ν, $\nu_d \ll k\bar{v}$, the first nonvanishing correction to the intensity in the center of the line $(\Omega = 0)$ has the form

$$I(0) \approx \frac{1}{\sqrt{\pi}k\bar{v}}\left\{1 + \frac{\pi-2}{\sqrt{\pi}}\frac{\nu}{k\bar{v}} + \frac{2}{3\sqrt{\pi}}\frac{\nu_d}{k\bar{v}}\right\}. \tag{14.31}$$

A comparison of (14.31) with formulas (14.4) and (14.22) shows that hard and soft collisions make an additive contribution with the same coefficients as when they are considered separately. In the far wing the intensity is given by the sum $\nu + \nu_d$:

$$I(\Omega) = \frac{1}{\sqrt{\pi}k\bar{v}}\frac{(\nu+\nu_d)(k\bar{v})^3}{2\sqrt{\pi}\,\Omega^4}. \tag{14.32}$$

*Expression (14.30) shows that we have the following asymptotic representation of the hypergeometric function in terms of the probability integral: $\frac{1}{\gamma-1}\Phi(1, \gamma; z) \approx \sqrt{\frac{\pi}{2z}}\, w\left(\frac{\gamma-1-z}{\sqrt{2z}}\right)$, if γ, $z \to \infty$, and $(\gamma-1-z)/(2z)^{\frac{1}{2}}$ is finite. Function ω is given by formula (14.20) or (14.21).

The same occurs if $\nu^2 + \Omega^2 \gg (k\bar{v})^2$ or $\nu_d^2 \gg (k\bar{v})^2$. In the first case, for instance, $I(\Omega)$ is given by the formula

$$I(\Omega) \approx \frac{1}{\sqrt{\pi} k\bar{v}} \frac{1}{2\sqrt{\pi}} \frac{(\nu + \nu_d)(k\bar{v})^3}{\left[\Omega^2 + \frac{(k\bar{v})^4}{4(\nu+\nu_d)^2}\right][\Omega^2 + (\nu + \nu_d)^2]}, \tag{14.33}$$

which is similar to formula (14.23). As was to be expected on the basis of general considerations of the statistics of the variables **r** and **v**, the total effective collision frequency in these limiting cases is significant.

4. Thus, the hard- and soft-collision models and joint consideration of soft and hard collisions lead to very similar results. The decisive factor for the correct description of the narrowing of the Doppler profile is the choice of the characteristic quantity t_v. What determines the value of t_v – soft or hard collisions or both kinds together – is of secondary importance. In the case of hard collisions t_v is the time between two successive collisions. In the case of soft collisions the atom undergoes a large number of collisions in time t_v, each of which causes an insignificant change in velocity [by a value $(1-\gamma)\mathbf{v}$, where $1-\gamma \ll 1$]. In both cases, however, a considerable time is required for a significant change in velocity.

§ 15. Broadening Due to Interactions and Doppler Effect

1. The general case where changes of both phase and velocity take place during collisions is described by the system of equations (13.17)-(13.20). However, it will be more convenient henceforth to work with the Fourier transformation of the distribution function with respect to φ. That is, we put

$$\tilde{f} = \int f(\mathbf{r}, \mathbf{v}, \varphi, t) e^{i\varphi} d\varphi. \tag{15.1}$$

Then formulas (13.17)-(13.20) can be written as follows:

$$\begin{gathered} \Phi(t) = \int d\mathbf{v} \int e^{-i\mathbf{kr}} \tilde{f}(\mathbf{r}, \mathbf{v}, t)\, d\mathbf{r}, \\ I(\Omega) = \frac{1}{\pi} \operatorname{Re} \int d\mathbf{v} \int e^{i(\Omega t - \mathbf{kr})} f(\mathbf{r}, \mathbf{v}, t)\, d\mathbf{r}\, dt, \\ \frac{\partial \tilde{f}}{\partial t} + \mathbf{v}\nabla\tilde{f} = -\tilde{f} \int A(\mathbf{v}, \mathbf{v}', \psi)\, d\mathbf{v}' d\psi + \int \tilde{A}(\mathbf{v}', \mathbf{v}) \tilde{f}(\mathbf{r}, \mathbf{v}', t)\, d\mathbf{v}', \\ \tilde{f}(\mathbf{r}, \mathbf{v}, \theta) = W(\mathbf{v})\, \delta(\mathbf{r}), \end{gathered} \tag{15.2}$$

where we have introduced the function

$$\tilde{A}(\mathbf{v}', \mathbf{v}) = \int A(\mathbf{v}', \mathbf{v}, \psi) e^{i\psi} d\psi. \tag{15.3}$$

Formally (15.2) is the same as formulas (13.23), which relate to purely Doppler broadening. The difference lies in the facts that (15.2) contains f instead of $\tilde{f}$, and in one of the terms of the collision integral the kernal $A(\mathbf{v}', \mathbf{v}, \psi)$ is replaced by the function $\tilde{A}(\mathbf{v}', \mathbf{v})$.

If changes in phase and velocity occur during different collisions, then

$$A(\mathbf{v}, \mathbf{v}', \varphi' - \varphi) = A_1(\mathbf{v}, \mathbf{v}')\, \delta(\varphi - \varphi') + B(\mathbf{v}, \varphi' - \varphi)\, \delta(\mathbf{v} - \mathbf{v}') \tag{15.4}$$

and the equation for $\tilde{f}$ has the form

$$\begin{gathered} \frac{\partial \tilde{f}}{\partial t} + \mathbf{v}\nabla\tilde{f} = -(\Gamma + i\Delta)\tilde{f} - \tilde{f} \int A_1(\mathbf{v}, \mathbf{v}')\, d\mathbf{v}' + \int A_1(\mathbf{v}', \mathbf{v}) \tilde{f}(\mathbf{r}, \mathbf{v}', t)\, d\mathbf{v}', \\ \Gamma + i\Delta = \int B(\mathbf{v}, \psi)[1 - e^{i\psi}]\, d\psi. \end{gathered} \tag{15.5}$$

The right side of equation (15.5) contains terms similar to those which occur in the description of purely Doppler broadening and broadening due to interaction [compare with (13.21) and (13.23)].

If changes in φ and $\mathbf{v}$ occur in the same collisions, then on the right side of the kinetic equation (15.2) we can also take out the term $-(\Gamma + i\Delta)\tilde{f}$. The remainder, however, is dependent not only on the change in velocity, but also on the change of phase:

$$\frac{\partial f}{\partial t} + \mathbf{v}\nabla\tilde{f} = -(\Gamma + i\Delta)\tilde{f} - \tilde{f}(\mathbf{r}, \mathbf{v}, t)\int A(\mathbf{v}, \mathbf{v}', \psi')\,d\mathbf{v}'\,d\psi' + \int \tilde{A}(\mathbf{v}', \mathbf{v})\tilde{f}(\mathbf{r}, \mathbf{v}', t)\,d\mathbf{v}'; \tag{15.6}$$

$$\tilde{A}(\mathbf{v}, \mathbf{v}') = \int A(\mathbf{v}, \mathbf{v}', \psi)\,e^{i\psi}\,d\psi; \tag{15.7}$$

$$\Gamma + i\Delta = \int A(\mathbf{v}, \mathbf{v}', \psi)\,[1 - e^{i\psi}]\,d\psi. \tag{15.8}$$

Equations (15.6) and (15.5) have the same structure. Equation (15.5), however, contains the real probability $A_1(\mathbf{v}, \mathbf{v}')$ of a change in velocity $\mathbf{v} \to \mathbf{v}'$ quite unconnected with the nature of the change in phase φ. In equation (15.7) the kernel of the collision integral is the complex function $\tilde{A}$, the form of which, generally speaking, depends on the distribution of the phase jumps. In addition, in contrast to equation (15.5) the different terms of the right side of (15.6) contain the same quantity $\tilde{A}$. In other words, the two terms of the collision integral in (15.6) do not describe different causes of broadening. This is evidently a direct reflection of the statistical dependence of the changes in velocity and phase.

There is no difficulty either in considering the case where only the phase φ, or the velocity $\mathbf{v}$, change in some collisions, and both φ and $\mathbf{v}$ change in others. Since the corresponding generalization of the kinetic equation is so obvious we will not write it out here.

2. We now determine the line profile in case (15.4). We begin with the hard-collision model, i.e., we use expression (14.15) for $A_1(\mathbf{v}', \mathbf{v})$. We then obtain the following equation for f:

$$\frac{\partial \tilde{f}}{\partial t} + \mathbf{v}\nabla\tilde{f} = -(\Gamma + i\Delta)\tilde{f} - \nu\left[\tilde{f} - W_M(\mathbf{v})\int\tilde{f}(\mathbf{r}, \mathbf{v}', t)\,d\mathbf{v}'\right]. \tag{15.9}$$

Using the same method as in § 14, we easily find from (15.9) the intensity distribution $I(\Omega)$ in the spectral line:

$$I(\Omega) = \frac{1}{\pi}\,\mathrm{Re}\left\{\int\frac{W(\mathbf{v})\,d\mathbf{v}}{\Gamma+\nu-i(\Omega-\Delta-\mathbf{kv})}\Bigg/\left[1 - \nu\int\frac{W_M(\mathbf{v})\,d\mathbf{v}}{\Gamma+\nu-i(\Omega-\Delta-\mathbf{kv})}\right]\right\}. \tag{15.10}$$

This expression differs from formula (14.18) (purely Doppler broadening) by the terms $\Gamma + i\Delta$ in the resonance denominators. If Γ, Δ are independent of the velocity of the emitting atom, then (15.10) can be put in the form of the convolution of the dispersion curve and the function (14.18), which gives the line profile when $\Gamma = \Delta = 0$:

$$I(\Omega) = \int\frac{\frac{\Gamma}{\pi}\,d\Omega'}{\Gamma^2 + (\Omega - \Delta - \Omega')^2}\,\mathrm{Re}\left\{\frac{\frac{1}{\pi}\int\frac{W(\mathbf{v})\,d\mathbf{v}}{\nu - i(\Omega' - \mathbf{kv})}}{1 - \nu\int\frac{W_M(\mathbf{v})\,d\mathbf{v}}{\nu - i(\Omega' - \mathbf{kv})}}\right\}. \tag{15.11}$$

In fact, the integrand in the integral with respect to Ω' has a single pole in the upper half-plane of the complex frequency: $\Omega' = \Omega - \Delta + i\Gamma$. Using the theorem of residues we can satisfy ourselves that (15.11) and (15.10) are identical.

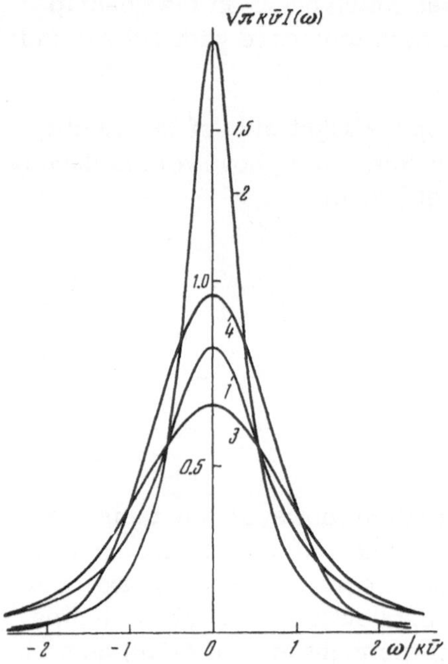

Fig. 17. Profile of the line broadened by Doppler effect and interactions. 1) $\nu = k\bar{v}$, $\Gamma = 0.4k\bar{v}$; 2) $\nu = k\bar{v}$, $\Gamma = 0$; 3) $\nu = 0$, $\Gamma = 0.4k\bar{v}$; 4) $\nu = \Gamma = 0$.

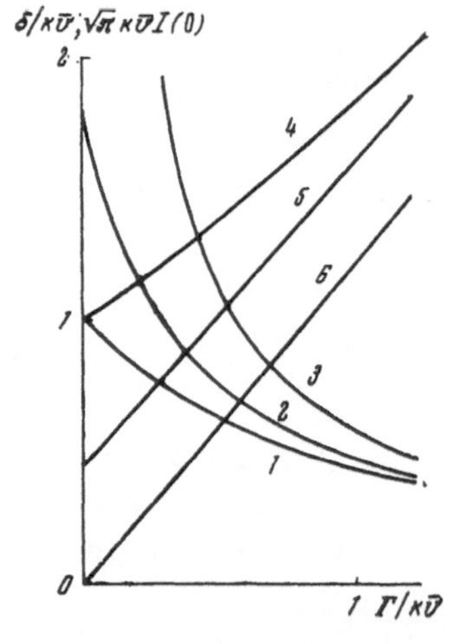

Fig. 18. Intensity (curves 1, 2, 3) at maximum and width (4, 5, 6) of line as functions of $\Gamma/k\bar{v}$ in the case of statistical independence of broadening due to interaction and the Doppler effect. 1, 4) $\nu = 0$; 2, 5) $\nu = k\bar{v}$; 3, 6) $\nu \to \infty$.

We note particularly that the above arguments make considerable use of the independence of Γ and Δ on $\mathbf{v}$. If this was not so the line profile would not be expressed in the form of a convolution. Thus, the broadenings due to interaction and the Doppler effect are statistically independent [even in the case (15.4) of collisions at different times] only if Γ, Δ are independent of the velocity of the emitting atom. This is quite natural, since if $\Gamma = \Gamma(\mathbf{v})$, $\Delta = \Delta(\mathbf{v})$, the increment of phase φ in a collision depends on the velocity acquired by the atom as a result of the previous elastic collision.

It is clear from considerations of symmetry that $\Gamma(\mathbf{v})$ and $\Delta(\mathbf{v})$ must be even functions of $\mathbf{v}$. Hence, it is easy to see that line (15.10) will be asymmetric.

Thus, correction for collisions leads to the conclusion that broadening due to interaction and the Doppler effect are statistically dependent even in the case where the changes in velocity and phase occur at different times. This gives rise to a new effect (asymmetry of the line), which is absent in the old versions of collision theory.

If Γ and Δ are independent of the velocity, the line profile is symmetric relative to the point $\Omega = \Delta$, i.e., the value of Δ determines the shift in the maximum of the line. It is clear from formula (15.10) that changes in phase due to collisions reduce the narrowing of the line due to elastic collisions. This follows formally from the fact that the integral term in the denominator of formula (15.10) is reduced by a factor $1 + \Gamma/\nu$ in comparison with (14.18). In the limiting case $\Gamma \gg \nu$ we can neglect it altogether and formula (15.10) becomes the usual convolution of the Gaussian and dispersion distributions:

$$I(\Omega) = \frac{\Gamma}{\pi^{3/2}\bar{v}} \int \frac{e^{-\mathrm{v}^2/\bar{v}^2}\, d\mathrm{v}}{\Gamma^2 + (\Omega - \Delta - \mathrm{kv})^2}\,. \qquad (15.12)$$

Figures 17, 18 give the results of numerical calculations of the line profile, the intensity at the maximum, and the line width δ for some values of the parameters. These figures show that when $\nu \gtrsim k\bar{v}$ the intensity at the maximum of the line rapidly decreases with increase in Γ, even if Γ is a small fraction (of the order of a few tenths) of ν. Thus, broadening due to interactions very effectively masks the narrowing of the Doppler profile. Nevertheless, when $\nu \sim \Gamma$ the narrowing is still quite appreciable (compare curves 1 and 3 in Fig. 17).

Very similar results are given by the soft-collision model. Here equation (15.5) for function $\tilde{f}$ has the form

$$\frac{\partial \tilde{f}}{\partial t} + \mathbf{v}\nabla_{\mathbf{r}}\tilde{f} = -(\Gamma + i\Delta)\tilde{f} + \nu_d\left[\operatorname{div}_{\mathbf{v}}(\mathbf{v}\tilde{f}) + \frac{\bar{v}^2}{2}\Delta_{\mathbf{v}}\tilde{f}\right]. \quad (15.13)$$

Solving this equation, we find the correlation function (for constant Γ, Δ, ν_d):

$$\Phi(t) = \exp\left\{-(\Gamma + i\Delta)t - \frac{(k\bar{v})^2}{2\nu_d^2}\left[\nu_d t - 1 + e^{-\nu_d t}\right]\right\}. \quad (15.14)$$

Thus, $\Phi(t)$ is the product of the correlation functions encountered in the separate analysis of broadening due to interaction and the Doppler effect [compare formulas (13.5) and (13.6)]. Since purely Doppler broadening in the hard- and soft-collision models is very similar, the resulting line shapes will also be similar. Hence, we will not make a thorough analysis of formula (15.14).

3. We now consider equation (15.7) with kernel $A(\mathbf{v}, \mathbf{v}', \psi)$ of the following form:

$$A(\mathbf{v}, \mathbf{v}', \psi) = A(\mathbf{v}, \mathbf{v}')\, B(\psi). \quad (15.15)$$

This expression means that a change in velocity and phase occurs in the same collision, but the magnitudes of these changes are quite independent of one another. Without loss of generality we can put

$$\int B(\psi)\, d\psi = 1. \quad (15.16)$$

In this case function $A(\mathbf{v}, \mathbf{v}')$ has the sense of the probability of a change in velocity $\mathbf{v} \to \mathbf{v}'$ in unit time, irrespective of the change in phase. For this function we make the same assumptions as in § 14. Then equation (15.7) for $\tilde{f}$ in the hard-collision model will be

$$\frac{\partial \tilde{f}}{\partial t} + \mathbf{v}\nabla\tilde{f} = -(\Gamma + i\Delta)\tilde{f} - \tilde{\nu}\left[\tilde{f} - W_M(\mathbf{v})\int \tilde{f}(\mathbf{r}, \mathbf{v}', t)\, d\mathbf{v}'\right], \quad (15.17)$$

where

$$\tilde{\nu} = \nu\tilde{B} = \nu\int B(\psi)\, e^{i\psi}\, d\psi = \nu - \Gamma - i\Delta, \quad (15.18)$$

and in the soft-collision model

$$\frac{\partial \tilde{f}}{\partial t} + \mathbf{v}\nabla_{\mathbf{r}}\tilde{f} = -(\Gamma + i\Delta)\tilde{f} + \tilde{\nu}_d\left[\operatorname{div}_{\mathbf{v}}(\mathbf{v}\tilde{f}) + \frac{\bar{v}^2}{2}\Delta_{\mathbf{v}}\tilde{f}\right], \quad (15.19)$$

$$\tilde{\nu}_d = \nu_d\left[1 - \frac{\Gamma + i\Delta}{\nu}\right].$$

Thus, in both models the effect of phase shifts in collisions reduces to replacement of the effective collision frequency ν, ν_d by the complex quantities $\tilde{\nu}$, $\tilde{\nu}_d$.

We consider first the hard-collision model. Repeating the calculations of § 14, we find

$$I(\Omega) = \frac{1}{\pi}\operatorname{Re}\left\{\int \frac{W(\mathbf{v})\, d\mathbf{v}}{\nu - i(\Omega - \mathbf{kv})} \Big/ \left[1 - \tilde{\nu}\int \frac{W_M(\mathbf{v})\, d\mathbf{v}}{\nu - i(\Omega - \mathbf{kv})}\right]\right\}. \quad (15.20)$$

As distinct from formula (15.10), the resonance denominators in expression (15.20) contain only ν, and not Γ and Δ. This is quite understandable, since the real part of the resonance factors depends on the mean duration of the train of waves into which the emission of the oscillator is divided by collisions [see the discussion of (14.18)]. In the case (15.10) the individual trains are produced both by phase jumps and changes in velocity. In this case changes in phase and velocity occur in the same collision, so that the mean number of trains in unit time is $1/\nu$,

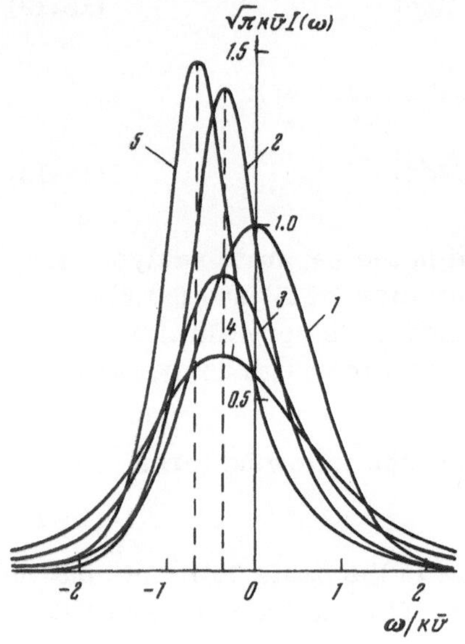

Fig. 19. Line profile in the case of statistical dependence of broadening due to interaction and the Doppler effect: 1) $\nu = \Gamma = \Delta = 0$; 2) $\nu = \frac{1}{2}k\bar{v}$, $\Gamma = 0$, $\Delta = \frac{1}{4}k\bar{v}$; 3) $\nu = \frac{1}{2}k\bar{v}$, $\Gamma = \Delta = \frac{1}{4}k\bar{v}$; 4) $\nu = \Gamma = \frac{1}{2}k\bar{v}$, $\Delta = \frac{1}{4}k\bar{v}$; 5) $\nu = \Delta = \frac{1}{2}k\bar{v}$, $\Gamma = 0$.

as in the absence of phase jumps. Hence, the resonance factors in (15.20) are the same as in (14.18). The effect of phase changes in collisions was manifested in the factor $\tilde{\nu}$ in (15.20). We recall that the denominator of formula (15.20) reflects the role of interference of the trains, which leads to narrowing of the profile of the Doppler-broadened line. On the other hand, it is easy to see from (15.16) and (15.18) that $|\tilde{\nu}| < \nu$. Hence, the appearance of $\tilde{\nu}$ in (15.20) can be interpreted as a reduction of coherence between the different trains.

We assume that $\tilde{\nu}$ is a real quantity. This is the case when the function $B(\psi)$ is symmetric. Then a simple redefinition of the parameters reduces formula (15.20) to (15.10). In fact, if we put $\nu - \Gamma$ in (15.20), then (15.20) is exactly the same as (15.10) with ν replaced by $\nu - \Gamma$. Thus, even $\Delta = \text{Im}\,\tilde{\nu} = 0$, the line profile, is qualitatively the same, irrespective of whether the changes in phase and velocity occur simultaneously or in different collisions.

If $\Delta \neq 0$, a much more interesting situation arises. In this case the line profile is not merely displaced [as occurred in the case of nonsimultaneous jumps in phase and velocity, see formula (15.10)], but is also asymmetric. This is clearly illustrated by Fig. 19, where the intensity distributions for different values of parameters are given.

It is known from the general theory of the spectral intensity of random steady processes that the centroid of the intensity distribution $I(\Omega)$ is connected with the derivative of the correlation function (see, for instance, [86]):

$$\langle\Omega\rangle = \int_{-\infty}^{\infty} \Omega I(\Omega)\, d\Omega = i\left(\frac{d\Phi}{dt}\right)_{t\to 0}. \tag{15.21}$$

Using the general expression (13.2) for $\Phi(t)$, we obtain

$$\langle\Omega\rangle = \left(\frac{d}{dt}\langle -\varphi + \mathbf{kr}\rangle\right)_{t\to 0}. \tag{15.22}$$

Since

$$\frac{d}{dt}\langle\mathbf{r}\rangle = \langle\mathbf{v}\rangle = 0; \tag{15.23}$$

$$\frac{d}{dt}\langle\varphi\rangle = -\Delta, \tag{15.24}$$

then formula (15.22) gives

$$\langle\Omega\rangle = \Delta. \tag{15.25}$$

Hence, the shift of the centroid of the line is quite independent of the elastic-scattering model and is due entirely to the broadening due to interaction. On the other hand, for the case $\nu \ll k\bar{v}$ it is easy to show that the shift of the maximum of the profile given by formula (15.20) is

$$\Omega_{max} = 2\Delta. \tag{15.26}$$

Thus, the asymmetry of the line is due to the fact that the maximum is shifted more than the centroid of the line.

We will show that the asymmetry of the line depends on the fact that the distribution of the random quantities **r** and φ is not normal. We assume the opposite. Then, from the definition of normally distributed random quantities we have [80]

$$\Phi(t) = \langle \exp\{-i\mathbf{kr} + i\varphi\}\rangle = \exp\left\{i\langle\varphi\rangle - \frac{1}{2}[\langle(\mathbf{kr})^2\rangle + 2\langle\mathbf{kr}\varphi\rangle + \langle\varphi^2\rangle]\right\}. \quad (15.27)$$

It is obvious that in our problem $\langle \mathbf{r}\varphi\rangle = 0$, since equal phase shifts will correspond to opposite displacements. Thus, the assumption of a normal distribution of **r** and φ automatically means the statistical independence of broadening due to interaction and the Doppler effect and, hence, symmetry of the line profile relative to the frequency $\Omega = \Delta$. It follows from the above that asymmetry of the line may be due entirely to the deviation of the statistics of **r** and φ from normal.

The expressed statement is well illustrated by the soft-collision model. It was shown above [see formula (13.27) and the discussion of formulas (14.24)] that in this model the shift of **r** is a normally distributed quantity. This is still true when phase changes are taken into account. However, the two-dimensional probability distribution for **r** and φ will no longer be normal. This is shown by the expression for the correlation function which can be found from equation (15.19):

$$\Phi(t) = \exp\left\{-(\Gamma + i\Delta)t - \frac{(k\bar{v})^2}{2\tilde{\nu}_d^2}[\tilde{\nu}_d t - 1 + e^{-\tilde{\nu}_d t}]\right\}, \quad (15.28)$$

$$\tilde{\nu}_d = \nu_d\left[1 - \frac{\Gamma + i\Delta}{\nu}\right].$$

The definition (13.1) of function $I(\Omega)$ shows that it will be a symmetric function of Ω if the correlation function is complex only due to the factor $\exp\{-i\Delta t\}$. This is not so for expression (15.28) if $\Delta \neq 0$, since $\tilde{\nu}_d$ is a complex number. Hence, the line profile will be asymmetric and this means that the distribution for **r**, φ is not normal. It should be noted, however, that since $\nu_d \ll \nu$ [see the discussion of model (14.26)] the effect of phase jumps in this case will be slight.

If the number of collisions is large, the statistics for **r**, φ will tend to normal. Hence, the line profile will become symmetric when $\nu \gg k\bar{v}$. Confining ourselves to the region of not very large frequencies $\Omega < \nu$, where the greater part of the energy is concentrated, we obtain from (15.20) the following expression for $I(\Omega)$:

$$I(\Omega) = \frac{1}{\pi}\frac{\Gamma + \frac{(k\bar{v})^2}{2\nu}}{(\Omega - \Delta)^2 + \left[\Gamma + \frac{1}{2}\frac{(k\bar{v})^2}{\nu}\right]^2}. \quad (15.29)$$

Thus, in this case the line is, in fact, symmetric.

We consider in conclusion the relation between the above theory and experiment. As regards the narrowing of the lines due to elastic collisions, we will be concerned, of course, not with establishing the effect itself (it has been observed in many investigations), but with determining its role in different conditions characteristic of the optical region of the spectrum, where it has never been taken into consideration. As was shown above, there may be narrowing when $\nu \sim \Gamma$, and in several cases such a situation can occur.

However, we think that the most interesting consequence of the theory is the appearance of asymmetry of the line profile when the statistical dependence of broadening due to interaction and the Doppler effect is taken into account. In some experiments asymmetry of the line has been observed at pressures for which there are no grounds for expecting a high intensity of the statistical wing. We have in mind [87], where asymmetry of the neon line $\lambda = 3.3913\ \mu$, $5s'[\frac{1}{2}]_1^0-4p'[\frac{3}{2}]_2$, was observed at pressures of about 1 mm Hg, and also [31, 61], which dealt with the nonlinear effects discovered in the neon transition $\lambda = 1.1523\ \mu$, $4s'[\frac{1}{2}]_1^0-3p'[\frac{3}{2}]_2$. It is possible that the line asymmetry found in [87, 31, 61] is due to the above-considered effect, although we cannot assert this conclusively without further investigation. The results of [31, 61] will be discussed more fully below, after the analysis of nonlinear effects.

§ 16. On the Relaxation Terms in the Equation for the Density Matrix

1. We write the Hamiltonian of the atom in the following form:

$$\hat{H} = \hat{H}_0 + \hbar\hat{V} + \hbar\hat{W}, \tag{16.1}$$

where $\hat{H}_0$ is the Hamiltonian of the isolated atom, and $\hbar\hat{V}$ and $\hbar\hat{W}$ describe the interaction of the atom with the external field and with the surrounding particles, respectively. In the collision approximation the matrix elements of $\hat{W}$ differ from zero only for a very short time ($\sim 10^{-13}$ sec), small in comparison with the lifetime ($\sim 10^{-8}$ sec). We also assume that during the lifetime the atom does not collide with any particle more than once. In addition, we will assume that collisions do not mix the states m, n between which the electromagnetic field causes transitions. For the optical region of the spectrum this assumption is usually valid. The fact is that the inelastic collision cross sections increase strongly with reduction in the difference of the energy levels. Hence, a pair of levels for which the transition lies in the optical region is usually perturbed by other, closer levels.

These conditions enable us to write the equation for the probability amplitudes in the folowing form:

$$i\dot{\hat{a}} = \hat{\alpha}\hat{a} + \hat{V}\hat{a}, \quad \hat{a} = \begin{pmatrix} a_m \\ a_n \end{pmatrix}, \tag{16.2}$$

where $\hat{\alpha}$ is a diagonal matrix. An explicit expression for $\hat{\alpha}$ can be obtained, for instance, from perturbation theory, as was done in [88]

$$\alpha_{ik} = \delta_{ik}\left[W_{ii} - i\sum_j{}' W_{ij}\int W_{ji}dt\right] \quad (i, k = m, n), \tag{16.3}$$

or can be assigned from model considerations [89]. All that matters to us is that $\hat{\alpha}$ is a random function of time, and the change of $\hat{\alpha}$ in time has the form of short "pulses," the overlap of which is neglected in the collision approximation.

The matrix $\hat{V}$ will also be a random function of time owing to the change in the velocity of the atom in collisions. For the reasons discussed in § 13, 15, matrices α and $\hat{V}$ in the general case are statistically dependent.

Decay due to spontaneous transitions can be taken into consideration if we add a term $-i\hat{\gamma}'a$ to the right side of (16.2), as in equation (3.5).

For what follows it is convenient to convert to the matrix

$$\hat{\sigma} = a\hat{a}^{\dagger},$$

the equation for which is easily obtained from (16.2):

$$i\dot{\hat{\sigma}} = \hat{\alpha}\hat{\sigma} - \hat{\sigma}\hat{\alpha}^{+} + \hat{V}\hat{\sigma} - \hat{\sigma}\hat{V}. \tag{16.4}$$

In the problems of interest to us we need to know σ_{ij}, averaged over the collisions. An equation of the type (16.4) is usually used for the mean values of σ_{ij}, but its random matrix $\hat{\alpha}$ is replaced by a matrix which is independent of time (see [58, 60], for instance). In the introduction and at the end of § 15 we mentioned the experimental data of [31, 87], which suggest that in some cases the introduction of such relaxation terms is invalid even in the collision model. The results of § 15 show that one of the possible restrictions may be due to the statistical dependence of $\hat{\alpha}$ and $\hat{V}$. This question is considered below.

2. We convert to a system of integral equations for σ_{ij}, which is easily obtained from (16.4):

$$\begin{aligned} \sigma_{mm}(t) &= \sigma_{mm}(t_0)\, e^{-i\delta_{mm}(t_0,\, t)} + 2\int_{t_0}^{t} \mathrm{Re}\{iV^{*}_{mn}(t')\,\sigma_{mn}(t')\}\, e^{-i\delta_{mm}(t',\, t)}\, dt', \\ \sigma_{nn}(t) &= \sigma_{nn}(t_0)\, e^{-i\delta_{nn}(t_0,\, t)} - 2\int_{t_0}^{t} \mathrm{Re}\{iV^{*}_{mn}(t')\,\sigma_{mn}(t')\}\, e^{-i\delta_{nn}(t',\, t)}\, dt', \\ \sigma_{mn}(t) &= i\int_{t_0}^{t} V_{mn}(t')\,[\sigma_{mm}(t') - \sigma_{nn}(t')]\, e^{-i\delta_{mn}(t',\, t)}\, dt'. \end{aligned} \tag{16.5}$$

Here $\delta_{ij}(t', t)$ denotes the integrals

$$\delta_{ij}(t', t) = \int_{t'}^{t} (\alpha_{ii} - \alpha^{*}_{jj})\, dt''. \tag{16.6}$$

The diagonal elements δ_{mm}, δ_{nn} are purely imaginary and obviously describe the decay of states m, n. The nondiagonal elements δ_{mn}, δ_{nm} are complex. Formula (16.3) shows that the imaginary parts of δ_{mn}, δ_{nm} are due to inelastic processes and the real parts to the shift in levels m, n during collisions. The quantity δ_{mn} is the same as the complex phase φ, which figured in § 13-15.

The factors $\exp\{-i\delta_{ij}(t', t)\}$ in (16.5) can be averaged independently of the averaging of the other factors. In fact, $\delta_{ij}(t', t)$ is determined by collisions at $t > t'$; the functions $V_{mn}(t')$ and $\sigma_{ij}(t')$, however, depend on the collisions which took place at $t_0 < t < t'$. It is easy to establish the explicit dependence on the time for $\exp\{-i\delta_{ij}(t', t)\}$. The quantities α_{ii} have the form of short, nonoverlapping "pulses." Hence, their number is given by a Poisson law. Using this, we can arrive at the following formulas:

$$\begin{aligned} \langle \exp\{-i\delta_{mm}(t_1, t_2)\}\rangle &= e^{-\Gamma_m (t_2 - t_1)}; & \Gamma_m &= \int P(g)\,[1 - e^{-i\delta_{mm}(g)}]\, dg, \\ \langle \exp\{-i\delta_{nn}(t_1, t_2)\}\rangle &= e^{-\Gamma_n (t_2 - t_1)}; & \Gamma_n &= \int P(g)\,[1 - e^{-i\delta_{nn}(g)}]\, dg, \\ \langle \exp\{-i\delta_{mn}(t_1, t_2)\}\rangle &= e^{-(\Gamma + i\Delta)(t_2 - t_1)}; & \Gamma + i\Delta &= \int P(g)\,[1 - e^{-i\delta_{mn}(g)}]\, dg. \end{aligned} \tag{16.7}$$

Here g denotes the set of collision parameters; P(g)dg is the number of collisions of type g in unit time; the quantities $\delta_{ij}(g)$ determine the contribution of one collision of type g to $\delta_{ij}(t_1, t_2)$ [compare with formula (13.5)].

We denote by ρ_{ij} the values of σ_{ij} averaged over δ_{ij}. Then from (16.7) and (16.5) we obtain

$$\rho_{mm} = \rho_{mm}(t_0)\, e^{-\Gamma_m(t-t_0)} + 2\int_{t_0}^{t} \mathrm{Re}\{i\langle V_{mn}^{*}(t')\,\sigma_{mn}(t')\rangle\}\, e^{-\Gamma_m(t-t')}\, dt',$$

$$\rho_{nn} = \rho_{nn}(t_0)\, e^{-\Gamma_n(t-t_0)} - 2\int_{t_0}^{t} \mathrm{Re}\{i\langle V_{mn}^{*}(t')\,\sigma_{mn}(t')\rangle\}\, e^{-\Gamma_n(t-t')} dt', \quad (16.8)$$

$$\rho_{mn} = i\int_{t_0}^{t} \langle V_{mn}(t')\,[\sigma_{mm}(t') - \sigma_{nn}(t')]\rangle\, e^{-(\Gamma+i\Delta)(t-t')}\, dt'.$$

If V_{mn} and σ_{ij} are statistically independent, then

$$\langle V_{mn}\sigma_{ij}\rangle = \langle V_{mn}\rangle\langle\sigma_{ij}\rangle = V_{mn}\rho_{ij}$$

and (16.8) will be a system of equations for ρ_{ij}. The reverse conversion to a differential equation gives

$$\begin{aligned} \dot\rho_{mm} + \Gamma_m\rho_{mm} &= 2\mathrm{Re}\,\{iV_{mn}^{*}\rho_{mn}\}, \\ \dot\rho_{nn} + \Gamma_n\rho_{nn} &= -2\mathrm{Re}\,\{iV_{mn}^{*}\rho_{mn}\}, \\ \dot\rho_{mn} + (\Gamma + i\Delta) &= iV_{mn}[\rho_{mm} - \rho_{nn}], \\ \rho_{nm} &= \overset{*}{\rho}_{mn}. \end{aligned} \quad (16.9)$$

Thus, we have arrived at equations for ρ_{ij} with these relaxation constants. The constants Γ_m, Γ_n obviously determine the decay of the excited states m, n in the absence of an external field, $V_{mn} = 0$; the constants Γ and Δ are the width and shift of the spontaneous emission line when $V_{mn} = 0$. Decay due to spontaneous transitions leads to an increase in Γ_m, Γ_n and Γ by $2\gamma_m$, $2\gamma_n$, and $\gamma_m + \gamma_n$, respectively.

It is clear from formula (16.6) that the quantities δ_{ij} are connected with one another:

$$\mathrm{Im}\, 2\delta_{mn} = \mathrm{Im}\,[\delta_{mm} - \delta_{nn}].$$

Using this relationship, we can easily show that

$$2\Gamma \geqslant \Gamma_m + \Gamma_n. \quad (16.10)$$

The equality sign occurs when $\delta_{mm} = \delta_{nn}$, Re $\delta_{mn} = 0$ (i.e., when quenching has just taken place, and it is the same for both levels) in the Karplus-Schwinger model [59], according to which in each collision $i\delta_{mm}$, $i\delta_{nn} \gg 1$ and, finally, in the opposite case $i\delta_{mm}$, $i\delta_{nn} \ll 1$, Re $\delta_{mn} = 0$.

The system of equations (16.9) was obtained on the assumption of the statistical independence of the changes in velocity and δ_{ij}. In the opposite case the introduction of relaxation terms into the equation for the matrix density is, generally speaking, invalid.

3. In (16.9) the averaging was only over δ_{ij}. If the velocities of the atoms are unaltered in the collisions, the system of equations (16.9) is the final one. In the opposite case we can take two courses. We can solve system (16.4) [or (16.9), if the changes in velocity and δ_{ij} are statistically independent], regarding **r**(t), **v**(t) and, hence, $V_{mn}(t)$ as random functions of time, and then carry out the averaging. The other course is to average the equations (16.4) [or (16.9)] themselves. This gives a fairly complicated system of integro-differential equations, which can probably be solved only by the method of successive approximations (we note that the results of § 13-15 correspond to the first approximation). In this calculation procedure the second method of averaging has no advantages over the use of the initial systems (16.4), (16.9). In addition, in the analysis of (16.4), (16.9) the physical sense of the particular assumptions is clearer. Hence, in § 17 we will proceed from the systems of equations (16.4), (16.9).

§ 17. The Saturation Effect with Collisions Taken into Account

1. We consider first the cases in which for some reason or other we can neglect the change in velocity of the atom in collisions. In such conditions we can use the system of equations (16.9), regarding $V_{mn}(t)$ as a regular function of time

$$\dot{\rho}_{mm} + \Gamma_m \rho_{mm} = 2\mathrm{Re}\{iV^*_{mn}\rho_{mn}\}, \quad \rho_{nn} + \Gamma_n \rho_{nn} = -2\mathrm{Re}\{iV^*_{mn}\rho_{mn}\},$$
$$\dot{\rho}_{mn} + (\Gamma + i\Delta)\rho_{mn} = iV_{mn}[\rho_{mm} - \rho_{nn}], \quad \rho_{mn} = \overset{*}{\rho}_{nm}. \tag{17.1}$$

We assume that spontaneous decay is included in Γ_m, Γ_n, and Γ, i.e., $2\gamma_m$, $2\gamma_n$, and $\gamma_m + \gamma_n$, respectively, are added to the values given by formulas (16.7).

If the external field is monochromatic and the atoms are stationary, the solution of system (17.1) has the form of a linear combination of exponential functions

$$\rho_{mm} = Ae^{\alpha t}, \quad \rho_{nn} = Be^{\alpha t}, \quad \rho_{mn} = Ce^{[\alpha - i(\Omega - \Delta)]t}, \tag{17.2}$$

and α is found from the characteristic equation

$$(\alpha + \Gamma_m)(\alpha + \Gamma_n)[(\alpha + \Gamma)^2 + (\Omega - \Delta)^2] + 4(\alpha + \Gamma)\left(\alpha + \frac{\Gamma_m + \Gamma_n}{2}\right)G^2 = 0. \tag{17.3}$$

The probabilities of induced emission W_m and absorption W_n are equal:

$$W_m = \frac{2\Gamma}{\Gamma_m} \frac{G^2}{(\Omega - \Delta)^2 + \Gamma^2\left[1 + \frac{2(\Gamma_m + \Gamma_n)}{\Gamma\Gamma_m\Gamma_n} G^2\right]}; \tag{17.4}$$

$$W_n = \frac{\Gamma_m}{\Gamma_n} W_m. \tag{17.5}$$

The most significant difference between (17.4) and expression (4.5), which relates to the case of purely spontaneous decay, is due to the term $2(\Gamma_m + \Gamma_n)G^2/\Gamma_m\Gamma_n\Gamma$ in the denominator, which determines the saturation. It contains all three relaxation constants. It is of interest to note that the rates of decay Γ_m, Γ_n of the levels m, n are not in the form of a sum, as might be expected from general considerations relating to line widths. In this case the sum of the reciprocals $\Gamma_m^{-1} + \Gamma_n^{-1}$, i.e., the sum of the decay times, is significant. This is quite understandable, since it is $\Gamma_m^{-1} + \Gamma_n^{-1}$ which determines the lifetime of the atom at both levels m, n, i.e., the time during which the atom can interact with the field.

We consider the features of the relaxation kinetics due to the external field. If $G = 0$, each of the functions ρ_{ij} decays at the same rate:

$$\rho_{mm} \propto e^{-\Gamma_m t}, \quad \rho_{nn} \propto e^{-\Gamma_n t}, \quad \rho_{mn} \propto e^{-\Gamma t}. \tag{17.6}$$

In the presence of a field ρ_{ij} are determined by a linear combination of four exponents, the indices of which are found from equation (17.3). For simplicity we put $\Omega - \Delta = 0$. Then one root of equation (17.3), is $\alpha_4 = -\Gamma$, and the coefficient of $e^{\alpha_4 t}$ in formulas (17.2), as can easily be shown, becomes zero. Hence, in the resonance case there remains the sum of three exponential functions. The corresponding roots are found from the equation

$$(\alpha + \Gamma_m)(\alpha + \Gamma_n)(\alpha + \Gamma) + 4\left(\alpha + \frac{\Gamma_m + \Gamma_n}{2}\right)G^2 = 0. \tag{17.7}$$

Equation (17.7) is easily solved in the cases $\Gamma_m = \Gamma_n$ or $\Gamma = (\Gamma_m + \Gamma_n)/2$.

$$\alpha_{1,2} = -\frac{\Gamma + \Gamma_m}{2} \pm \sqrt{\left(\frac{\Gamma - \Gamma_m}{2}\right)^2 - 4G^2}, \quad \alpha_3 = -\Gamma_m, (\Gamma_m = \Gamma_n); \tag{17.8}$$

$$\alpha_{1,2} = -\frac{\Gamma_m + \Gamma_n}{2} \pm \sqrt{\left(\frac{\Gamma_n - \Gamma_m}{2}\right)^2 - 4G^2}, \quad \alpha_3 = -\Gamma, (2\Gamma = \Gamma_m + \Gamma_n). \tag{17.9}$$

The roots $\alpha_{1,2}$, which depend on G, are of most interest. A comparison of (17.8), (17.9) with formula (4.13) shows that here it is only a question of redefinition of the parameters. If the indicated conditions are not fulfilled, the roots of equation (17.6) have a relatively simple form only when one of the relaxation constants is much greater than the other two. The case of practical interest is

$$\Gamma \gg \Gamma_n, \ \Gamma_m. \tag{17.10}$$

Here the approximate formulas for the roots are

$$\alpha_{1,2} \approx -\frac{\Gamma_m + \Gamma_n}{2} - \frac{2G^2}{\Gamma} \pm \sqrt{\left(\frac{\Gamma_n - \Gamma_m}{2}\right)^2 + \left(\frac{2G^2}{\Gamma}\right)^2}, \quad \alpha_3 = -\Gamma + \frac{4G^2}{\Gamma}. \tag{17.11}$$

The situation becomes even more simple when

$$\Gamma_n \gg \frac{G^2}{\Gamma}, \ \Gamma_m. \tag{17.12}$$

Then

$$\alpha_1 = -\Gamma_m - \frac{2G^2}{\Gamma}, \quad \alpha_2 = -\Gamma_n - \frac{2G^2}{\Gamma}, \quad \alpha_3 = -\Gamma + \frac{4G^2}{\Gamma}. \tag{17.13}$$

In the case of initial conditions $\rho_{mm}(t_0) = 1$ the solution of system (17.1) will be

$$\begin{gathered} \rho_{mm} = e^{\alpha_1 (t - t_0)}, \quad \rho_{nn} = + \frac{2G^2}{\Gamma \Gamma_n} [1 - e^{-\Gamma_n (t - t_0)}], \\ \rho_{mn} = i \frac{G}{\Gamma} \rho_{mm} [1 - e^{-\Gamma (t - t_0)}]. \end{gathered} \tag{17.14}$$

This case is perfectly analogous to that which occurs when $\gamma_n \gg \gamma_m + (G^2/\gamma_n)$ and which was fully investigated in Chapters I-III [see formulas (4.16) and (4.17)]. Induced transitions lead to an increase in the decay rate $(-\alpha_1)$ of the upper level, and the population of the lower level "manages to follow" the population of the upper owing to condition (17.12).

Germogenova [51] suggested a method of solving the system (17.1) for an arbitrary time dependence of the field when inequalities (17.10) and (17.12) are satisfied. In this method ρ_{ij} are expressed in quadratures, as in (4.26).

In calculation of the emission and absorption of a "weak" field acting on the atom simultaneously with a "strong" field the system of equations (17.1) also leads to qualitatively the same results as in Chapters II and III. For instance, the expression for the probability of emission of a weak-field photon (frequency ω_μ) in the presence of a strong monochromatic field has the form

$$\begin{aligned} W_{\mu m} = \frac{2G_\mu^2}{\Gamma_m} \frac{1}{1 + \dfrac{2(\Gamma_m + \Gamma_n)}{\Gamma \Gamma_m \Gamma_n} \dfrac{G^2 \Gamma^2}{\Gamma^2 + \Omega^2}} \operatorname{Re} \left\{ \frac{1}{\Gamma + i\Omega_\mu} \left[1 - \frac{2G^2}{\Gamma - i\Omega} \times \right.\right. \\ \left.\left. \times \frac{[2\Gamma + i(\Omega_\mu - \Omega)][\Gamma + i(\Omega_\mu - 2\Omega)]/[\Gamma + i(\Omega_\mu - \Omega)]}{\left[\Gamma + i(\Omega_\mu - \Omega) + \dfrac{\Omega^2}{\Gamma + i(\Omega_\mu - \Omega)}\right] \left[\dfrac{\Gamma_m + \Gamma_n}{2} + i(\Omega_\mu - \Omega) - \dfrac{(\Gamma_n - \Gamma_m)^2/4}{\dfrac{\Gamma_m + \Gamma_n}{2} + i(\Omega_\mu - \Omega)}\right] + 4G^2} \right] \right\}; \end{aligned} \tag{17.15}$$

$$\Omega_\mu = \omega_\mu - \omega_{mn} - \Delta, \quad \Omega = \omega - \omega_{mn} - \Delta.$$

The first line in (17.15) gives a profile of dispersion form, the magnitude of which is reduced by the saturation effect. The second line in (17.15) leads to a considerable complication of the relationship between $W_{\mu m}$ and Ω_μ. In particular, when $\Omega_\mu \to \Omega$, we obtain

$$W_{\mu m} = \frac{2G_\mu^2}{\Gamma_m} \frac{\Gamma}{\Gamma^2 + \Omega_\mu^2} \frac{1}{\left[1 + \frac{2(\Gamma_m + \Gamma_n)}{\Gamma\Gamma_m\Gamma_n} \frac{G^2\Gamma^2}{\Gamma^2 + \Omega^2}\right]^2}. \quad (17.16)$$

From (17.4) and (17.16) we have

$$\frac{\alpha_\mu}{\alpha} = \frac{W_{\mu m}/G_\mu^2}{W_m/G^2} = \frac{1}{1 + \frac{2(\Gamma_m + \Gamma_n)}{\Gamma\Gamma_m\Gamma_n} \frac{G^2\Gamma^2}{\Gamma^2 + \Omega^2}} < 1. \quad (17.17)$$

This relationship is analogous to (11.14). Thus, when collisions are taken into account the amplification of the weak field for $\Omega_\mu \to \Omega$ is less than for the strong field, as was the case for purely spontaneous relaxation.

2. We turn now to the induced emission of atoms moving in the field of a standing monochromatic wave. A similar problem was solved in § 5 with collisions neglected. In this section we will take into account the change in velocity in collisions and also the change in phase of the atomic oscillator, and we will give due regard to the statistical dependence of these perturbations. As a starting point we will use the system of equations (16.4) for functions σ_{ij} which have not been averaged over collisions.

We will solve the system (16.4) by the method of successive approximations, taking V_{mn} small, and retain only the first nonvanishing corrections which are nonlinear with respect to $|V_{mn}|^2$. For initial conditions $\sigma_{mn}(t_0) = 1$ the standard procedure leads to the following expression for the emission probability

$$W_m = -2\mathrm{Re}\left\{i\int_{t_0}^{\infty} V_{mn}^*\sigma_{mn}\,dt\right\} = \int_{t_1}^{\infty} dt_1 \int_{t_1}^{t_1} dt_2 f(t_2, t_1)\, e^{-i\delta_{mm}(t_1, t_2)} -$$

$$-\int_{t_0}^{\infty}\int_{t_0}^{t_1}\int_{t_0}^{t_2}\int_{t_0}^{t_3} f(t_4, t_3)\, f(t_2, t_1)\, [e^{-i\delta_{mm}(t_3, t_2)} + e^{-i\delta_{nn}(t_3, t_2)}]\, e^{-i\delta_{mm}(t_0, t_4)}\, dt_1\, dt_2\, dt_3\, dt_4. \quad (17.18)$$

Here we introduce the symbol

$$f(t, t') = 2\mathrm{Re}\{V_{mn}^*(t)\, V_{mn}(t')\, e^{-i\delta_{mn}(t', t)}\}. \quad (17.19)$$

In correspondence with the above the matrix element $V_{mn}(t)$ has the form

$$V_{mn}(t) = Ge^{-i\Omega t}\cos[\mathbf{kr}(t) + \mu], \quad \Omega = \omega - \omega_{mn}. \quad (17.20)$$

Expression (17.18) must be averaged over the initial coordinates of the atoms (i.e., over μ) and over the collisions. Functions $\mathbf{r}(t)$ and $\delta_{ij}(t, t')$ are random functions. Averaging over μ gives

$$\overline{W}_m = \frac{G^2}{\Gamma_m} I_1(\Omega) - \frac{G^4}{2\Gamma_m} I_2(\Omega); \quad (17.21)$$

$$I_1(\Omega) = \Gamma_m \int_{t_0}^{\infty} dt_1 \int_{t_1}^{\infty} dt_2\, e^{-i\delta_{mm}(t_0, t_1)} \cos\mathbf{kr}(t_1, t_2)\, \mathrm{Re}\{e^{i\Omega(t_2-t_1)-i\delta_{mn}(t_1, t_2)}\},$$

$$I_2(\Omega) = \Gamma_m \int_{t_0}^{\infty} dt_1 \int_{t_1}^{\infty} dt_2 \int_{t_2}^{\infty} dt_3 \int_{t_3}^{\infty} dt_4\, e^{-i\delta_{mm}(t_0, t_1)} [e^{-i\delta_{mm}(t_2, t_3)} + e^{-i\delta_{nn}(t_2, t_3)}] \mathrm{Re}\{e^{i\Omega(t_2-t_1)-i\delta_{mn}(t_1, t_2)}\}\, \mathrm{Re}\{e^{i\Omega(t_4-t_3)-i\delta_{mn}(t_3, t_4)}\} \times$$

$$\times \{\cos\mathbf{k}[\mathbf{r}(t_3, t_4) - \mathbf{r}(t_1, t_2)] + \cos\mathbf{k}[\mathbf{r}(t_3, t_4) + \mathbf{r}(t_1, t_2)] + \cos\mathbf{k}[\mathbf{r}(t_1, t_4) + \mathbf{r}(t_2, t_3)]\}. \quad (17.22)$$

Here $\mathbf{r}(t', t)$ denotes the displacement of the atom in time $t - t'$. If the Doppler width $k\bar{v}$ is much greater than all the relaxation constants, then in the fourfold integral in (17.22) we can retain only $\cos\mathbf{k}[\mathbf{r}(t_3, t_4) - \mathbf{r}(t_1, t_2)]$, and drop the other two cosines. In fact, when the collision frequencies are small the displacements of the atom $\mathbf{r}(t', t)$ are strongly correlated in different lengths of time. This means that $\cos \mathbf{k}[\mathbf{r}(t_3, t_4) + \mathbf{r}(t_1, t_2)]$ and $\cos \mathbf{k}[\mathbf{r}(t_1, t_4) + \mathbf{r}(t_2, t_3)]$ will oscillate strongly in the whole region of integration with respect to t_1, t_2, t_3, t_4, excluding an interval of width of the order $1/k\bar{v}$ near the origin of coordinates. In the same conditions $\cos \mathbf{k}[\mathbf{r}(t_3, t_4) - \mathbf{r}(t_1, t_2)]$ is close to unity near the hyperplane $t_4 - t_3 = t_2 - t_1$. Thus,

$$I_1(\Omega) = \Gamma_m \int_{t_0}^{\infty} e^{-i\delta_{mm}(t_0, t_1)}\, dt_1 \int_{t_1}^{\infty} \operatorname{Re}\{e^{i\Omega(t_2-t_1)-i\delta_{mn}(t_1, t_2)-i\mathbf{kr}(t_1, t_2)}\}\, dt_2,$$

$$I_2(\Omega) = \frac{\Gamma_m}{2} \int_{t_0}^{\infty} e^{-i\delta_{mm}(t_0, t_1)} dt_1 \int_{t_1}^{\infty} dt_2 \int_{t_2}^{\infty} dt_3 \int_{t_3}^{\infty} dt_4\, [e^{-i\delta_{mm}(t_2, t_3)} + e^{-i\delta_{nn}(t_2, t_3)}] \times$$

$$\times \operatorname{Re}\{e^{i\Omega(t_4-t_3+t_2-t_1)-i\delta_{mn}(t_3, t_4)-i\delta_{mn}(t_1, t_2)-i\mathbf{kr}(t_3, t_4)+i\mathbf{kr}(t_1, t_2)} + e^{i\Omega(t_4-t_3-t_2+t_1)-i\delta_{mn}(t_3, t_4)+i\delta_{mn}(t_1, t_2)-i\mathbf{kr}(t_3, t_4)+i\mathbf{kr}(t_1, t_2)}\}. \quad (17.23)$$

Here the cosines are represented by exponential functions and the parity of cos z is used.

3. We proceed to average over the collisions. We note that the factor $\exp\{-i\delta_{mm}(t_0, t_1)\}$ depends on the collisions which have taken place in the interval (t_0, t_1) and the rest of the integrands include collisions for $t > t_1$. Hence, the factor $\exp\{-i\delta_{mm}(t_0, t_2)\}$ can be averaged independently after using formula (16.7). It is clear from considerations of stationarity in time that the remaining part of the integrands depends on the differences $t_2 - t_1$, $t_3 - t_1$, etc., and not on t_1. This enables us to carry out integration with respect to t_1:

$$I_1(\Omega) = \int_{t_1}^{\infty} \langle \operatorname{Re}\{e^{i\Omega(t_2-t_1)-i\delta_{mn}(t_1, t_2)-i\mathbf{kr}(t_1, t_2)}\}\rangle\, dt_2; \quad (17.24)$$

$$I_2(\Omega) = \frac{1}{2} \int_{t_1}^{\infty} dt_2 \int_{t_2}^{\infty} dt_3 \int_{t_3}^{\infty} dt_4 \langle [e^{-i\delta_{mm}(t_2, t_3)} + e^{-i\delta_{nn}(t_2, t_3)}] \times$$

$$\times \operatorname{Re}\{e^{i\Omega(t_4-t_3+t_2-t_1)-i\delta_{mn}(t_3, t_4)-i\delta_{mn}(t_1, t_2)-i\mathbf{kr}(t_3, t_4)+i\mathbf{kr}(t_1, t_2)} + e^{i\Omega(t_4-t_3-t_2+t_1)-i\delta_{mn}(t_3, t_4)+i\delta_{mn}(t_1, t_2)-i\mathbf{kr}(t_3, t_4)+i\mathbf{kr}(t_1, t_2)}\}\rangle. \quad (17.25)$$

It is easy to see that $(1/\pi)I_1(\Omega)$ is the same as function $I(\Omega)$ considered in § 13-15. This was to be expected, since in the linear approximation with respect to $|V_{mn}|^2$ the profiles of the spontaneous and induced emission lines are identical. Hence,

$$I_1(\Omega) = \pi I(\Omega). \quad (17.26)$$

Thus, to calculate $I_1(\Omega)$ we can use the results of § 14, 15.

In the calculation of integral $I_2(\Omega)$ we will assume that during inelastic collisions there is no change in the velocity of the atom. This assumption is usually valid, since quenching of excited states occurs only in collisions with electrons. Thus, the factors $\exp\{-i\delta_{mn}(t', t)\}$ can be averaged separately by using formulas (16.7). For the same reason we can extract the factor associated with inelastic collisions from $\exp\{-i\delta_{mn}(t', t)\}$. With these assumptions we obtain the expressions

$$I_2(\Omega) = \frac{1}{2} \int_{t_1}^{\infty} dt_2 \int_{t_2}^{\infty} dt_3 \int_{t_3}^{\infty} dt_4\, [e^{-\Gamma_m(t_3-t_2)} + e^{-\Gamma_n(t_3-t_2)}]\, e^{-\Gamma(t_4-t_3+t_2-t_1)} \times$$

$$\times \operatorname{Re}\{\langle \exp[(\Omega - \Delta)(t_4 - t_3 + t_2 - t_1) - i\delta_{43} - i\delta_{21} - i\mathbf{kr}_{43} + i\mathbf{kr}_{21}]\rangle +$$

$$+ \langle \exp[(\Omega - \Delta)(t_4 - t_3 - t_2 + t_1) - i\delta_{43} + i\delta_{21} - i\mathbf{kr}_{43} + i\mathbf{kr}_{21}]\rangle\}, \quad (17.27)$$

where we have introduced the symbols

$$\delta_{j+1, j} = \delta_{mn}(t_j, t_{j+1}), \quad \mathbf{r}_{j+1, j} = \mathbf{r}(t_j, t_{j+1}), \quad (17.28)$$

and Γ, Δ are the width and shift due to statistically independent collisions. Radiation damping is included in Γ_m, Γ_n, Γ.

We introduce the probability

$$W \equiv W(\mathbf{r}_{43}, \mathbf{r}_{32}, \mathbf{r}_{21}; \delta_{43}, \delta_{32}, \delta_{21}; t_4, t_3, t_2, t_1)$$

that within intervals of time $t_2 - t_1$, $t_3 - t_2$, $t_4 - t_3$ changes $\mathbf{r}_{43}$, $\mathbf{r}_{32}$,..., δ_{21} in the coordinates of the atom and phase occur. This probability can be put in the following form:

$$W = \int f_4(\mathbf{r}_{43}, \mathbf{v}_4, \delta_{43}, t_4 - t_3; \mathbf{v}_3) f(\mathbf{r}_{32}, \mathbf{v}_3, \delta_{32}, t_3 - t_2; \mathbf{v}_2) f_2(\mathbf{r}_{21}, \mathbf{v}_2, \delta_{21}, t_2 - t_1; \mathbf{v}_1) f(\mathbf{v}_1)\, d\mathbf{v}_1 d\mathbf{v}_2 d\mathbf{v}_3 d\mathbf{v}_4. \tag{17.29}$$

Here f_i are the arbitrary probabilities that an atom with velocity $\mathbf{v}_{i-1}$ in time $t_i - t_{i-1}$ acquires a velocity $\mathbf{v}_i$ and is shifted through $\mathbf{r}_{i,i-1}$ receiving an increment of phase $\delta_{i,i-1}$ (i = 4, 3, 2). Functions f_i satisfy the kinetic equation

$$\frac{\partial f_i}{\partial t} + \mathbf{v}_i \nabla f_i = S \tag{17.30}$$

and initial conditions

$$f_i(\mathbf{r}_{i,i-1}, \mathbf{v}_i, \delta_{i,i-1}, 0; \mathbf{v}_{i-1}) = \delta(\mathbf{r}_{i,i-1})\, \delta(\mathbf{v}_i - \mathbf{v}_{i-1})\, \delta(\delta_{i,i-1}). \tag{17.31}$$

In other words, functions f_i are analogous to the distribution function f introduced in § 13 and differ from the latter only in the initial conditions (17.31) [compare (13.14)]. The velocity distribution of the atoms at time t_1, which is the origin for the whole problem, is given by function $f(\mathbf{v}_1)$ in (17.29).

Substitution of (17.29) in (17.27) gives

$$\begin{aligned} I_2(\Omega) = \frac{1}{2} \operatorname{Re} \int \{F(0, \mathbf{v}_3, 0, \Gamma_m; \mathbf{v}_2) + F(0, \mathbf{v}_3, 0, \Gamma_n; \mathbf{v}_2)\} \times \\ \times F(\mathbf{k}, \mathbf{v}_4, 1, \Gamma - i(\Omega - \Delta); \mathbf{v}_3) \{F(-\mathbf{k}, \mathbf{v}_2, 1, \Gamma - i(\Omega - \Delta); \mathbf{v}_1) + \\ + F(-\mathbf{k}, \mathbf{v}_2, -1, \Gamma + i(\Omega - \Delta); \mathbf{v}_1)\} f(\mathbf{v}_1)\, d\mathbf{v}_1 d\mathbf{v}_2 d\mathbf{v}_3 d\mathbf{v}_4, \end{aligned} \tag{17.32}$$

where the functions F are Fourier transforms of f_i:

$$F(\mathbf{k}, \mathbf{v}_i, \eta, p, \mathbf{v}_{i-1}) = \int e^{-i\mathbf{kr} - i\eta\delta - pt} f(\mathbf{r}, \mathbf{v}_i, \delta, t, \mathbf{v}_{i-1})\, d\mathbf{r}\, d\mathbf{v}\, d\delta. \tag{17.33}$$

The functions F satisfy an integral equation similar to (14.14):

$$(\nu + p + i\mathbf{k}\mathbf{v}_i) F(\mathbf{k}, \mathbf{v}_i, \eta, p, \mathbf{v}_{i-1}) = \delta(\mathbf{v}_i - \mathbf{v}_{i-1}) + \int \widetilde{A}(\mathbf{v}', \mathbf{v}_i, \eta) F(\mathbf{k}, \mathbf{v}', \eta, p, \mathbf{v}_{i-1})\, d\mathbf{v}',$$

$$\widetilde{A}(\mathbf{v}', \mathbf{v}_i, \eta) = \int A(\mathbf{v}', \mathbf{v}, \psi)\, e^{-i\psi\eta}\, d\psi, \tag{17.34}$$

where $A(\mathbf{v}', \mathbf{v}, \psi)$ is the kernel of the collision integral S.

Thus, the averaging of the induced emission probability can be reduced to the solution of the same kinetic equation as in linear theory and to calculation of the integral (17.32) of the products of the Fourier transforms of the distribution functions. In the subsequent approximations of perturbation theory only the number of cofactors in (17.32) and the multiplicity of the integral are altered. When terms of the order of $|V_{mn}|^{2l}$ are taken into account (17.32) will have a $2l$-fold integral of productions, each of which contains $2l - 1$ functions F and a function $f(\mathbf{v}_1)$. In this sense the theory expounded in § 13 is the special case corresponding to $l = 1$.

We recall that in formula (17.21) we dropped two terms: $\cos \mathbf{k}(\mathbf{r}_{43}+\mathbf{r}_{21})$ and $\cos \mathbf{k}(\mathbf{r}_{41}+\mathbf{r}_{32})$. It was noted in the discussion of formula (17.21) that these terms are small if the collision frequency is much less than $k\bar{v}$. This assumption is not a fundamental restriction of the theory, since the contribution of these terms to $I_2(\Omega)$ can be expressed by integrals of the same functions F. Appropriate calculations showed that these integrals are actually $k\bar{v}/\nu$ times less than the remaining expression.

4. To make formula (17.32) more specific we use the hard-collision model considered in § 15:

$$A(\mathbf{v}', \mathbf{v}, \psi) = \nu W_M(\mathbf{v}) B(\psi), \tag{17.35}$$

where $W_M(\mathbf{v})$ is the equilibrium velocity distribution. The solution of equation (17.34) then has the form

$$F(\mathbf{k}, \mathbf{v}_i, \eta, p; \mathbf{v}_{i-1}) = \frac{1}{\nu + p + ik\mathbf{v}_i}\left\{\delta(\mathbf{v}_i - \mathbf{v}_{i-1}) + \frac{\nu \widetilde{B}(\eta) W_M(v_i)}{[\nu + p + ik\mathbf{v}_{i-1}]\left[1 - \sqrt{\pi}\frac{\nu}{k\bar{v}}\widetilde{B}(\eta) w\right]}\right\}, \tag{17.36}$$

where

$$w = \frac{k\bar{v}}{\sqrt{\pi}} \int \frac{W_M(\mathbf{v})\, d\mathbf{v}}{\nu + p + ik\mathbf{v}}, \quad \widetilde{B}(\eta) = \int B(\psi) e^{-i\eta\psi}\, d\psi. \tag{17.37}$$

Substituting (17.36) in (17.32), integrating with respect to $\mathbf{v}_1$, $\mathbf{v}_2$, $\mathbf{v}_3$, and $\mathbf{v}_4$, and using formulas (17.26) and (15.20), we obtain the following expression for $\langle \overline{W}_m \rangle$:

$$\langle \overline{W}_m \rangle = \frac{G^2}{\Gamma_m}\frac{\sqrt{\pi}}{k\bar{v}}\left\{\mathrm{Re}\frac{w}{1 - \frac{\nu\sqrt{\pi}}{k\bar{v}}\widetilde{B}(1) w} - \frac{G^2}{4}\left[\frac{1}{\nu + \Gamma_m} + \frac{1}{\nu + \Gamma_n}\right] \times \right.$$
$$\left. \times \frac{1}{\nu + \Gamma}\mathrm{Re}\left[\frac{\nu + \Gamma}{\nu + \Gamma - i(\Omega - \Delta)}\frac{w}{\left[1 - \frac{\nu\sqrt{\pi}}{k\bar{v}}\widetilde{B}(1) w\right]^2} + \frac{w}{\left|1 - \frac{\nu\sqrt{\pi}}{k\bar{v}}\widetilde{B}(1) w\right|^2}\right]\right\}. \tag{17.38}$$

In function w we must put $p = \Gamma - i(\Omega - \Delta)$. If the elastic collision frequency $\nu = 0$ and, in addition, $\Gamma_m, \Gamma_n, \Gamma \ll k\bar{v}$, then $w \approx \exp\{-(\Omega - \Delta)^2/(k\bar{v})^2\}$ and

$$\langle \overline{W}_m \rangle = \frac{G^2}{\Gamma_m}\frac{\sqrt{\pi}}{k\bar{v}} e^{-(\Omega-\Delta)^2/(k\bar{v})^2}\left\{1 - \frac{G^2(\Gamma_m + \Gamma_n)}{4\Gamma\Gamma_m\Gamma_n}\left[1 + \frac{\Gamma^2}{\Gamma^2 + (\Omega - \Delta)^2}\right]\right\}. \tag{17.39}$$

This expression becomes (5.21), which was obtained by neglecting collisions, when $\Gamma_m = 2\gamma_m$, $\Gamma_n = 2\gamma_n$, $\Gamma = \gamma_m + \gamma_n$, $\Delta = 0$. Thus, in the absence of elastic collisions a consideration of the perturbation of the emitting atom by other particles reduces to a redefinition of the parameters, and the general form of the expression for $\langle \overline{W}_m \rangle$ is unaltered. The depth of the dip on the profile of $\langle W_m(\Omega) \rangle$, which depends on the term $\Gamma^2/[\Gamma^2 + (\Omega - \Delta)^2]$, will decrease with increase in the concentration of perturbing particles. We note also that $\langle \overline{W}_m \rangle$ as a function of frequency is symmetric relative to $\Omega = \Delta$.

Now let $\nu \neq 0$, and $\widetilde{B}(1)$ be a real number; this means that in collisions in which the velocity of the atom changes the mean value of the phase jumps is zero. In this case, as can easily be seen from (17.38) and (17.37), the probability $\langle \overline{W}_m \rangle$ will also be symmetric relative to $\Omega - \Delta$, but the width of the dip will be increased and its depth reduced. If $\mathrm{Im}\,\widetilde{B}(1) \neq 0$, then $\langle W(\Omega) \rangle$ is an asymmetric function of the frequency. The asymmetry is due to two factors. As

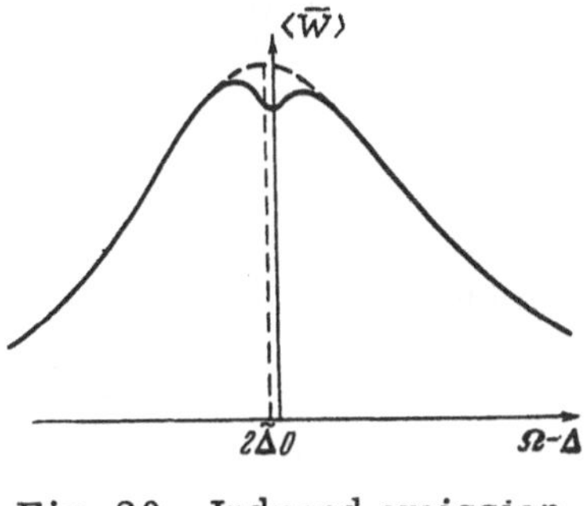

Fig. 20. Induced emission probability as a function of frequency in the case of statistical dependence of broadening due to the Doppler effect and interactions.

was shown in § 15, function $I_1(\Omega)$ is asymmetric. In addition, and this is more important, the maximum of function $I_1(\Omega)$ does not coincide with that of $I_2(\Omega)$. In fact, according to (15.25), the maximum of $I_1(\Omega)$ occurs when

$$\Omega - \Delta = -2\nu \operatorname{Im} B(1) = +2\tilde{\Delta} \tag{17.40}$$

[we note that the quantity Δ in (15.25) corresponds to $\tilde{\Delta}$ in (17.40)]. The maximum of function $I_2(\Omega)$ depends on the "sharp" factor $1/[\Gamma + \nu - i(\Omega - \Delta)]$, i.e., it occurs when $\Omega - \Delta = 0$. Hence, in the region of small frequencies ($\Omega - \Delta$, $\tilde{\Delta} \ll k\bar{v}$), we have

$$\langle \overline{W}_m \rangle = \frac{G^2}{\Gamma_m} \frac{\sqrt{\pi}}{k\bar{v}} \left\{ 1 - \frac{(\Omega - \Delta - 2\tilde{\Delta})^2}{(k\bar{v})^2} - \frac{G^2}{4(\Gamma + \nu)} \left[\frac{1}{\Gamma_m + \nu} + \frac{1}{\Gamma_n + \nu} \right] \times \right.$$
$$\left. \times \left[1 + \frac{(\Gamma + \nu)^2}{(\nu + \Gamma)^2 + (\Omega - \Delta)^2} \right] \right\}. \tag{17.41}$$

Thus, the maxima of functions $I_1(\Omega)$ and $I_2(\Omega)$ are separated from one another by $2\tilde{\Delta}$, i.e., by twice the displacement of the line which would be produced by elastic conditions if the statistical dependence of the changes in velocity and phase is neglected (Fig. 20).

5. An obvious explanation of the increase in width of the dip with increase in ν is that owing to elastic collisions the atoms "diffuse" within the velocity distribution and the effect of the external field on this distribution is reduced [see formulas (5.8), (5.20), and the subsequent discussions]. It is physically clear that the effect of this diffusion will depend greatly on the type of elastic collision. To clarify this question we consider the soft-collision model. In this case jumps will be neglected for the sake of simplicity.

In the case of interest to us the kinetic equation (17.30) has the form

$$\frac{\partial f}{\partial t} + \mathbf{v}\nabla_r f = \nu_d \left[\operatorname{div}_v (\mathbf{v} f) + \frac{\bar{v}^2}{2} \Delta_v f \right]. \tag{17.42}$$

Function F, defined in (17.33), is equal to

$$F(\mathbf{k}, \mathbf{v}_i, 0, p; v_{i-1}) = \int d\varkappa \int_0^\infty dt\, e^{-i\varkappa v_i - pt} X(\mathbf{k}, \varkappa, t, v_{i-1}), \tag{17.43}$$

$$X(k, \varkappa, t, v_{i-1}) = e^{i v_{i-1}\left[k \frac{1 - e^{-\nu_d t}}{\nu_d} + \varkappa e^{-\nu_d t}\right] - \frac{1}{2}[G'\varkappa^2 + 2H'\varkappa k + P' k^2]},$$

$$G' = \frac{\bar{v}^2}{2}[1 - e^{-2\nu_d t}], \quad H' = \frac{\bar{v}^2}{2\nu_d}[1 - e^{-\nu_d t}]^2,$$

$$P' = \frac{\bar{v}^2}{\nu_d^2}\left[\nu_d t - 1 + e^{-\nu_d t} - \frac{1}{2}(1 - e^{-\nu_d t})^2\right].$$

The substitution of (17.43) in (17.32) and integration with respect to $\varkappa$ and v_i leads to the following expression for $I_2(\Omega)$:

$$I_2(\Omega) = \frac{1}{2} \int_0^\infty \int_0^\infty \int_0^\infty d\tau_1 d\tau_2 d\tau_3 [e^{-\Gamma_m \tau_2} + e^{-\Gamma_n \tau_3}] e^{-\Gamma(\tau_1 + \tau_3)} \{\cos(\Omega - \Delta)(\tau_1 + \tau_3) + \cos(\Omega - \Delta)(\tau_1 - \tau_3)\} \times$$
$$\times \exp\left\{ -\frac{(k\bar{v})^2}{2\nu_d^2} [\nu_d(\tau_1 + \tau_3) - 2 + e^{-\nu_d \tau_1} + e^{-\nu_d \tau_3} - e^{-\nu_d \tau_2}(1 - e^{-\nu_d \tau_1})(1 - e^{-\nu_d \tau_3})] \right\}. \tag{17.44}$$

The last line of the integrand in (17.44) represents the role of motion of the atom and collisions involving a change of velocity. It is easy to see that the index of the exponent is

$$\frac{1}{2}\langle[\mathbf{kr}_{43}-\mathbf{kr}_{21}]^2\rangle=\frac{1}{2}[\langle(\mathbf{kr}_{43})^2\rangle+\langle(\mathbf{kr}_{21})^2\rangle-2\langle(\mathbf{kr}_{43})(\mathbf{kr}_{21})\rangle]$$

in accordance with the fact that in the soft-collision model the displacement of the atom is a normally distributed random quantity. The variables τ_i in (17.44) are connected with t_i by the relationships $\tau_i = t_{i+1}-t_i$. When $\nu_d = 0$ the index of the exponent is $(kv/2)^2(\tau_3-\tau_1)^2$, which corresponds to complete correlation of the displacements of the atom during different lengths of time. When $\nu_d \neq 0$ this correlation will be disturbed, but when $\nu_d \ll \Gamma_m\Gamma_n$ the disturbances will be relatively small. We can show that when

$$\frac{\nu}{\Gamma}\lesssim 4\frac{\min(\Gamma_m,\Gamma_n)}{k\bar{v}}\ll 1 \qquad (17.45)$$

we need only take into account the reduction of $\langle(\mathbf{kr}_{43})(\mathbf{kr}_{21})\rangle$ with increase in $\tau_2 = t_3-t_2$; then $I_2(\Omega)$ can be put in the following form:

$$I_2(\Omega)=\frac{\sqrt{\pi}}{2k\bar{v}}\int_0^\infty\left[\cos(\Omega-\Delta)z+e^{-\left(\frac{\Omega-\Delta}{k\bar{v}}\right)^2}\right]\left[\frac{1}{\Gamma_m+\frac{(k\bar{v})^2\nu_d}{8}z^2}+\frac{1}{\Gamma_n+\frac{(k\bar{v})^2\nu_d}{8}z^2}\right]e^{-\Gamma z}\,dz. \qquad (17.46)$$

The integral in (17.46) can be expressed by integral power functions, but we will not write out the corresponding formulas, since an analysis of expressions of this type was made in [90].

We note first of all that $I_2(\Omega)$ is symmetric relative to $\Omega=\Delta$. We can say the same about $I_1(\Omega)$. In approximation (17.45) it will be

$$I_1(\Omega)=\frac{\sqrt{\pi}}{k\bar{v}}e^{-\left(\frac{\Omega-\Delta}{k\bar{v}}\right)^2}. \qquad (17.47)$$

Hence, $\langle\overline{W}_m\rangle$ as a whole is a symmetric function of the frequency.

The function $I_2(\Omega)$ can, of course, be divided into two parts, one of which varies smoothly with the frequency and the other much more sharply. The first part is due to $\exp\{-(\Omega-\Delta)^2\times(k\bar{v})^{-2}\}$ (width $k\bar{v}$). The second part, due to $\cos(\Omega-\Delta)z$, contains two integrals of the same type, differing only in the values of the parameters Γ_m, Γ_n. The frequency dependence of each of these terms is, according to [90], close to a dispersion curve with a width given by the following approximate formula:

$$\beta_i\approx\Gamma\left\{1+\frac{\frac{3}{8}\left(\frac{k\bar{v}}{\Gamma}\right)^2\frac{\nu_d}{\Gamma_i}}{\sqrt{1+\frac{13}{8}\left(\frac{k\bar{v}}{\Gamma}\right)^2\frac{\nu_d}{\Gamma_i}}}\right\},\quad i=m,n;\ \frac{k\bar{v}}{\Gamma}\sqrt{\frac{\nu_d}{\Gamma_i}}<16. \qquad (17.48)$$

Thus,

$$\langle\overline{W}_m\rangle=\frac{G^2}{\Gamma_m}\frac{\sqrt{\pi}}{k\bar{v}}e^{-\left(\frac{\Omega-\Delta}{k\bar{v}}\right)}\left\{1-\frac{G^2}{4}\left[\frac{1}{\beta_m\Gamma_m}\left(1+\frac{\beta_m^2}{\beta_m^2+(\Omega-\Delta)^2}\right)+\frac{1}{\beta_n\Gamma_n}\left(1+\frac{\beta_n^2}{\beta_n^2+(\Omega-\Delta)^2}\right)\right]\right\}. \qquad (17.49)$$

It is easy to show that β_m, $\beta_n \ll kv$ when condition (17.45) and the applicability criterion (17.48) are satisfied. Hence, as in other models, the function $I_2(\Omega)$ has a relatively sharp dip close to frequency $\Omega=\Delta$.

The values of the parameters β_m, β_n are given by the ratios of the combinations $k\bar{v}(\nu_d/\Gamma_m)^{\frac{1}{2}}$, $k\bar{v}(\nu_d/\Gamma_n)^{\frac{1}{2}}$ to the collision width Γ. The physical sense of these ratios is readily understandable from the following considerations. According to formula (17.43), the variance of the velocity of the atom is

$$\sqrt{\langle [v_t - v_{t-1}e^{-\nu_d t}]^2 \rangle} = \sqrt{G'} = \frac{\bar{v}}{\sqrt{2}}\sqrt{1-e^{-2\nu_d t}} \approx \bar{v}\sqrt{\nu_d t} \qquad (17.50)$$

(if $\nu_d t \ll 1$), i.e., the "wandering" of the atom in velocity space conforms to a diffusion law (diffusion coefficient $\bar{v}^2\nu_d/2$). During the lifetime $t \sim 1/\Gamma_m$ (or $1/\Gamma_n$) the velocity changes, hence, by $\bar{v}(\nu_d/\Gamma_m)^{\frac{1}{2}}$ (or $\bar{v}(\nu_d/\Gamma_n)^{\frac{1}{2}}$), which is $k\bar{v}(\nu_d/\Gamma_m)^{\frac{1}{2}}$ ($k\bar{v}(\nu_d/\Gamma_n)^{\frac{1}{2}}$) in spectral units. Thus, the parameters $(k\bar{v}/\Gamma)\ (\nu_d/\Gamma_m)^{\frac{1}{2}}$, $(k\bar{v}/\Gamma)\ (\nu_d/\Gamma_m)^{\frac{1}{2}}$ are none other than the ratios of the Doppler shift due to diffusion of the atom during its sojourn at levels m, n, to the total collision width of the line.

In the linear problem (§ 13-15) soft collisions led to an appreciable change in the line profile only when ν_d was comparable with $k\bar{v}$. In the nonlinear problem under consideration it is sufficient that $\nu_d \sim k\bar{v}\Gamma^2\Gamma_m/(k\bar{v})^3 \ll k\bar{v}$ for the effect of collisions to become quite appreciable. This difference is due to the fact that in the particular case the velocity distribution is nonequilibrium (in contrast to the linear approximation), and the role of collisions consists primarily in removing this nonequilibrium. The width of the nonequilibrium addition is Γ/k and it is quite understandable that the collisions will be effective if during the lifetime Γ_m^{-1}, Γ_n^{-1} the variance of the velocity $\bar{v}(\nu_d/\Gamma_m)^{\frac{1}{2}}$, $\bar{v}(\nu_d/\Gamma_n)^{\frac{1}{2}}$ is of the order of, or greater than, Γ/k. Hence, it is obvious that formula (17.48) will also contain the parameters $(k\bar{v}/\Gamma)(\nu_d/\Gamma_m)^{\frac{1}{2}}$, $(k\bar{v}/\Gamma)\times(\nu_d/\Gamma_n)^{\frac{1}{2}}$. On the other hand, conditions (17.45) mean that the variance of the velocity in time $1/\Gamma_m$, $1/\Gamma_n$ does not exceed $\bar{v}$, although it may be greater than Γ/k.

It is of interest to compare hard and soft collisions from this viewpoint. According to formula (17.41), hard collisions will be effective in the sense interesting us, if $\nu \sim \Gamma_m, \Gamma_n, \Gamma$. It is easy to see that the change in the characteristic parameter is due to the specific nature of hard collisions. In fact, during the lifetime $1/\Gamma_m$, $1/\Gamma_n$ there will be ν/Γ_m, ν/Γ_n collisions, which will alter the velocity of the atom by an amount of the order of $\bar{v}$. Since the whole velocity distribution of the atoms has width $\bar{v}$, the relative number of atoms arriving in the velocity interval of interest to us, will be $(\nu/\Gamma_m)(\Gamma/k\bar{v})$, $(\nu/\Gamma_n)(\Gamma/k\bar{v})$. When these collisions were neglected the relative number of atoms was $\Gamma/k\bar{v}$. Hence, hard collisions will increase the number of atoms by a factor of $1+\nu/\Gamma_m$, $1+\nu/\Gamma_n$, which was reflected in formulas (17.39) and (17.4·) by the replacement of Γ_m^{-1}, Γ_n^{-1} by $(\nu+\Gamma_m)^{-1}$, $(\nu+\Gamma_n)^{-1}$. Thus, as distinct from linear theory, the hard- and soft-collision models lead to significantly different results in the nonlinear problem — hard and soft collisions will have the same effect on the dip if $\nu_d \sim \nu(\Gamma/k\bar{v})^2 \ll \nu$.

We consider the expression for β_m, β_n from the viewpoint of dependence on the concentration of perturbing particles, on which only ν_d depends. At sufficiently small ν_d only unity will remain under the root sign in (17.48).

$$\beta_i \approx \Gamma + \frac{3}{8}\frac{(k\bar{v})^2}{\Gamma}\frac{\nu_d}{\Gamma_i}, \quad i = m, n, \ \frac{13}{8}\frac{(k\bar{v})^2}{\Gamma}\frac{\nu_d}{\Gamma_i} \ll 1. \qquad (17.51)$$

It is easy to see, however, that in this case the addition due to collisions will be about 10% of Γ. Hence, the experimental detection of collision effects will be possible only in the opposite limiting case, when we can neglect unity under the root sign in (17.48):

$$\beta_i \approx \Gamma + \frac{3}{2\sqrt{26}}k\bar{v}\sqrt{\frac{\nu_d}{\Gamma_i}}; \quad \frac{kv}{\Gamma}\sqrt{\frac{\nu_d}{\Gamma_i}} > 1; \quad i = m, n. \qquad (17.52)$$

Thus, in cases of practical interest the addition to Γ is proportional to the square root of the concentration of the perturbing particles.

We note, finally, another detail which may be of interest in comparison with experiment. In the two considered collision models the "sharp" term is half of the value $I_2(\Omega)$ at the maximum $\Omega = \Delta$ [see formulas (17.41) and (17.46)]. As Perel' reported, this property of $I_2(\Omega)$ is presumably independent of the elastic-collision model, since it reflects the fact that at $\Omega = \Delta$ the two traveling waves forming the standing wave interact effectively with the same atoms ($|kv| \simeq \Gamma$), whereas at $|\Omega - \Delta| \gg \Gamma$ for each of the traveling waves different groups of atoms are significant ($|\Omega\Delta - \pm kv| \lessgtr \Gamma$).

6. The effect of elastic collisions of the atoms on the nonlinear part of the induced emission probability was considered in [26, 31, 91]. From qualitative considerations it is stated in [26] that when $\nu \sim \Gamma$ hard collisions can be neglected. This conclusion contradicts our results, according to which hard collisions are insignificant only if $\nu \ll \Gamma_m, \Gamma_n, \Gamma$. Since the initial premises of the interpretation expounded at the end of the preceding section **5** essentially agree with the conception of the author of [26], the disagreement is due simply to the inconsistency of the arguments in [26]. It is also stated in [26] that soft collisions will lead to an increase in the width of the dips. This assertion agrees with our results, but the quantitative assessment of the effect in [26] is incorrect. The author of [26] ignored the diffusive nature of the change of velocity in soft collisions, with the result that the broadening of the dip would, according to [26], be proportional to the collision frequency ν_d, and not $(\nu_d)^{\frac{1}{2}}$, as is the case according to (17.52).

In [31] the following formula was given without deduction for the induced emission probability:

$$\langle W\rangle = \frac{\sqrt{\pi}}{k\bar{v}}\frac{G^2}{\Gamma_m} e^{-\left(\frac{\Omega}{k\bar{v}}\right)^2}\left\{1 - \beta G^2\left[1 + \frac{\gamma\gamma'}{\gamma'^2 + \Omega^2}\right]\right\}. \tag{17.53}$$

According to [31], γ' includes the rate of radiation damping, broadening due to interactions, and the frequency of weak elastic collisions.

The quantity γ differs from γ' only by the frequency of soft elastic collisions. Hence, in our symbols

$$\gamma' = \Gamma + a\nu_d, \quad \gamma = \Gamma, \quad \Delta = 0. \tag{17.54}$$

It is pointed out in [31] that in the analysis of elastic scattering the wave function of the optical electron was regarded as unperturbed. In other words, it was assumed that the phase jumps $\delta_{nm} = 0$.

Formula (17.53) differs in several respects from our results contained in (17.49) and (17.48). Firstly, formula (17.49) has two dispersion terms, while (17.53) has only one. This difference disappears only if $\Gamma_m = \Gamma_n$. Secondly, as in [26], the frequency of the collisions, and not the square root of it, is introduced into the width of the dip γ'. Thirdly, the contribution of the "sharp" term to I_2 when $\Omega = 0$ is, according to (17.53), less than half:

$$1\Big/\left[1 + \frac{\gamma'}{\gamma}\right] < \frac{1}{2}.$$

It was pointed out in section **5** that this fraction is exactly a half. Since (17.53) was published without deduction it is difficult to indicate the reason for the discrepancy.

Information about [91] can be obtained only from short abstracts of the paper. According to these, the authors of [91] found a broadening and shift of the dip of the curve of $\langle W\rangle$ when collisions were taken into account. The specific physical mechanism giving rise to these effects is not indicated in the abstracts and it is impossible to form any definite opinion about [91].

7. It was mentioned at the end of § 16 that the role of collisions could be analyzed by two methods. The first method, used above, consisted in the solution of equations (16.4) for $\hat{\sigma}$ and subsequent averaging over the collision parameters. The second method consists in first obtaining the equation for the averaged density matrix $\hat{\rho}$ and then solving it for the particular form of V(t, z).

We give without deduction the equations for ρ_{ij}, which correspond to the averaging system adopted in this section:

$$\left(\frac{\partial}{\partial t}+\mathbf{v}\nabla\right)\rho_{jj}=-\Gamma_j\rho_{jj}+S_j\pm 2\mathrm{Re}\,[iV^{*}_{mn}\rho_{mn}],\quad j=m,\,n,$$
$$\left(\frac{\partial}{\partial t}+\mathbf{v}\nabla\right)\rho_{mn}=-(\Gamma+i\Delta)\,\rho_{mn}+S+iV_{mn}(\rho_{mm}-\rho_{nn}), \tag{17.55}$$

$$S_j=-\nu\rho_{jj}+\int A\,(\mathbf{v}',\,\mathbf{v},\,\psi)\,\rho_{jj}(\mathbf{v}')\,d\mathbf{v}'d\psi',$$
$$S=-\tilde{\nu}\rho_{mn}+\int\tilde{A}\,(\mathbf{v}',\,\mathbf{v})\,\rho_{mn}(\mathbf{v}')\,d\mathbf{v}'. \tag{17.56}$$

The collision integrals (17.56) are similar to those in equation (15.2) for the Fourier transform $\tilde{f}$ of the distribution function with respect to the phase variable. In (17.56) it is easy to include the quenching due to collisions involving a change in velocity. For this it is sufficient to put

$$\tilde{\Gamma}_j\equiv\nu-\int A\,(\mathbf{v},\,\mathbf{v}',\,\psi')\,d\mathbf{v}'d\psi'\neq 0. \tag{17.57}$$

The quantity $\tilde{\Gamma}_j$ obviously characterizes quenching in the considered collisions. Equations (17.55) lead to the same results as those obtained above.

Part II

APPLICATION TO LASER THEORY

CHAPTER V

Monochromatic Generation

§ 18. Self-Consistent Problem

The electrodynamics of lasers must be fundamentally nonlinear. In fact, the medium of a laser is in such a thermodynamically nonequilibrium state that it does not absorb, but emits electromagnetic waves. Hence, for such a medium solutions which do not increase in time are impossible in the linear approximation. The limitation on the increase in field may be due to the fact that the current induced by the field depends nonlinearly on the field. Physically nonlinearity may occur when the probabilities of induced and relaxation transitions are comparable. Calculation of the current is a quantum-mechanical problem, to which Part I was devoted.

It is obvious that in general nonlinear theory the electrodynamic and quantum-mechanical problems are inseparable. The Maxwell and Schrödinger equations, supplemented with rules for averaging over the particles of the medium, constitute a system of equations which must be solved jointly.

Below we will consider only monochromatic generation. In this practically important case the electrodynamic and quantum-mechanical problems can be separated to some extent. This simplification is due to the fact that the nonlinear field dependence of such characteristics as the mean current, polarization, and emission probability is not determined by the instantaneous value of the field, but by the mean value (during the oscillation period $2\pi/\omega$) of the square of the field amplitude. Thus, in the case in which we are interested we can introduce a dielectric constant which depends on the field amplitude and formulate the problem as follows. The Schrödinger equation is solved for some field external to the atom. The general form of this field (standing wave, one or several traveling waves, etc.) and its parameters (amplitude and phase at different points in the laser) are found from the solution of the electrodynamic problem. The problem as a whole will be self-consistent in the sense that the Schrödinger equation and the Maxwell equations will contain the same field. This is the formulation of the problem which will be adopted below. The results of Part I will be used in the quantum-mechanical part of the problem. In the following sections we will be concerned mainly with the electrodynamic part of the problem.

Thus, we will proceed from the equation

$$\Delta E + \frac{\omega^2}{c^2}\varepsilon E = 0, \tag{18.1}$$

where the dielectric constant ε consists of two terms:

$$\varepsilon = \varepsilon_0 + \Delta\varepsilon, \quad \Delta\varepsilon = \Delta\varepsilon' - i\Delta\varepsilon''. \tag{18.2}$$

The term $\Delta\varepsilon$ describes the contribution to the dielectric constant of the levels m, n, the transition between which is used for generation. The term ε_0 is determined by all the other atoms of the laser medium and is assumed to be independent of the field, the coordinates, and the time.

It was shown in § 6 that because of the saturation effect a laser medium is optically inhomogeneous, i.e., $\Delta\varepsilon = \Delta\varepsilon(\mathbf{r})$. The inhomogeneity is relatively slight. Knowing the gain α, we can easily evaluate $\Delta\varepsilon''$:

$$\Delta\varepsilon'' = \frac{\lambda}{2\pi}\alpha. \tag{18.3}$$

Since $\lambda \sim 10^{-4}$ cm, $\alpha \sim 10^{-1}$-10^{-4} cm^{-1}, then $\Delta\varepsilon'' \sim 10^{-6}$-$10^{-9}$. Hence, it appears at first sight that we can use geometrical optics, adapted for the case of weakly inhomogeneous media (see, for instance, [92, 93]). In fact, however, this is not so. The fact is that wave reflection occurs on inhomogeneities of the medium and this is ignored in the geometrical-optics approximation. We assume that the field established in the laser consists of standing wave. Then the laser medium will be periodically inhomogeneous with a period of $\lambda/2$ (see § 6). The reflection coefficient for each period can be roughly evaluated from the Fresnel formulas. Assuming that the change in ε is of the order of $\Delta\varepsilon''$, we find that the reflection coefficient of one period is approximately $\Delta\varepsilon''/4\varepsilon_0$ or, according to (18.3), $\lambda\alpha/8\pi\varepsilon_0$. The relative change in field amplitude due to induced emission will be $\lambda\alpha/2$. Thus, reflection from an inhomogeneity and amplification are effects of the same order and, hence, the geometrical-optics approximation cannot be used.

§ 19. One-Dimensional Laser Model

1. Below we will consider the one-dimensional laser model. We will assume that a plane-parallel layer of active substance is enclosed between planes x = 0 and x = l. Inside the layer we have

$$\frac{d^2E}{d\xi^2}+\left(1+\frac{\Delta\varepsilon}{\varepsilon_0}\right)E=0, \tag{19.1}$$

where we have introduced the dimensionless coordinates

$$\xi=2\pi\frac{x}{\lambda}=\frac{\omega}{c}\sqrt{\varepsilon_0}\,x. \tag{19.2}$$

The solution of equation (19.1) will be sought in the form

$$E=p_1(\xi)e^{-i\xi}+p_2(\xi)e^{i\xi}. \tag{19.3}$$

It is easy to verify that (19.3) is the solution of equation (19.1) if functions $p_1(\xi)$, $p_2(\xi)$ satisfy the first-order system of equations

$$\frac{dp_1}{d\xi}=\frac{\Delta\varepsilon}{2i\varepsilon_0}[p_1+p_2e^{2i\xi}],\quad \frac{dp_2}{d\xi}=-\frac{\Delta\varepsilon}{2i\varepsilon_0}[p_1e^{-2i\xi}+p_2]. \tag{19.4}$$

The boundary conditions for functions $p_1(\xi)$, $p_2(\xi)$ have the form

$$p_2(0)=\rho_1p_1(0),\quad p_1(L)=\rho_2p_2(L),\quad L=\frac{\omega}{c}\sqrt{\varepsilon_0}\,l, \tag{19.5}$$

where ρ_1 and ρ_2 are the coefficients of reflection of the plane wave from the half-spaces $\xi < 0$ and $\xi > L$, respectively. If on the planes $\xi = 0$ and $\xi = L$ the dielectric constant changes abruptly, then ρ_1, ρ_2 are Fresnel reflection coefficients. If the reflectors are of some complex form, as is usually the case (multilayer coatings, metal mirrors or selective systems of the Fabry–Perot type), then ρ_1, ρ_2 are the reflection coefficients for the reflectors as a whole.

The radiation fluxes through unit surface of the boundaries of the layer at $\xi = 0$ and $\xi = L$ are

$$\Pi_1=\frac{c\sqrt{\varepsilon_0}}{8\pi}t_1|p_1(0)|^2,\quad \Pi_2=\frac{c\sqrt{\varepsilon_0}}{8\pi}t_2|p_2(L)|^2, \tag{19.6}$$

where t_1 and t_2 are the transmission coefficients (for intensities) of the reflectors.

We will show that $p_1(\xi)$, $p_2(\xi)$ vary slowly in the intervals $\xi_2 - \xi_1 \sim 1$. Integrating (19.4) in the interval ξ_1, ξ_2, we find

$$|p_{1,2}(\xi_2)-p_{1,2}(\xi_1)|<\max\left|\frac{\Delta\varepsilon}{2\varepsilon_0}\right|\{\max|p_1|+\max|p_2|\}[\xi_2-\xi_1].$$

Since $\Delta\varepsilon/\varepsilon_0 \sim 10^{-6}$–$10^{-9} \ll 1$, this estimate means that when ξ changes by several units (i.e., when the coordinate x changes by a value of the order of the wavelength) the functions $p_1(\xi)$, $p_2(\xi)$ change very little. On the other hand, the functions $\exp\{\pm 2i\xi\}$ and $\Delta\varepsilon(\xi)$ in (19.4) vary rapidly. Hence, it is reasonable to convert from (19.4) to equations averaged over the period of oscillations of the functions $\exp\{\pm 2i\xi\}$. Integrating (19.4) in the interval $\xi-(\pi/2)$, $\xi+(\pi/2)$, we obtain

$$\begin{aligned} p_1\left(\xi+\frac{\pi}{2}\right)-p_1\left(\xi-\frac{\pi}{2}\right)&=\frac{1}{2i\varepsilon_0}\int\limits_{\xi-\pi/2}^{\xi+\pi/2}\Delta\varepsilon(\xi')[p_1(\xi')+p_2(\xi')e^{2i\xi'}]\,d\xi',\\ p_2\left(\xi+\frac{\pi}{2}\right)-p_2\left(\xi-\frac{\pi}{2}\right)&=\frac{i}{2\varepsilon_0}\int\limits_{\xi-\pi/2}^{\xi+\pi/2}\Delta\varepsilon(\xi')[p_1(\xi')e^{-2i\xi'}+p_2(\xi')]\,d\xi'.\end{aligned} \tag{19.7}$$

In these expressions the argument ξ' of functions $p_1(\xi')$ and $p_2(\xi')$ can be replaced by ξ, i.e., in the integration we can regard $p_1(\xi)$ and $p_2(\xi)$ as constant. We also introduce the moduli and phases of the functions $p_1(\xi)$ and $p_2(\xi)$:

$$p_1(\xi) = P_1(\xi)\, e^{i\varphi_1(\xi)},\ p_2(\xi) = P_2(\xi)\, e^{i\varphi_2(\xi)}.$$

The dielectric constant $\Delta\varepsilon$ is a function of

$$|E|^2 = |p_1 e^{-i\xi} + p_2 e^{i\xi}|^2 = P_1^2 + P_2^2 + 2P_1P_2\cos(2\xi + \varphi_2 - \varphi_1).$$

We expand $\Delta\varepsilon$ in a Fourier series in the interval $n\pi + (\varphi_2 - \varphi_1)/2$, $(n+1)\pi + (\varphi_2 - \varphi_1)/2$:

$$\Delta\varepsilon = \overline{\Delta\varepsilon} - 2\sum_{m=1}^{\infty} c_m \cos m\,(2\xi + \varphi_2 - \varphi_1), \tag{19.8}$$

$$c_m = -\frac{1}{\pi}\int\limits_{n\pi+(\varphi_2-\varphi_1)/2}^{(n+1)\pi+(\varphi_2-\varphi_1)/2} \Delta\varepsilon(\xi)\cos m\,(2\xi + \varphi_2 - \varphi_1)\, d\xi.$$

The quantities $\overline{\Delta\varepsilon}$, c_m are also obviously slowly varying functions of ξ. Hence, when (19.8) is substituted in the integrand in (19.7) they can be regarded as constant. If, finally, the differences $p_{1,2}[\xi + (\pi/2)] - p_{1,2}[\xi - (\pi/2)]$ in (19.7) are replaced by the approximate values $\pi dp_{1,2}/d\xi$, then we arrive at the following system of equations for p_1, p_2:

$$\frac{dp_1}{d\xi} = \frac{\overline{\Delta\varepsilon}}{2i\varepsilon_0} p_1 - \frac{c_1}{2i\varepsilon_0} p_2 e^{-i(\varphi_2-\varphi_1)},\quad \frac{dp_2}{d\xi} = \frac{c_1}{2i\varepsilon_0} p_1 e^{i(\varphi_2-\varphi_1)} - \frac{\overline{\Delta\varepsilon}}{2i\varepsilon_0} p_2. \tag{19.9}$$

Thus, the field inside the laser consists of two traveling waves with slowly varying amplitudes (over the length of λ). The changes in the complex amplitudes $p_1(\xi)$, $p_2(\xi)$ are made up, according to (19.9), of two terms. The terms $(\Delta\bar{\varepsilon}/2i\varepsilon_0)p_{1,2}$ are the same as in the consideration of waves in a homogeneous medium with averaged (over a length of $\lambda/2$) characteristics ($\overline{\Delta\varepsilon}$). These terms correspond to the geometrical-optics approximation. The terms proportional to c_1 are due to the fact that on inhomogeneities of the medium due to saturation and separated by a distance π from one another, the second (first) wave is reflected and interferes with the first (second).*

We will later require equations for the real amplitudes $P_1(\xi)$, $P_2(\xi)$ and phases $\varphi_1(\xi)$ and $\varphi_2(\xi)$. These equations are easily found from (19.9):

$$\begin{gathered}\frac{dP_1}{d\xi} = -\frac{\overline{\Delta\varepsilon''}}{2\varepsilon_0} P_1 + \frac{c_1''}{2\varepsilon_0} P_2,\quad \frac{dP_2}{d\xi} = -\frac{c_1''}{2\varepsilon_0} P_1 + \frac{\overline{\Delta\varepsilon''}}{2\varepsilon_0} P_2,\\ \frac{d\varphi_1}{d\xi} = -\frac{\overline{\Delta\varepsilon'}}{2\varepsilon_0} + \frac{c_1'}{2\varepsilon_0}\frac{P_2}{P_1},\quad \frac{d\varphi_2}{d\xi} = \frac{\overline{\Delta\varepsilon'}}{2\varepsilon_0} - \frac{c_1'}{2\varepsilon_0}\frac{P_1}{P_2}.\end{gathered} \tag{19.10}$$

If ρ_1 and ρ_2 are written in terms of the reflection coefficients r_1, r_2 for the intensities and jump of the phases

$$\rho_1 = \sqrt{r_1}\, e^{i\alpha_1},\ \rho_2 = \sqrt{r_2}\, e^{i\alpha_2}, \tag{19.11}$$

the boundary conditions (19.5) take the form

$$P_{2,0} = \sqrt{r_1}\, P_{1,0},\ P_{1L} = \sqrt{r_2}\, P_{2L}, \tag{19.12}$$

$$\varphi_{20} - \varphi_{10} = \alpha_1,\ \varphi_{1L} - \varphi_{2L} = \alpha_2 + 2L + 2\pi m, \tag{19.13}$$

*Equations (19.9) were obtained independently in [94].

where the subscripts 0 and L relate the quantities to the boundaries $\xi = 0$ and $\xi = L$. We also write out equations for $P_2 \pm P_1$, which are of interest in some questions:

$$\frac{d}{d\xi}(P_2+P_1)=\frac{\overline{\Delta\varepsilon''}+c_1'}{2\varepsilon_0}(P_2-P_1),\quad \frac{d}{d\xi}(P_2-P_1)=\frac{\overline{\Delta\varepsilon''}-c_1''}{2\varepsilon_0}(P_2+P_1). \tag{19.14}$$

The derivatives $d\varphi_1/d\xi$, $d\varphi_2/d\xi$ determine the contribution of $\Delta\varepsilon$ to the phase velocities of the waves $p_1e^{-i\xi}$, $p_2e^{i\xi}$. The terms $\pm\overline{\Delta\varepsilon'}/2\varepsilon_0$ in equations (19.10) correspond to the geometrical-optics approximation. The terms proportional to c_1' are due to reflection of waves on inhomogeneities of the medium. The resultant wave is the result of interference of two waves – the direct one and the return one due to reflection. The change in phase velocity due to interference of these waves will depend on the ratio of the amplitudes of the components. This also, obviously, accounts for the appearance of the factors P_2/P_1, P_1/P_2 in (19.10).

2. We consider some general properties of the solutions of system (19.10). We note that $\Delta\varepsilon'' > 0$. Hence, as can easily be seen from expression (19.8) for c_1, the inequalities $\overline{\Delta\varepsilon''} > c_1$ is satisfied. Since $P_1, P_2 > 0$, then $d/d\xi\,(P_2 - P_1) > 0$. Hence, the function $P_2 - P_1$ increases monotonically. It follows from inequalities $r_1, r_2 < 1$ and (19.12) that $P_{20} - P_{10} < 0$, $P_{2L} - P_{1L} > 0$. Thus, function $P_2 - P_1$ on the segment 0, L increases monotonically and changes sign. Hence, it becomes zero at one (and only one) point, which we denote by $\tilde{\xi}$. On the left of $\tilde{\xi}$ we have $P_2 - P_1 < 0$, and on the right $P_2 - P_1 > 0$. The same applies to the sign of the derivative $d/d\xi\,(P_2 + P_1)$. Hence, function $P_2 + P_1$ has a single minimum at the point $\tilde{\xi}$. The greatest value of the sum $P_2 + P_1$ is attained on one of the boundaries of the layer ($\xi = 0$ or L). Function $P_2 - P_1$, according to the boundary conditions and the above-demonstrated monotonicity, varies in the limits

$$-\frac{1-\sqrt{r_1}}{1+\sqrt{r_1}}(P_{20}+P_{10})<P_2-P_1<\frac{1-\sqrt{r_2}}{1+\sqrt{r_2}}(P_{2L}+P_{1L}). \tag{19.15}$$

Some general statements can be made about each of the amplitudes P_1, P_2. Having in mind the inequality $\overline{\Delta\varepsilon''} > |c_1''|$, we can see from equations (19.10) that $dP_1/d\xi < 0$, at least when $P_1 \geq P_2$, i.e., when $\xi \leq \tilde{\xi}$. Similarly, $dP_2/d\xi > 0$ when $\xi \geq \tilde{\xi}$. Hence, the amplitude P_1 decreases monotonically on the segment 0, $\tilde{\xi}$, while P_2 increases monotonically on the segment $\tilde{\xi}$, L (continuous curves in Fig. 21). In the general case we cannot draw any conclusions about the behavior of P_1 and P_2 on the supplementary segments. On them P_1 and P_2 can vary as on segments 0, $\tilde{\xi}$; $\tilde{\xi}$, L, respectively, and can have derivatives of different sign (broken curves in Fig. 21). Whatever case occurs depends on the nature of the saturation effect. In particular, if the dielectric constant $\Delta\varepsilon$ depends on $|E|^2$ according to the formula

$$\Delta\varepsilon=\frac{\beta}{1+\sigma^2|E|^2}, \tag{19.16}$$

which corresponds to the expression (17.4) for the induced emission probability, then $dP_1/d\xi < 0$, $dP_2/d\xi > 0$ throughout the layer. We can easily verify this by using formulas (20.3). In the case of moving atoms this will be all the more valid, since the inhomogeneity of the medium due to saturation will be less in this case than for stationary atoms (see Par. 3, § 6).

Physically the monotonic increase in the amplitudes P_1, P_2 means, obviously, that for any ratio of P_2 and P_1 amplification always dominates over reflection from inhomogeneities. Even if the "reflected" wave [i.e., the one in the terms $c_1''P_{1,2}/2\varepsilon_0$ in (19.10)] is much more intense, then its role is less than the "amplified" one, since when $P_2/P_1 \gg 1$ or $P_2/P_1 \ll 1$ the reflection coefficient decreases accordingly, because the "standing" component of the field is reduced and the medium becomes more homogeneous.

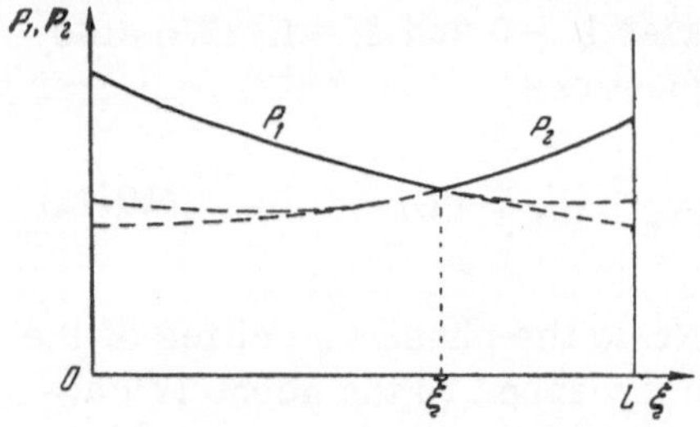

Fig. 21. Change in amplitudes P_1, P_2 along layer.

We recall that ε_0, according to hypothesis, has no imaginary part. If this is not so, the conclusions regarding the change in P_1, P_2 become invalid. In particular, it was shown in [95] that in the presence of nonmodulated absorption the amplitudes P_1, P_2 vary nonmonotonically even in model (19.16).

3. We assume that the ratio of the real and imaginary parts of $\Delta\varepsilon$ is independent of $|E|^2$. This is the case, in particular, in the model described by the system of equations (17.1):

$$\frac{\Delta\varepsilon'}{\Delta\varepsilon''} = \frac{\omega - \omega_{mn} - \Delta}{\Gamma}. \tag{19.17}$$

In this case we can find the eigenvalues of the frequency for arbitrary values of the field. In fact, when (19.17) is satisfied $\overline{\Delta\varepsilon'}$, $\overline{\Delta\varepsilon''}$ and c_1', c_1'' will be proportional in pairs and equations (19.10) for the phases can be written as

$$\frac{d\varphi_1}{d\xi} = \frac{\Delta\varepsilon'}{\Delta\varepsilon''}\frac{1}{P_1}\frac{dP_1}{d\xi}, \quad \frac{d\varphi_2}{d\xi} = \frac{\Delta\varepsilon}{\Delta\varepsilon''}\frac{1}{P_2}\frac{dP_2}{d\xi}, \tag{19.18}$$

from which it follows that

$$\varphi_1 = \frac{\Delta\varepsilon'}{\Delta\varepsilon''}\ln P_1 + \text{const}, \quad \varphi_2 = \frac{\Delta\varepsilon'}{\Delta\varepsilon''}\ln P_2 + \text{const}. \tag{19.19}$$

Using the boundary conditions (19.14), (19.13) and having in mind the definition (19.5) of parameter L, we easily find the following equation for the eigenvalues of the frequencies:

$$\omega = \frac{c}{l\sqrt{\varepsilon_0}}\pi m\left[1 - \frac{\alpha_1 + \alpha_2}{\pi m}\right] - \frac{\Delta\varepsilon'}{2\Delta\varepsilon''}\frac{c}{l\sqrt{\varepsilon_0}}\ln\frac{1}{\sqrt{r_1 r_2}}, \tag{19.20}$$

where m is a whole number. In this equation, $\alpha_{1,2}$ and $r_{1,2}$, as well as $\Delta\varepsilon'/\Delta\varepsilon''$, may also depend on ω. It is clear from (19.20) that the natural frequencies do not depend on the field intensity. This is entirely due, obviously, to the assumption that the ratio $\Delta\varepsilon'/\Delta\varepsilon''$ is independent of the field. For the same reason equation (19.20) is satisfied in the case where the inhomogeneity of the medium is ignored. If $\Delta\varepsilon'/\Delta\varepsilon''$ is given by formula (19.17), and $\alpha_{1,2}$, $r_{1,2}$ are independent of ω, the natural frequency is given by the expression

$$\omega - \omega_{mn} - \Delta = (\omega_c - \omega_{mn} - \Delta)\frac{\Gamma}{\Gamma + \Delta\omega_c}; \tag{19.21}$$

$$\omega_c = \frac{c}{l\sqrt{\varepsilon_0}}\pi m\left[1 - \frac{\alpha_1 + \alpha_2}{\pi m}\right], \quad \Delta\omega_c = \frac{c}{4l\sqrt{\varepsilon_0}}\ln\frac{1}{r_1 r_2}, \tag{19.22}$$

where ω_c is the natural frequency of the layer at $\Delta\varepsilon = 0$, and $\Delta\omega_c$ is the resonance width of the layer. Formula (19.21) was obtained in [5, 27, 96], where the medium of the layer was assumed to be homogeneous.

4. We compare our initial equations with the theory based on the transport equations [97, 98]. In this theory the main quantities are the fluxes associated with the waves $P_1e^{-i\xi}$, $P_2e^{i\xi}$:

$$S_1(\xi) = \frac{c\sqrt{\varepsilon_0}}{8\pi}P_1^2(\xi), \quad S_2(\xi) = \frac{c\sqrt{\varepsilon_0}}{8\pi}P_2^2(\xi). \tag{19.23}$$

The real flux of electromagnetic energy is $S_2 - S_1$. For the functions S_1, S_2 we derive from (19.10) the following equations:

$$\frac{dS_1}{d\xi} = -\frac{\overline{\Delta\varepsilon''}}{\varepsilon_0} S_1 + \frac{c_1''}{\varepsilon_0}\sqrt{S_1 S_2}, \quad \frac{dS_2}{d\xi} = \frac{\overline{\Delta\varepsilon''}}{\varepsilon_0} S_2 - \frac{c_1''}{\varepsilon_0}\sqrt{S_1 S_2}. \tag{19.24}$$

The transport equations used in [97, 98] have, in our symbols, the form

$$\frac{dS_1}{d\xi} = -\frac{\Delta\varepsilon''(S_1 + S_2)}{\varepsilon_0} S_1, \quad \frac{dS_2}{d\xi} = \frac{\Delta\varepsilon''(S_1 + S_2)}{\varepsilon_0} S_2. \tag{19.25}$$

Equations (19.25) and (19.24) differ in two points. Firstly, (19.24) contains the mean value of $\Delta\varepsilon''$, while (19.25) contains the dielectric constant as a function of the mean values of $S_1 + S_2$. It is clear that in the case of an inhomogeneous field $\overline{\Delta\varepsilon\,[|E(x)|^2]} \neq \Delta\varepsilon\,[\overline{|E|^2}]$. Secondly, the right sides of equations (19.25) do not contain the terms $\pm (c_1''/\varepsilon_0)(S_1S_2)^{\frac{1}{2}}$, which are present in (19.24) and describe the reflection of the waves on inhomogeneities of the medium. Thus, the transport equations (19.25) ignore the interference effects which cause inhomogeneity of the medium or result from this inhomogeneity.

5. In the practically important special case of high reflection coefficients

$$1 - r_1 \ll 1, \quad 1 - r_2 \ll 1 \tag{19.26}$$

we can obtain an approximate expression for the amplitudes and phases $P_{1,2}$, $\varphi_{1,2}$. When conditions (19.26) are satisfied the amplitudes P_1 and P_2 are almost equal and vary little over the extent on the layer. This can be seen directly from inequalities (19.15). Hence, in a first approximation in $1 - r_1$, $1 - r_2$ equations (19.14) take the form

$$\frac{d(P_2 + P_1)}{d\xi} = 0, \quad \frac{d(P_2 - P_1)}{d\xi} = \frac{\overline{\Delta\varepsilon''} - c_1''}{2\varepsilon_0}(P_2 + P_1),$$

and $\overline{\Delta\varepsilon''}$, c_1'' can be regarded as independent of ξ. Hence,

$$P_2 + P_1 = A, \quad P_2 - P_1 = \frac{A}{2\varepsilon_0}(\overline{\Delta\varepsilon''} - c_1'')\xi + B. \tag{19.27}$$

From equations (19.10) in the same approximation we find the phases

$$\varphi_1 = -\frac{\overline{\Delta\varepsilon'} - c_1'}{2\varepsilon_0}\xi + C, \quad \varphi_2 = \frac{\overline{\Delta\varepsilon'} - c_1'}{2\varepsilon_0}\xi + D. \tag{19.28}$$

We determine the constants of integration in (19.27), (19.28) from the boundary conditions (19.12), (19.13) and as a result we obtain

$$E = e^{i\varphi} A\left\{\cos\left(\xi + \frac{\alpha_1}{2}\right) - \right.$$
$$\left. - \frac{i}{4}\left[1 - r_1 - (2 - r_1 - r_2)\left(1 + i\frac{\overline{\Delta\varepsilon'} - c_1'}{\overline{\Delta\varepsilon''} - c_1''}\right)\frac{\xi}{L}\right]\sin\left(\xi + \frac{\alpha_1}{2}\right)\right\}; \tag{19.29}$$

$$\omega = \omega_c - \frac{\overline{\Delta\varepsilon'} - c_1'}{\overline{\Delta\varepsilon''} - c_2''}\Delta\omega_c; \tag{19.30}$$

$$\frac{\overline{\Delta\varepsilon''} - c_1''}{2\varepsilon_0} = \frac{\Delta\omega_c}{\omega}, \tag{19.31}$$

where $e^{i\varphi}$ is an unessential phase factor. Thus, in the case of high reflection coefficients the field inside the layer is almost a standing wave with slowly varying amplitude. From the system of equations (19.30) and (19.31) we can determine the natural frequency and the amplitude A of the steady field.

Relationship (6.2) between the imaginary part of the dielectric constant and the emission probability reveals the physical sense of condition (19.31). Integration of (6.2) with respect to ξ over the whole width of the layer when (19.29) and (19.31) are taken into account gives

$$[Q_m W_m - Q_n W_n]\hbar\omega_{mn} l = \frac{\omega_{mn}}{16\pi} A^2 [\overline{\Delta\varepsilon''} - c_1''] l = \frac{c\sqrt{\varepsilon_0}}{16\pi}\left[1 - \frac{r_1 + r_2}{2}\right] A^2 = 4\pi^2 \frac{c\hbar}{\lambda^3 \gamma_{mn}}\left[1 - \frac{r_1 + r_2}{2}\right] G_0^2. \tag{19.32}$$

Thus, (19.31) indicates the equality of the flux emerging from the layer and the power of the radiation originating in the layer.

Expression (19.6) for the flux which has passed through the reflectors at $\xi = 0$ and $\xi = L$, can be presented in the following way:

$$\Pi_{1,2} = \frac{c\sqrt{\varepsilon_0}}{32\pi} t_{1,2} A^2 = \hbar\omega\left[1 - \frac{a_{1,2}}{1 - \frac{r_1 + r_2}{2}}\right] l\,[Q_m W_m - Q_n W_n], \tag{19.33}$$

where a_1, a_2 are the coefficients of loss inside the reflecting systems (absorption and scattering).

Relationships (19.29)-(19.31) can be obtained from the direct solution of (19.1) by the method of successive approximations, with $(\Delta\varepsilon/\varepsilon_0)E$ regarded as a small quantity. In fact, in the zero approximation we have ($\Delta\varepsilon = 0$)

$$E^{(0)} = A\left[\cos(\xi + \psi) - \frac{i}{4}(1 - r_1)\sin(\xi + \psi)\right].$$

To calculate the next approximation we introduce $E \approx A\cos(\xi + \psi)$ into the term with $\Delta\varepsilon/\varepsilon_0$, since $1 - r_1 \ll 1$. Then

$$E^{(1)} = A\left\{\cos(\xi + \psi) - \frac{i}{4}\left[1 - r_1 + \frac{\overline{\Delta\varepsilon} - c_1}{2\varepsilon_0}\xi\right]\sin(\xi + \psi)\right\}. \tag{19.34}$$

Using the boundary conditions for E

$$\frac{dE}{d\xi} = -i\frac{1 - \rho_1}{1 + \rho_1}E \quad (\xi = 0); \qquad \frac{dE}{d\xi} = i\frac{1 - \rho_2}{1 + \rho_2}E \quad (\xi = L), \tag{19.35}$$

we again arrive at (19.29)-(19.31).

6. Throughout the treatment above we considered a one-dimensional laser model. The three-dimensional and two-dimensional problems give rise to new difficulties. The fact is that in these cases the field amplitude varies along the y and z axes too, and owing to the saturation effect the dielectric constant will depend on y, z. This fact, which has already been reported in [99], greatly complicates the analysis. Kuznetsova [100], however, showed that when condition

$$\left(\frac{k}{k_{y,z}}\right)^2 \Delta\varepsilon \ll 1 \tag{19.36}$$

is satisfied, the dependence of the eigenfunctions of the resonator on y, z is practically the same as in the case of a homogeneous medium. Hence, in the indicated approximation, we can use the results of [99, 101-103], where the form of the eigenfunctions was determined with saturation neglected.

In the case of small relative losses we can obtain a relationship similar to (19.31) for the multidimensional problem. In steady conditions we have

$$\hbar\omega \int_V [W_m Q_m - W_n Q_n]\, dV = \frac{c\sqrt{\varepsilon_0}}{8\pi} \oint |E_s|^2 dS. \tag{19.37}$$

Dividing the right and left sides of this equation by $\frac{\varepsilon_0}{8\pi}\int_V |E|^2 dV$, we obtain

$$\frac{8\pi\hbar\omega}{\varepsilon_0} \frac{\overline{W_m Q_m} - \overline{W_n Q_n}}{\overline{|E|^2}} = \frac{\frac{c\sqrt{\varepsilon_0}}{8\pi}\oint |E_s|^2 dS}{\frac{\varepsilon_0}{8\pi}\int_V |E|^2 dV} = \Delta\omega_c, \tag{19.38}$$

where the stroke above denotes averaging over the whole volume of the resonator. This relationship is similar to (19.31). According to the above, when condition (19.36) is satisfied the averaging in (19.38) can be carried out with the eigenfunctions of the resonator obtained with saturation neglected.

§ 20. Stationary Atoms

In the model described by the system of equations (17.1) the dielectric constant is given by the formula

$$\Delta\varepsilon = \frac{\beta' - i\beta''}{1 + \sigma^2 |E|^2}, \quad \frac{\beta'}{\beta''} = \frac{\Delta\varepsilon'}{\Delta\varepsilon''} = \frac{\Omega - \Delta}{\Gamma}. \tag{20.1}$$

$$\sigma^2 = \frac{\Gamma(\Gamma_m + \Gamma_n) p_{mn}^2}{\Gamma_m \Gamma_n [\Gamma^2 + (\Omega - \Delta)^2]}, \quad \beta'' = \frac{\lambda^3}{8\pi^2} 2\gamma_{mn}\left(\frac{Q_m}{\Gamma_m} - \frac{Q_n}{\Gamma_n}\right) \frac{\Gamma}{\Gamma^2 + (\Omega - \Delta)^2}. \tag{20.2}$$

In this model the generation frequency can be found from formula (19.20) or (19.21) and below we will deal with the power output.

The mean quantities $\overline{\Delta\varepsilon''}$ and c_1'' are

$$\begin{aligned} \overline{\Delta\varepsilon''} &= \frac{\beta''}{\sqrt{1 + \sigma^2 (P_1 - P_2)^2}\,\sqrt{1 + \sigma^2 (P_1 + P_2)^2}}, \\ c_1'' &= \overline{\Delta\varepsilon''}\, \frac{4\sigma^2 P_1 P_2}{[\sqrt{1 + \sigma^2 (P_1 - P_2)^2} + \sqrt{1 + \sigma^2 (P_1 + P_2)^2}]^2}. \end{aligned} \tag{20.3}$$

The solution of equations (19.10) with these $\overline{\Delta\varepsilon''}$ and c_1'' can be put in the following way:

$$\sqrt{1 + \sigma^2 (P_1 + P_2)^2} - \sqrt{1 + \sigma^2 (P_1 - P_2)^2} = \text{const}; \tag{20.4}$$

$$\begin{aligned} &-\frac{\beta''}{\varepsilon_0}\xi - (P_1^2 - P_2^2)\sigma^2 + \ln\left[\frac{\sigma(P_1 + P_2)}{\sqrt{1 + \sigma^2 (P_1 + P_2)^2}} + \frac{\sigma(P_2 - P_1)}{\sqrt{1 + \sigma^2 (P_1 - P_2)^2}}\right] - \\ &-\ln\left[\frac{\sigma(P_1 + P_2)}{\sqrt{1 + \sigma^2 (P_1 + P_2)^2}} - \frac{\sigma(P_2 - P_1)}{\sqrt{1 + \sigma^2 (P_2 - P_1)^2}}\right] = \text{const}. \end{aligned} \tag{20.5}$$

The constants of integration in (20.4) and (20.5) can conveniently be related to the radiation fluxes Π_1, Π_2 through the boundaries of the layer [see (19.6)], or to the dimensionless quantities proportional to them

$$F_1 = \sigma^2 (1 - r_1) P_{10}^2, \quad F_2 = \sigma^2 (1 - r_2) P_{2L}^2. \tag{20.6}$$

The values of F_1, F_2 can be determined from the system of equations obtained from (20.4), (20.5) with the aid of boundary conditions (19.12):

$$\sqrt{1 + n_1 F_1} - \sqrt{1 + F_1/n_1} = \sqrt{1 + n_2 F_2} - \sqrt{1 + F_2/n_2},$$
$$n_{1,2} = \frac{1 + \sqrt{r_{1,2}}}{1 - \sqrt{r_{1,2}}}; \tag{20.7}$$

$$2\zeta \equiv \frac{\beta''}{\varepsilon_0} L = F_1 + F_2 + \ln \frac{\left\{n_1\left[1 + \sqrt{1 + \frac{F_1}{n_1}}\right] + 1 + \sqrt{1 + n_1 F_1}\right\}\left\{n_2\left[1 + \sqrt{1 + \frac{F_2}{n_2}}\right] + 1 + \sqrt{1 + n_2 F_2}\right\}}{\left\{n_1\left[1 + \sqrt{1 + \frac{F_1}{n_1}}\right] - 1 - \sqrt{1 + n_1 F_1}\right\}\left\{n_2\left[1 + \sqrt{1 + \frac{F_2}{n_2}}\right] - 1 - \sqrt{1 + n_2 F_2}\right\}}. \tag{20.8}$$

Thus, by solving equations (20.7), (20.8) we can find F_1, F_2 and from them the fluxes Π_1 and Π_2 and the boundary values of the amplitudes P_{10}, P_{2L}. Using the latter we can determine the constants of integration in formulas (20.4), (20.5) and calculate $P_1(\xi)$, $P_2(\xi)$. Phases $\varphi_1(\xi)$, $\varphi_2(\xi)$ are then found from (19.19).

Before proceeding to the analysis of formulas (20.7), (20.8), we consider two limiting cases. When the reflection coefficients are high ($n_{1,2} \gg 1$), we can discard the terms F_1/n_1 and F_2/n_2 in the logarithm. In fact, they play a substantial role only for $F_{1,2} > n_{1,2}$. It is easy to show, however, that in equation (20.8) the main role is played by the term $F_1 + F_2$ and the logarithmic term gives a correction of the order $1/n_{1,2}$. In addition, it follows from equation (20.7) that $n_1F_1 = n_2F_2$ when n_1, $n_2 \gg 1$. Taking these considerations into account we can easily obtain

$$2\zeta = \frac{\beta''}{\varepsilon_0} L = \left[\frac{1}{n_1} + \frac{1}{n_2}\right] \sqrt{1 + n_1 F_1}\left[1 + \sqrt{1 + n_1 F_1}\right] = \frac{1}{2}\left[1 - \frac{r_1 + r_2}{2}\right]\left[1 + \sigma^2 A^2 + \sqrt{1 + \sigma^2 A^2}\right]. \tag{20.9}$$

Solving the equation for A^2, we find

$$\sigma^2 A^2 = 2\eta - \frac{1}{2}\left[1 + \sqrt{1 + 8\eta}\right], \quad \eta = \frac{\beta'' L}{\varepsilon_0 \left[1 - \frac{r_1 + r_2}{2}\right]}. \tag{20.10}$$

These equations can also be obtained directly from the approximate formula (19.31).

The second limiting case which we require corresponds to neglect of the inhomogeneity of the medium. Here the expression for the flux has the form [5, 97] (in our symbols)

$$2\zeta = \frac{\beta''}{\varepsilon_0} L = F_1 + F_2 + \frac{1}{2} \ln \frac{1}{r_1 r_2}. \tag{20.11}$$

This relationship can be derived from (20.8) if we put $F_1 = F_2 = 0$ in the logarithm.

We turn now to an analysis of formulas (20.7), (20.8) in the general case. The value of the parameter ζ at which F_1, F_2 becomes zero is

$$2\zeta_{\text{thresh}} = \frac{1}{2} \ln \frac{1}{r_1 r_2}. \tag{20.12}$$

The same expression for the threshold value of ζ follows also from (20.11). However, the inhomogeneity of the medium affects the curve of increase of the flux. In the simplest case of $F_1 = F_2 = F$ we have for the derivative $d\zeta/dF$

$$\frac{d\zeta}{dF} = 1 + \frac{1}{2\sqrt{1 + nF}\sqrt{1 + F/n}}. \tag{20.13}$$

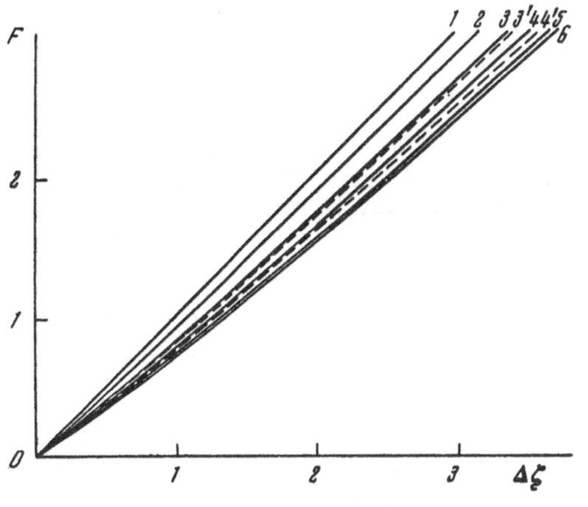

Fig. 22. Dependence of F on $\Delta\zeta = \zeta - \zeta_{\text{thresh}}$; $r_1 = r_2 = r$. 1) From (20.11); 2) r = 0.96; 3) r = 0.75; 4) r = 0.5; 5) r = 0.11; 6) r = 0.03, coincides with graphical accuracy with (20.15); the curves 3' and 4' are plotted from (20.10).

Near the threshold $d\zeta/dF$ is independent of n (i.e., is the same for any reflection coefficients on the boundaries of the layer) and is 3/2. In the case (20.11) the derivative $d\zeta/dF$ is everywhere unity. Thus, the rate of increase of the flux depends significantly on the inhomogeneity.

Figure 22 gives the values of the radiation fluxes in relation to $\zeta - \zeta_{\text{thresh}}$ for different values of the reflection coefficient. The straight line 1 corresponds to formula (20.11). The broken curves are plotted from the approximate formula (20.10). As Fig. 22 shows, these expressions are suitable in practice up to $r_1, r_2 \approx 50\%$. With increase in ζ the relative difference of the graphs from line 1 becomes less and less. With reduction in the reflection coefficients r the graphs deviate from the line 1 and when $r \to 0$ they approach a limiting curve denoted in Fig. 22 by the number 6. We consider this case in more detail, since it is of interest from the viewpoint of clarifying the role of the inhomogeneity of the medium.

Thus, let

$$n_{1,2} - 1 \ll 1. \tag{20.14}$$

Expanding expression (20.8) in powers of $n_1 - 1$, $n_2 - 1$ and confining ourselves to the first approximation, we can obtain the following formula:

$$(n_1 - 1)\frac{F_1}{\sqrt{1+F_1}} = (n_2 - 1)\frac{F_2}{\sqrt{1+F_2}}, \tag{20.15}$$

$$2\zeta = F_1 + F_2 + \frac{1}{2}\ln\frac{1}{r_1 r_2} + \frac{1}{2}\ln(1+F_1)(1+F_2).$$

As distinct from (20.11), formula (20.15) contains another logarithmic term $\frac{1}{2}\ln(1+F_1)(1+F_2)$, which, hence, reflects the role of inhomogeneity. Curve 6 in Fig. 22 corresponds to formulas (20.15). Hence, formulas (20.15) give a good description of the true state of affairs for $r_1, r_2 <$ 10%. Thus, even at the lowest r_1, r_2, modulation of the dielectric constant affects the field within the layer and the radiation flux. The effect of this is greatest near the threshold, when F_1, F_2 are small and the logarithmic term is half of the main term $F_1 + F_2$. However, the latter is valid in the general case of arbitrary values of r_1, r_2, as was already noted in connection with formula (20.13).

The conducted analysis of the general equations and the individual special cases shows that the inhomogeneity of the medium (over the extent of $\lambda/2$) due to the saturation effect must always be taken into account. It is easy to show that in equations (19.24) we cannot neglect the terms $\pm(c_1''/\varepsilon_0)(S_1S_2)^{\frac{1}{2}}$, which describe reflection, except, perhaps, at the point of threshold excitation. In fact, these terms could be neglected if the inequalities

$$\begin{aligned} \frac{c_1''}{\overline{\Delta\varepsilon''}}\sqrt{\frac{S_2}{S_1}} &= \frac{2\alpha S_2}{1+\alpha(S_1+S_2)+\sqrt{[1+\alpha(S_1+S_2)]^2 - 4\alpha^2 S_1 S_2}} \ll 1, \\ \frac{c_1''}{\overline{\Delta\varepsilon''}}\sqrt{\frac{S_1}{S_2}} &= \frac{2\alpha S_1}{1+\alpha(S_1+S_2)+\sqrt{[1+\alpha(S_1+S_2)]^2 - 4\alpha^2 S_1 S_2}} \ll 1. \end{aligned} \tag{20.16}$$

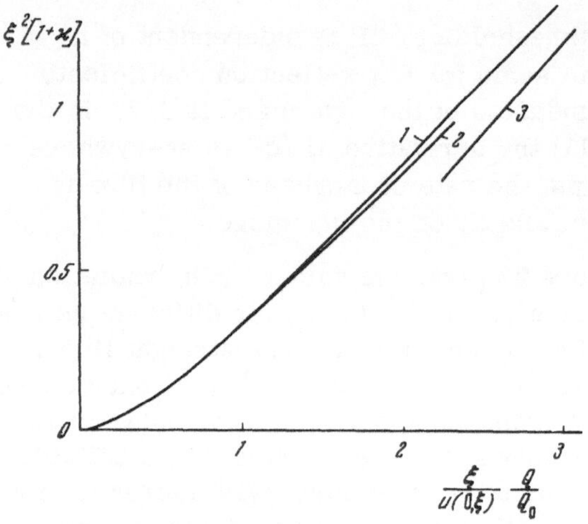

Fig. 23. Field amplitude as a function of excitation level.

were simultaneously satisfied. It is obvious that (20.16) is satisfied only at low saturation αS_1, $\alpha S_2 \ll 1$. It may appear at first sight that reflections should not play a role in less stringent conditions, viz., in all cases where modulation of the dielectric constant is small. Weak modulation may also occur at high saturation, if the field consists of two traveling waves, the amplitudes of which differ greatly in magnitude. It is physically clear, however, that this small inhomogeneity cannot be neglected, since if the amplitudes of the two waves differ considerably the weak reflections of the wave with larger amplitude considerably after the wave with lower amplitude.

Returning to inequalities (20.16) we point out the following. If αS_1, $\alpha S_2 \ll 1$, the role of reflections of inhomogeneities is actually small. But in these conditions, as can easily be seen from formulas (20.1), (20.3), $\overline{\Delta\varepsilon''}$ differs from the constant β'' by a value of the same order of smallness as the term describing reflection. Hence, if we ignore reflection, we should in general neglect saturation, i.e., confine ourselves to linear theory. Using the latter, however, we can determine only the threshold point, but the output power cannot be obtained.

§ 21. Moving Atoms

1. We calculate the flux produced by a gas laser in cases where the induced emission probability is given by a formula of type (5.37), which we modify by introducing three relaxation constants in accordance with (17.4). We will assume that the reflection coefficients of the laser mirrors are high, so that in (5.37) we can put $G_1 = G_2$.

We consider first the case $\omega = \omega_{mn} + \Delta$. Substitution of (5.37) in (19.32) gives the following equation for G^2:

$$\frac{\xi}{u(0,\xi)}\frac{Q}{Q_0} = \frac{\xi\sqrt{1+\varkappa}}{u(0,\xi\sqrt{1+\varkappa})}. \tag{21.1}$$

Here we put

$$\xi = \frac{\Gamma}{k\bar{v}},\quad Q = Q_m - \frac{\Gamma_m}{\Gamma_n}Q_n,\quad \varkappa = G^2\frac{\Gamma_m+\Gamma_n}{\Gamma\Gamma_m\Gamma_n},$$
$$Q_0 = \frac{2\sqrt{\pi}}{\lambda^2}\frac{\Gamma_m k\bar{v}}{\gamma_{mn}u(0,\xi)}\frac{1-r}{l},\quad u(0,z) = e^{z^2}[1-\Phi(z)], \tag{21.2}$$
$$2r = r_1 + r_2,$$

where $\Phi(z)$ is the probability integral. The value of Q_0 is the threshold level of excitation of the system at which generation begins. Thus, Q/Q_0 indicates the excess over the threshold excitation.

Determining $\varkappa$ from equation (21.1) we can calculate the radiation flux from unit area of the end face:

$$\Pi_{1,2} = \frac{\hbar\omega}{2}\frac{t_{1,2}}{1-r}lQ\langle\overline{W}_m\rangle = \frac{\hbar\omega}{2}\frac{t_{1,2}}{1-r}\frac{\Gamma_n}{\Gamma_n+\Gamma_m}lQ\sqrt{\pi}\,\xi u(0,\xi)\varkappa. \tag{21.3}$$

The solution of equation (21.1) is shown in Fig. 23 (curve 1). The quantities $\xi Q/u(0,\xi)Q_0$, $\xi^2(1 + x)$ are plotted on the axes. In these variables, as can be seen from (21.1), the graph is universally applicable for arbitrary values of $\xi = \Gamma/k\bar{v}$. The threshold of generation ($\varkappa = 0$, $Q = Q_0$) corresponds to the point $\xi/u(0, \xi), \xi^2$. Since $\varkappa \propto E^2$, then with accuracy to the constant term ξ^2 the graph gives the relationship between the radiation flux and the excitation level of the system. We note that near the threshold, $\xi Q/Q_0 < 1$, and the graph is essentially nonlinear. In the approximation used the medium is regarded as inhomogeneous (see Par. 3, § 6). Hence, the nonlinearity of the graph is of different origin from that in the case of Fig. 22. Here it is obviously due to Doppler broadening of the line ($\xi \ll 1$).

On different parts of curve 1 (Fig. 23) we can obtain suitable analytical formulas for the solution of equation (21.1). Taking the first term of the expansion of $\Phi(z)$ in a power series and slightly altering the coefficient in this term to obtain a good approximation in a wide range of Q/Q_0 we can obtain the formulas

$$\varkappa = \frac{(Q/Q_0)^2}{1+2\xi\left(\frac{Q}{Q_0}-1\right)} - 1,\ \langle \overline{W}_m\rangle = \frac{\Gamma_n}{\Gamma_n+\Gamma_m}\sqrt{\pi}\,\xi\left\{\frac{Q/Q_0}{1+2\xi\left(\frac{Q}{Q_0}-1\right)} - \frac{Q_0}{Q}\right\},$$

$$\Pi = \Pi_1 + \Pi_2 = \hbar\omega\left[1-\frac{a}{1-r}\right] l\,\frac{\Gamma_n}{\Gamma_n+\Gamma_m}\cdot\frac{\sqrt{\pi}\,\xi}{1-\xi}\left\{\frac{(Q/Q_0)^2}{1+2\xi\left(\frac{Q}{Q_0}-1\right)} - 1\right\} Q_0, \tag{21.4}$$

$$2a = a_1 + a_2.$$

Curve 2 in Fig. 23 corresponds to these expressions. For $\xi Q/Q_0 < 1.5$ the error is less than 1.5%. Near the threshold the relationship between Π and Q/Q_0 is quadratic and it is only when $\xi Q/Q_0 \sim 1$ that the relationship becomes linear owing to the term $2\xi\,[(Q/Q_0) - 1]$ in the denominator obtained from the expansion of $\Phi(z)$. In the limiting case $\xi Q/Q_0 \gg 1$ we find

$$\varkappa = \frac{1}{\sqrt{\pi}\,\xi}\left[\frac{Q}{Q_0} - \frac{\sqrt{\pi}}{2\xi}\right],\ \langle W_m\rangle = \frac{\Gamma_n}{\Gamma_n+\Gamma_m}\left[1-\frac{\sqrt{\pi}}{2\xi}\,\frac{Q_0}{Q}\right],$$

$$\Pi = \hbar\omega\left[1-\frac{a}{1-r}\right]\frac{l\Gamma_n}{\Gamma_n+\Gamma_m}\left[Q - \frac{\sqrt{\pi}}{2\xi}\,Q_0\right]. \tag{21.5}$$

Line 3 in Fig. 23 is drawn from this formula. This figure shows that (21.5) gives a good approximation only when $\xi Q/Q_0 \gtrsim 3$.

We will discuss the comparison of the theoretical deductions and experimental results. In this comparison we should have in mind that the theory ignores some properties of real lasers. It is known that the concentration of excited atoms varies across the discharge tube [27, 33, 104]. In addition, the field amplitude also depends on the radius, which can easily be taken into account. At sufficiently high levels of excitation there is simultaneous generation on several modes. Unfortunately, there do not appear to be any investigations in which special measures to exclude these factors have been taken. Nevertheless, we give the results of a comparison of (21.4) with the data of [105], where the relationship between the flux and the length l of the active part of the laser was measured.

For convenience of discussion we rewrite expression (21.3) for the power output by replacing Q/Q_0 by the ratio l/l_0, where l_0 is the threshold value of l at which generation begins:

$$P = \sigma\Pi = c\left\{\frac{(l/l_0)^2}{1+2\xi\left(\frac{l}{l_0}-1\right)} - 1\right\},\ c = \sigma(1-r-a)\frac{2\pi\hbar\omega\Gamma\Gamma_m\Gamma_n}{\lambda^2\gamma_{mn}(\Gamma_m+\Gamma_n)}. \tag{21.6}$$

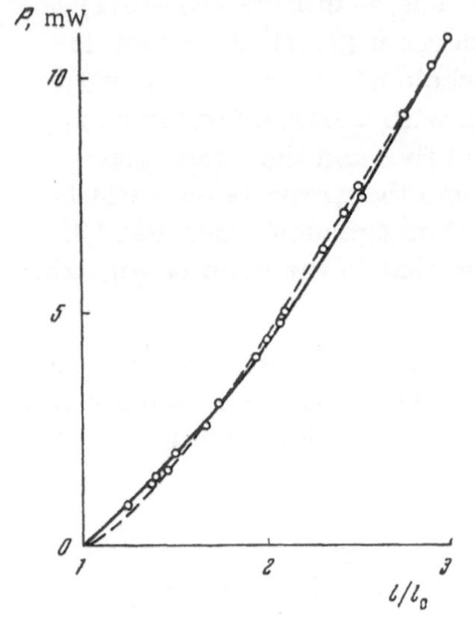

Fig. 24. Comparison of experimental data [105] (points) with theoretical curves (21.6) (continuous curve) and [105] (broken curve).

Here σ is the cross-sectional area of the beam. It is relatively simple to compare theory and experiment for the relationship between P and l/l_0, given by the expression in braces in (21.6), which contains only $\xi = \Gamma/kv$ in addition to l/l_0. If Γ is unknown, then we can consider the converse formulation of the problem: to choose ξ so that the theoretical curve satisfies the experimental values of $P(l/l_0)$. The factor c can be chosen from the experimental value of the derivative of the flux with respect to l at the threshold point

$$l_0 \frac{dP}{dl}\bigg|_{l=l_0} = 2c(1-\xi). \tag{21.7}$$

The main parameters of the laser in [105] and the working transition $Ne5s'[\frac{1}{2}]_1^0-3p'[\frac{3}{2}]_2$ were: $\lambda = 6328$ Å, $\gamma_n = 4.2 \cdot 10^7\ sec^{-1}$, $k\bar{v} = 5.7 \cdot 10^9\ sec^{-1}$ (T = 400°K), distance between mirrors 4 m, $1 - r - a = 0.015$. Figure 24 shows the experimental values of P from [105] (points). The continuous curve is obtained from (2.16) with $\xi = 0.057$ and 2c = 3.5 MW. There is a good agreement between the theoretical and experimental data.

The interpretation given in [105] differs from that given above. It was assumed in [105] that the radiation spectrum consists of many lines situated at distances much less than the natural line width. It was also assumed that the real distribution of energy in the modes could be replaced by some continuous distribution, and for each Fourier component a condition of generation of the type (19.32) is satisfied. These ideas were used in [105] to obtain a formula for the relationship between the power output and the length of the laser and other parameters. The broken curve in Fig. 24 corresponds to this formula. As this figure shows, the graph of [105] does not differ much from ours. An exception is the region close to the threshold, where the discrepancy is greater than the error of measurement (5% [105]). Thus, it appears that only this region is affected by the theoretical model, and special attention must be paid to this region in testing the theory.* It was specially noted in [105] that the region close to the threshold is not described satisfactory by the formulas obtained in that paper. Our curve, however, is in good agreement with the experimental data. Physically this is quite understandable. When $l/l_0 - 1$ is small generation occurs on one or a few modes. Hence, in this case the errors due to substitution of a continuous distribution for the real one will be particularly large. This is reflected, in particular, in the fact that, according to the formulas of [105], at the threshold point $dP/dl = 0$. On the other hand, at large values of l/l_0 both models lead to an enlargement of the range of velocities of atoms which interact effectively with the field. Hence, at large l/l_0 the power output is not affected by the model.

2. To investigate the region immediately adjacent to the threshold of generation we use the results of sections 2-5, § 17. Substituting expression (17.21) in formula (19.32) we find the field amplitude:

$$G^2 = 2\frac{I_1(\Omega)}{I_2(\Omega)}\left[\frac{Q}{Q_0} - 1\right]\frac{Q_0}{Q}, \tag{21.8}$$

---

*Measurements of the power output in a He–Ne ($\lambda = 1.15\ \mu$) laser in (106, 107) were made far from the threshold and, hence, we cannot use these data for comparison with the theory.

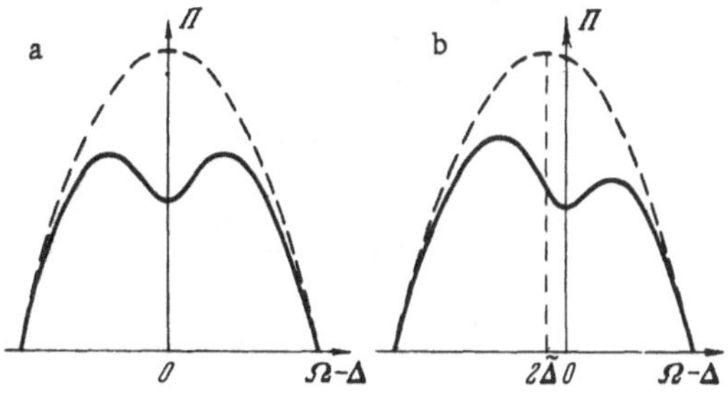

Fig. 25. Emission flux Π as a function of output frequency. a) Constructed from formula (21.12); b) from (21.13).

where the threshold excitation probability is

$$Q_0 = \frac{2\pi\Gamma_m}{\lambda^2\gamma_{mn} I_1(\Omega)} \frac{1-r}{l}. \tag{21.9}$$

Near the threshold the factor Q_0/Q in (21.8) can be taken as unity. Then the emission flux will be

$$\Pi = [1-r-a] \frac{4\pi\hbar\omega}{\lambda^2\gamma_{mn}} \frac{I_1(\Omega)}{I_2(\Omega)} \left[\frac{Q}{Q_0} - 1\right]. \tag{21.10}$$

Thus, the approximation used to obtain formula (17.21) enables us to determine only the derivative $d\Pi/dQ$ at the threshold point, which, naturally, differs from zero.

Of greatest interest in (21.10) is the dependence on the output frequency Ω. I_2, I_1, and, hence, Q_0 depend on Ω. We put

$$\eta = \frac{Q}{Q_{0\,\max}}, \quad Q_0 = Q_{0\,\max} \frac{I_{1\,\max}}{I_1(\Omega)}, \quad Q_{0\,\max} = \frac{2\pi\Gamma_m}{\lambda^2\gamma_{mn}} \frac{1-r}{l} \frac{1}{I_{1\,\max}};$$

$Q_{0\max}$ is the value of Q_0 at the point where $I_1(\Omega)$ is a maximum. Then

$$\Pi = A \frac{I_1(\Omega)}{I_2(\Omega)} \left[\eta - \frac{I_{1\,\max}}{I_1(\Omega)}\right], \tag{21.11}$$

where the factor A is independent of Ω.

First we use expression (17.39). In view of the smallness of $\eta - 1$ formula (21.11) gives

$$\Pi = A_1 \cdot \frac{\eta - 1 - \left(\frac{\Omega - \Delta}{k\bar{v}}\right)^2}{1 + \frac{\Gamma^2}{\Gamma^2 + (\Omega - \Delta)^2}}. \tag{21.12}$$

In Fig. 25a the continuous curve is plotted from formula (21.12) and the broken curve corresponds to the frequency dependence of the numerator in (21.12). Curve (21.12) is characterized by symmetry relative to $\Omega = \Delta$ and halving of Π at the point $\Omega = \Delta$ due to the denominator.

When the statistical dependence of Doppler broadening and broadening due to interaction are taken into account, the symmetry disappears, since $I_1(\Omega)$ and $I_2(\Omega)$ have their maximum values at different points. In the hard-collision model the value of $\langle \overline{W}_m \rangle$ at low $\nu/k\bar{v}$ is given by formula (17.41). Using it we obtain

$$\Pi = A_2 \frac{\eta - 1 - \left(\frac{\Omega - \Delta - 2\tilde{\Delta}}{k\bar{v}}\right)^2}{1 + \frac{(\Gamma + \nu)^2}{(\Gamma + \nu)^2 + (\Omega - \Delta)^2}}, \tag{21.13}$$

where $\tilde{\Delta}$ is the shift of the line which would occur in the case of hard collisions with change of velocity, if the statistical dependence of the changes in phase and velocity is ignored. Figure 25b illustrates formula (21.13).

Asymmetry of the curve $\Pi(\Omega)$ has been observed by several authors (see [31, 61] and the citations of special papers in [91]). Initially no special significance was attached to the asymmetry, but it became clear later that it is due to collision processes. In cold versions of collision theory neither interactions nor changes in velocity gave asymmetry. Hence, it was not clear in what way collisions could lead to asymmetry, since the usual conditions of applicability of collision theory in [31, 61] were satisfied. In view of this the essential question raised in [62] is revision of the criterion of applicability of collision theory.

It was shown above that the asymmetry of $\Pi(\Omega)$ may be due to the asymmetry of $\langle \overline{W}_m \rangle$ which arises in the collision approximation if the statistical dependence of changes in velocity and phase are taken into account. Unfortunately, in [31] there was not enough experimental information to test the theory.

CHAPTER VI

Stability of Monochromatic Generation

§ 22. Formulation of Problem

Lasers generate simultaneously several quasi-monochromatic lines corresponding to different oscillation modes which have natural frequencies within the width of the spontaneous emission line. This applies to lasers using gases or solids. In the case of gases the theoretical interpretation of this effect raises no difficulties [26], since in gases the width of the spontaneous emission line depends mainly on the Doppler effect ("inhomogeneous" broadening). On the other hand, in solids, particularly ruby, the broadening of the line is due to relaxation processes ("homogeneous" broadening). In this case established theoretical ideas led to the conclusion that it is possible to produce only one line within the width of the spontaneous emission line [27, 64, 108, 109]. In these investigations, however, either the optical inhomogeneity of the medium [108, 109], or the change in the gain curve [27, 64], established in Chapter III, was ignored.

The following analysis of the question of the stability of operation is based on the following considerations. Steady-state generation is considered. We assume that a monochromatic standing wave of frequency ω is set up in the laser. This field causes saturation and makes the medium inhomogeneous. It is henceforth called strong. We will assume that, in addition to the strong field, the resonator cavity contains other waves, the frequency ω_μ of which differ from ω, and the amplitudes of which are so small that they lead to hardly any saturation. These fields, which are actually produced by the spontaneous emission of the laser medium, will be called weak. Our problem is to determine the conditions in which weak fields are decaying or increasing in time. In the first case the initial generation of monochromatic emission is stable, whereas in the second case it is unstable.

§ 23. General Formulas

1. We consider first of all a laser model in the form of a plane-parallel layer bounded by mirrors with reflection coefficients r_1, r_2. We will assume for the sake of simplicity that

$$1 - r_1 \ll 1, \quad 1 - r_2 \ll 1. \tag{23.1}$$

It is easy to show by the method of successive approximations that the plane solutions of the wave equation, which exponentially depend on the time within the layer, have the form

$$E_\mu(x, t) = A_\mu e^{-i(\omega_\mu + i\gamma_\mu)t}\left\{\cos\left(k_\mu x + \frac{\alpha_1}{2}\right) - \frac{i}{4}\left[1 - r_1 - (2 - r_1 - r_2)\right]\frac{x}{l}\sin\left(k_\mu x + \frac{\alpha_1}{2}\right)\right\}; \tag{23.2}$$

$$\omega_\mu = \frac{\pi c}{l\sqrt{\varepsilon_0}}\mu - \frac{c}{\sqrt{\varepsilon_0}}\frac{k_\mu}{l}\int_0^l \cos^2\left(k_\mu x + \frac{\alpha_1}{2}\right)\delta\varepsilon'(x, \omega_\mu)\,dx; \tag{23.3}$$

$$\frac{1}{\varepsilon_0} 2k_\mu \int_0^l \cos^2\left(k_\mu x + \frac{\alpha_1}{2}\right)\delta\varepsilon''(x, \omega_\mu)\,dx = 1 - \frac{r_1 + r_2}{2} + \frac{2\gamma_\mu l}{c}; \tag{23.4}$$

$$k_\mu = \frac{\omega_\mu}{c}\sqrt{\varepsilon_0},\quad \varepsilon = \varepsilon_0 + \delta\varepsilon(x, \omega_\mu),$$
$$\delta\varepsilon(x, \omega_\mu) = \delta\varepsilon'(x, \omega_\mu) - i\delta\varepsilon''(x, \omega_\mu). \tag{23.5}$$

Formulas (23.2)-(23.5) are similar to formulas (19.29)-(19.31), (18.2), which relate to a strong field.

When condition (23.1) is satisfied, the amplitude of the strong field varies almost periodically with x. Hence, $\delta\varepsilon$ will be a periodic function of x with a period $\lambda/2 = \pi/k$. Expanding $\delta\varepsilon$ in a Fourier series, similar to (19.8), we obtain, instead of (23.3), (23.4)

$$\omega_\mu = \frac{c}{2l\sqrt{\varepsilon_0}}\left\{2\pi\mu - \overline{\delta\varepsilon''} + 2\sum_{m=1}^{\infty} c'_{m\mu}\frac{1}{l}\int_0^l \cos(2k_\mu x + \alpha_1)\cos m(2kx + \alpha_1)\,dx\right\},$$

$$\frac{k_\mu l}{\varepsilon_0}\left\{\overline{\delta\varepsilon''} - 2\sum_{m=1}^{\infty} c''_{m\mu}\frac{1}{l}\int_0^l \cos(2k_\mu x + \alpha_1)\cos m(2kx + \alpha_1)\,dx\right\} = 1 - \frac{r_1 + r_2}{2} + \frac{2\gamma_\mu l}{c}. \tag{23.6}$$

It is easy to see that for whole $2k_\mu l$, $2kl$ and $k_\mu \neq k$ the integrals in (23.6) become zero. Hence,

$$\omega_\mu = \omega_c - \frac{c}{2l\sqrt{\varepsilon_0}}\overline{\delta\varepsilon'},\quad \frac{k_\mu l}{\varepsilon_0}\overline{\delta\varepsilon''} = 1 - \frac{r_1 + r_2}{2} + \frac{2\gamma_\mu l}{c}. \tag{23.7}$$

We also write similar equations for the strong field [i.e., equations (19.30), (19.31)]:

$$\omega = \omega_c - \frac{c}{2l\sqrt{\varepsilon_0}}(\overline{\Delta\varepsilon'} - c'_1),\quad \frac{kl}{\varepsilon_0}(\overline{\Delta\varepsilon''} - c''_1) = 1 - \frac{r_1 + r_2}{2}. \tag{23.8}$$

The last formula differs from (23.7) in the terms c'_1, c''_1, which are absent in (23.7) [the absence in (23.8) of a term similar to $2\gamma_\mu l/c$ is obviously due to the assumption that the strong field is monochromatic, $\gamma = 0$]. It is easy to understand the physical cause of this difference. Terms c'_1, c''_1 in (23.8) are due to reflection of the wave on periodical inhomogeneities of the medium due to saturation (see § 19). The weak fields are also reflected from inhomogeneities. However, the half weak-field wavelength $\lambda_\mu/2 = \pi/k_\mu$ differs from the period $\lambda/2$ of the inhomogeneity. Owing to this, the difference in path between successive reflected waves differs from λ_μ, and the reflected waves extinguish one another. This was expressed in the integrals in (23.6) becoming zero.

Another interpretation of the same effect is of interest. At the antinodes of the strong field the dielectric constant is a minimum and at the nodes it is a maximum. Near the nodes, however, the strong field itself is small and these parts of the volume make a small contribution to the total emission of the laser. In the region of an antinode, however, where the field is a maximum, $\Delta\varepsilon$ has a small value. The dielectric constant $\delta\varepsilon$ has the same period as $\Delta\varepsilon$. However, $\lambda_\mu \neq \lambda$, and the antinodes of the weak field on advance through the layer take all possible positions between the minima of $\delta\varepsilon$. Hence, the integrals in (23.3), (23.4) are determined by the mean value of $\delta\varepsilon$.

The condition for instability of monochromatic generation is $\gamma_\mu > 0$ or, according to formulas (23.7), (23.8),

$$\chi \equiv \frac{\overline{\delta\varepsilon''}}{\overline{\Delta\varepsilon''} - c_1''} = \frac{2l}{c} \frac{\gamma_\mu}{1 - \frac{r_1 + r_2}{2}} + 1 > 1. \tag{23.9}$$

2. In real lasers instability of monochromatic generation may be produced by the increase in time of weak fields, which differ in spatial structure from the strong field not only along the x axis, but also along the y, z axes ("transverse" or "nonaxial" oscillation modes). If the conditions assumed in section 6, § 19 (relatively low excitation levels and small relative losses) are satisfied, the stability criterion, analogous to (23.9), has the form

$$\chi \equiv \frac{\overline{W_{m\mu} Q / |E_\mu|^2}}{\overline{W_m Q / |E|^2}} > 1 + \frac{\Delta f}{f}, \quad \Delta f = f_\mu - f, \tag{23.10}$$

where f_μ, f are the relative energy losses for the oscillation modes of the weak and strong fields, respectively. Formula (23.10) follows from (19.38) and from the proportionality of $\Delta\omega_c$ to the losses.

§ 24. Stationary Atoms

1. The question of amplification of a weak field in the presence of a strong one was considered in § 11 and § 17. In this section we use the results obtained there to determine the stability of monochromatic generation.

Figures 12 and 13 ($\Omega = 0$) show that the gain for the weak field is less than for the strong one (α), with the exception of cases of very large G and high frequencies (the side maximum of curve 5 in Fig. 12). The same obviously applies to the quantities $\delta\varepsilon''$ and $\Delta\varepsilon''$, which are proportional to α_μ and α. Hence, if the laser medium is homogeneous (in this case the question of increase or decay is determined simply by the relationship between $\delta\varepsilon''$ and $\Delta\varepsilon''$), then all weak harmonic fields will be decaying. Only when $G > 4\Gamma$ [see formula (11.5)] at frequencies $\Omega_\mu = \pm(2G + \Gamma)$ we have $\Delta\varepsilon'' < \delta\varepsilon''$. We can easily find from formula (20.9) that $G > 4\Gamma$ when $\eta > 37$. Thus, on the assumption of homogeneity of the medium increasing solutions appear only at very high levels of excitation of the system at a distance of several widths from the line center. However, a consideration of the inhomogeneity of the medium radically alters the position.

2. We consider first the case where the weak-field frequency is so close to that of the strong field that in the expression for $\delta\varepsilon_\mu''$ we can put $\Omega = \Omega_\mu$. Here we use the most general model (§ 17). Using formulas (17.16), (17.4), and (10.1) we find

$$\chi = \frac{1}{2} [1 + \sqrt{1 + \varkappa}] \left[1 - \frac{\varkappa/2}{1 + \varkappa}\right], \quad \varkappa = 2G^2 \frac{\Gamma(\Gamma_m + \Gamma_n)}{[\Gamma^2 + \Omega^2] \Gamma_m \Gamma_n}. \tag{24.1}$$

At low $\varkappa$ it follows from (24.1) that

$$\chi \approx 1 - \frac{\varkappa}{4}, \tag{24.2}$$

i.e., in these conditions χ decreases with increase in $\varkappa$, which means greater amplification of the strong field and stability generation. With further increase in $\varkappa$, however, the quantity χ passes through a minimum at $\varkappa \approx 1.3$ and then becomes unity at

$$\varkappa = 2(1 + \sqrt{2}), \tag{24.3}$$

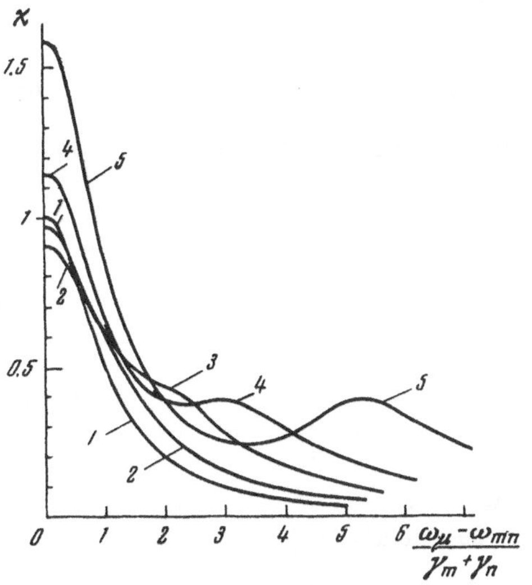

Fig. 26. Ratio of amplifications of weak and strong fields, averaged over the length of the laser, as a function of frequency; $\gamma_m = \gamma_n$, $\omega = \omega_{mn}$. 1) $G_0 = 0$; 2) $G_0 = \gamma_m$; 3) $G_0 = 2\gamma_m$; 4) $G_0 = 3\gamma_m$; 5) $G_0 = 5\gamma_m$.

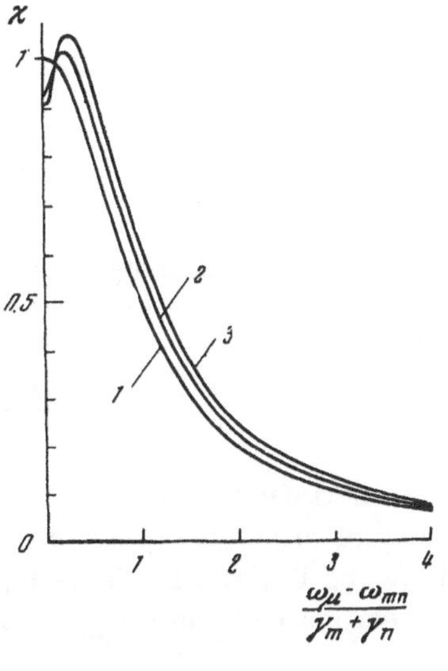

Fig. 27. Ratio of amplifications of weak and strong fields, averaged over the length of the laser, as a function of frequency; $\gamma_m = 0.05\,\gamma_n$, $\omega = \omega_{mn}$. 1) $G_0 = 0$; 2) $G_0^2 = 0.05\gamma_m\gamma_n$; 4) $G_0^2 = \gamma_m\gamma_n$.

while at greater fields it increases monotonically. We can find from (20.9) that the critical field value (24.3) exceeds the generation threshold by a factor of 4.1. Hence, despite the fact that at each point of the laser $\alpha_\mu < \alpha$, the inhomogeneity of the medium results in the mean amplification of the weak field becoming greater than that of the generated emission.

3. The behavior of $\delta\varepsilon''$ at $\Omega_\mu \neq \Omega$ depends significantly on the relationships between the parameters Γ_m, Γ_n, Γ, and G (compare Figs. 12 and 13). From formulas (17.15) and (17.4) we can calculate χ in the general case, but the obtained result is too complicated for analysis. Hence, we consider some special cases. Let

$$\Gamma_m = \Gamma_n, \quad \Gamma = (\Gamma_m + \Gamma_n)/2 = \Gamma_m, \quad \Omega = 0. \tag{24.4}$$

The calculation gives

$$\chi = \frac{1+\sqrt{1+\varkappa}}{4} \cdot \left\{\frac{1}{1+\xi^2} + \frac{2\sqrt{1+\varkappa}}{\sqrt{(1+\varkappa-\xi^2)^2}\,\sqrt{\sqrt{(1+\varkappa-\xi^2)^2+\xi^2}+1+\varkappa-\xi^2}}\right\}, \tag{24.5}$$

$$\xi = \Omega_\mu/\Gamma, \quad \varkappa = 4G^2/\Gamma^2.$$

Near $\Omega_\mu = 0$ the main role at large $\varkappa$ is played by the first term, since the second term is of the order of $1/\varkappa$. Hence, near the line center the frequency dependence of χ has the usual form of a dispersion curve. Amplification of the weak field will decrease with the frequency. This is quite understandable, since when $\Gamma_n = \Gamma_m = \Gamma$ near the line center there is always a maximum of $\delta\varepsilon''_\mu(\Omega_\mu)$ (compare Fig. 12). The value of χ, calculated from (24.5), is shown in Fig. 26. At large $\varkappa$ there is a subsidiary maximum close to $\Omega_\mu = \Gamma(1+\varkappa)^{\frac{1}{2}}$. For $\varkappa > 70$, as can easily be shown, $\chi > 1$ for $\Omega_\mu = \Gamma(1+\varkappa)^{\frac{1}{2}}$. However, χ begins to exceed unity at low Ω_μ for much smaller values of $\varkappa$.

The presented results of calculations illustrate the great effect of inhomogeneity on amplification of the weak field. Firstly, at all frequencies $\overline{\delta\varepsilon''} > 0$, whereas local values of $\delta\varepsilon''$ may be negative (see Figs. 12 and 13). Secondly, the greatest values of $\overline{\delta\varepsilon''}$ are attained at the line center, whereas at the antinodes of the strong field $\delta\varepsilon''$ has its greatest value on the wings of the line.

Figure 26 shows that inhomogeneity of the medium leads to instability of solutions with frequencies close to the transition frequency. The critical amplitude of the strong field at which the output becomes unstable is much lower; $\varkappa > 4.8$ from (24.3), instead of 64, according to (11.5).

4. We now consider a case which is typical for lasers using a crystal as an active medium. Here the inequalities

$$\Gamma \gg \Gamma_n \gg \Gamma_m. \tag{24.6}$$

are usually satisfied. We confine ourselves, in addition, to not too high excitation levels, so that the condition

$$\Gamma_n \gg \Gamma_m + \frac{G^2}{\Gamma}, \tag{24.7}$$

is satisfied, and we put $\Omega = 0$. Then, using formula (17.15), we can find

$$\chi = \frac{1}{2}\left[1 + \sqrt{1+\varkappa}\right]\left\{\frac{\Gamma^2}{\Gamma^2 + \Omega_\mu^2} - \frac{\varkappa}{2}\frac{\sqrt{2}\Gamma_m^2}{\sqrt{(1+\varkappa)^2\Gamma_m^2 + \Omega_\mu^2}\sqrt{\sqrt{[(1+\varkappa)^2\Gamma_m^2 + \Omega_\mu^2]^2 + (\varkappa\Omega_\mu\Gamma_m)^2} + (1+\varkappa)^2\Gamma_m^2 + \Omega_\mu^2}}\right\}. \tag{24.8}$$

The general nature of the relationship between χ and Ω_μ is the same as for $\delta\varepsilon''$ – the second term in the braces leads to a narrow dip with a width of the order of $\Gamma_m(1 + \varkappa)^{1/2}$. This is well illustrated by Fig. 27.

The most significant feature of (24.8) is that near $\Omega_\mu = 0$ the inequality $\chi > 1$ holds for much smaller $\varkappa$ than in the case (24.5). The role of the ratio Γ_m / Γ can be clearly seen in the limiting case $\varkappa \ll 1$. Then (24.8) will obviously take the form

$$\chi \approx \frac{1}{2}\left[1 + \sqrt{1+\varkappa}\right]\left\{\frac{\Gamma^2}{\Gamma^2 + \Omega_\mu^2} - \frac{\varkappa}{2}\frac{\Gamma_m^2}{\Gamma_m^2 + \Omega_\mu^2}\right\}. \tag{24.9}$$

The extreme value $\chi = \chi_m$ is attained when

$$\Omega_\mu^2 + \Gamma_m^2 \approx \Gamma\Gamma_m\sqrt{\varkappa}. \tag{24.10}$$

Substituting (24.10) in (24.9) we find that $\chi_m > 1$, if

$$\varkappa > 4(3 + \sqrt{8})\left(\frac{\Gamma_m}{\Gamma}\right)^2. \tag{24.11}$$

Finally, it follows from (20.9) that corresponding to such values of $\varkappa$ there is an excess over the excitation threshold

$$\eta - 1 = 3(3 + \sqrt{8})\left(\frac{\Gamma_m}{\Gamma}\right)^2. \tag{24.12}$$

If Γ_m/Γ is sufficiently small, then conditions (24.11) and (24.12) are satisfied when $\varkappa \ll 1$. Thus, the difference in relaxation constants increases the instability of monochromatic generation, and this increase is greater, the greater the difference between Γ_m and Γ. This is due both to specific amplification of the weak field in the presence of a strong one when $\Gamma_m \ll \Gamma$ and to inhomogeneity of the laser medium.

Such a situation is possible when the frequency interval defined by formula (24.10) does not contain the natural frequencies of the resonator. For instance, in resonators of length l = 10 cm the difference in frequencies of successive axial modes is $\sim 10^{10}$ sec^{-1}. The frequency from (24.10) at $\varkappa \sim (\Gamma_m/\Gamma)^2$ will be $\Omega_\mu \sim \Gamma_m$. In the real crystals used in actual lasers, $\Gamma_m \sim 10^3$-10^6 sec^{-1}. In such cases the second term in braces in (24.8) can be neglected and the critical values of $\varkappa$ and $\eta - 1$ will be

$$\varkappa = \frac{4\Omega_\mu^2}{\Gamma^2}, \quad \eta - 1 = 3\frac{\Omega_\mu^2}{\Gamma^2}. \tag{24.13}$$

Since the characteristic values of Γ are $(2-10) \cdot 10^{11}$ sec^{-1}, then here also the critical value of $\varkappa \sim 10^{-2} \ll 1$. Thus, in this case too, monochromatic generation is unstable at negligible $\eta - 1$ and this is entirely due to the inhomogeneity of the laser medium.

5. The conducted calculations show that in a one-dimensional laser model inhomogeneity of the medium due to saturation plays an exceptionally important role. Real systems differ from the considered model in many respects (three-dimensionality, inhomogeneity of the crystal and excitation, and so on). Nevertheless, some theoretical and experimental works published after the above analysis was made have shown that the effects established here are typical of many real lasers. In [110] an attempt was made to interpret "spiking" from essentially the same viewpoint as that adopted above. Similar effects have been found in SHF masers (see [111], for instance). It was shown experimentally in [112] that inhomogeneity due to saturation is significant in lasers with concentric resonators. The role of inhomogeneity of the medium was very clearly demonstrated in [113]. In this work conditions of operation with two methods of illuminating the ruby crystal with the exciting radiation were compared. In one case only the central part of the ruby was excited and then generation on several modes was observed. If, however, only the ends of the ruby adjoining the mirrors were illuminated, generation occurred on only one mode. This is quite understandable in the light of the ideas developed above, since in the center of the crystal the nodes of successive eigenfunctions of the resonator are maximally spaced.

§ 25. Moving Atoms. Gas Lasers

1. An analysis of conditions of breakdown of monochromatic generation will be made for the case considered in Par. 3, § 12. According to (5.27) and (12.5), we have

$$\langle W \rangle = \frac{\sqrt{\pi}\gamma_n}{k\bar{v}} \frac{\varkappa/2}{\sqrt{1+\varkappa/2}}\, u\left(0, \frac{\gamma_n}{k\bar{v}}\sqrt{1+\varkappa/2}\right), \quad \varkappa = \frac{G^2}{\gamma_m \gamma_n}; \tag{25.1}$$

$$\langle W_\mu \rangle = \frac{\sqrt{\pi}\gamma_n}{k\bar{v}} \frac{G_\mu^2}{2\gamma_m\gamma_n} e^{-\Omega_\mu^2/(k\bar{v})^2} \left\{1 - \frac{\varkappa/2}{1+\frac{\varkappa}{2}+\sqrt{1+\varkappa/2}} \frac{\Gamma_1^2}{\Gamma_1^2+\Omega_\mu^2} - \right.$$
$$\left. -\frac{\varkappa}{2}\operatorname{Re}\left[\frac{1}{(1+i\xi)\sqrt{1+\frac{\varkappa}{2}}\sqrt{1+\frac{\varkappa/2}{1+i\xi}}\left[\sqrt{1+\frac{\varkappa}{2}}+\sqrt{1+\frac{\varkappa/2}{1+i\xi}}\right]}\right]\right\}, \tag{25.2}$$
$$\xi = \frac{\Omega_\mu}{2\gamma_m}.$$

The subsequent calculations depend significantly on whether the weak field is an axial or transverse mode. We will now consider the first of these two cases.

2. The difference in the frequencies of two neighboring axial modes is $\pi c/l$, which is approximately 10^9 sec^{-1} when $l = 10^2$ cm. On the other hand, $\gamma_m, \gamma_n \sim 10^7$-$10^8$. Hence, in the case of axial modes we can drop the term in the second line of formula (25.2), which has an appreciable value only when $\Omega_\mu \lesssim 2\gamma_m$. Since, in addition, $\Omega_\mu, \gamma_n \ll k\bar{v}$, then, as will be clear, the critical value of $\varkappa \ll 1$. Hence, in the other terms of formula (25.2) we can confine ourselves to the first correction for $\varkappa$. Finally, we must bear in mind that the losses and transverse structure of the field in all axial modes are the same. Taking these considerations into account we can use (23.10), (25.1), and (25.2) to obtain the following expression for the critical strong-field amplitude at which generation of the neighboring axial mode occurs:

$$\varkappa_0 = \frac{G_0^2}{\gamma_m\gamma_n} = 4\frac{\bar{g}}{\bar{g}^2}\frac{(2\gamma_n)^2+\Omega_\mu^2}{(k\bar{v})^2}. \tag{25.3}$$

Here $\varkappa_0$ and $g(y, z) = g$ are defined as

$$\varkappa = \varkappa_0 g(y, z), \tag{25.4}$$

and the stroke above denotes averaging over y, z. Function g(y, z) gives the relationship between the square of the field and the transverse coordinates. If the mirrors are rectangular, then

$$g(y, z) = \left[1 - \cos\frac{\pi}{a}(m+1)y\right]\left[1 - \cos\frac{\pi}{b}(n+1)z\right]. \tag{25.5}$$

For axial modes m = n = 0. In this case formula (25.3) takes the form

$$\varkappa_0 = \frac{64}{9}\frac{(2\gamma_n)^2 + \Omega_\mu^2}{(k\bar{v})^2}, \quad \frac{\bar{g^2}}{\bar{g}} = \frac{9}{16}. \tag{25.6}$$

It is easy to show also that corresponding to the critical value of $\varkappa_0$ from (25.3) there is the following excess over the generation threshold

$$\frac{Q}{Q_0} - 1 = \frac{(2\gamma_n)^2 + \Omega_\mu^2}{(k\bar{v})^2}. \tag{25.7}$$

3. In the case of transverse modes $\Omega_\mu \sim 10^6\ \mathrm{sec}^{-1} \ll 2\gamma_m$. Hence, in (25.2) we can put

$$\langle W_\mu \rangle = \frac{\sqrt{\pi}\gamma_n}{k\bar{v}}\frac{G_\mu^2}{4\gamma_m\gamma_n}\left\{\frac{1}{\sqrt{1+\varkappa/2}} + \frac{1}{[1+\varkappa/2]^{3/2}}\right\}. \tag{25.8}$$

We introduce function g(y, z) in accordance with (25.5). Then the criterion of instability (23.10) can be written in the following way:

$$\chi = \overline{\frac{1}{2}\left\{\frac{g}{\sqrt{1+\varkappa/2}}\left[1 + \frac{1}{1+\varkappa/2}\right]\right\}} \Big/ \overline{\left\{\frac{g_0}{\sqrt{1+\varkappa/2}}\right\}} > 1 + \frac{\Delta f}{f}, \tag{25.9}$$

where g_0 relates to axial modes. For transverse modes the losses are greater than for axial modes, i.e., $\Delta f > 0$. On the other hand, if $g \equiv g_0$, then, as can easily be seen from (25.9), χ is always <1. Thus, condition (25.9) will be satisfied only when the transverse distribution of the fields in the axial and transverse modes is different, $g \not\equiv g_0$. We can show that for the modes, TEM_{0m}, TEM_{n0}, χ is always <1. For TEM_{nm} (m, n ≠ 0) the calculation gives $\chi > 1$. Figure 28 gives the results of calculation of $\chi - 1$, performed by an electronic digital computer,* from formula (25.9) with g, g_0 determined from (25.5), i.e., for rectangular mirrors. The abscissa of the point of intersection of the graphs in Fig. 28 with the straight line $\chi - 1 = \Delta f/f$ gives the value of $\varkappa$ at which generation of the particular mode is possible.

We recall that Δf increases rapidly with increase in the numbers m, n of the transverse modes. According to [101]

$$f_{\mathrm{diffr}} = 2\pi^2\left\{\frac{\beta(N_a+\beta)}{(N_a+\beta)^2+\beta^2}(m+1)^2 + \frac{\beta(N_b+\beta)}{(N_b+\beta)^2+\beta^2}(n+1)^2\right\}, \tag{25.10}$$

$$N_a = \frac{2\pi a^2}{l\lambda}, \quad N_b = \frac{2\pi b^2}{l\lambda}, \quad \beta \approx 0.824.$$

Here a, b are the dimensions of the mirrors in the directions of the y and z axes. For typical values of the parameters ($\lambda = 1\ \mu$, $l = 10^2$ cm, $a = b = 1$ cm) we have

$$\Delta f = f_\mu - f \approx 10^{-3}[m(m+2) + n(n+2)],$$

*The calculations were made by A. Konstantinova of the Computer Department of FIAN.

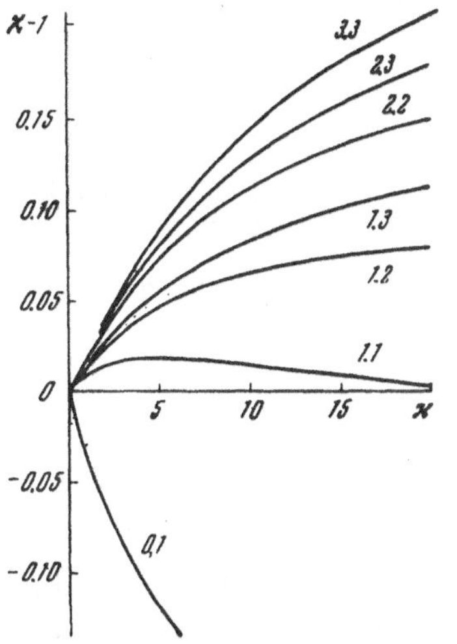

Fig. 28. Ratio of amplifications of weak and strong fields, averaged over the volume of the laser, as a function of frequency. The figures on the curves are the numbers m, n.

i.e., Δf is a few tenths of one per cent of the lowest transverse modes. Since

$$f = 1 - \frac{r_1 + r_2}{2} + f_{0\,\mathrm{diffr}} = 1 - \frac{r_1 + r_2}{2} + 2 \cdot 10^{-3},$$

then $\Delta f / f \sim 0.1$, since usually r_1, $r_2 = 0.99$. With the used values of the parameters there can be no generation of transverse modes with any subscripts, since Δf increases rapidly with increase in m and n. On the other hand, generation of transverse modes is observed in experiments. This may be due to several circumstances. Firstly, formula (25.10) describes the difference in diffraction losses in a resonator with plane mirrors. In resonators with concave mirrors Δf is usually much smaller. The value of f depends mainly on the reflection coefficients, so that in the case of concave mirrors $\Delta f / f$ is much less than in a resonator with plane mirrors. Secondly, in our calculations we ignored the radial inhomogeneity of the concentration of excited atoms. If the difference of populations is greatest on the axis of the discharge tube and decreases monotonically towards the walls (as was the case in [33], where the results of the corresponding measurements were given), the radial inhomogeneity hinders the generation of transverse modes. It was reported in [27] that in some discharge conditions there is a minimum amplification on the tube axis. In this case generation of transverse modes is facilitated. Finally, it is quite possible that generation of transverse modes is due to the fact that the assumptions made in regard to γ_m / γ_n in the deduction of formulas (25.1) and (25.2) are not valid. Unfortunately, the presently available published information is largely of a scrappy nature and it is difficult to make a sound choice between the possibilities.

APPENDIX

Spectroscopic Applications of Gas Lasers

1. Gas lasers are interesting systems from the viewpoint of a large number of purely spectroscopic problems which have largely exhausted the potentialities of old methods. In [114, 115], for instance, lines which had not been observed earlier in the spontaneous emission were detected. Hence, we have a new method of solving one of the basic problems of spectroscopy – the determination of the energies of the states of quantum systems. We think, however, that more interest attaches to the applications of lasers for other purposes, such as determination of the probabilities of radiation processes and collisions of excited atoms with other particles. We consider some of these possibilities more fully.

2. Ordinary spectroscopic methods enable us to determine either the product $A_{mn}N$ of the first Einstein coefficient for the transition m → n and the population of the upper ($N = N_m$) or lower ($N = N_n$) level, or $A_{mn}[(N_m/g_m) - (N_n/g_n)]$, where g_m, g_n are the statistical weights of the levels m, n. The determination of A_{mj}/A_{nk} of transitions with different upper and lower levels also requires a knowledge of N_m/N_n. Here laser systems may be useful, since there are various methods of maintaining a particular value of the ratio of the populations of two levels in such systems.

Let generation on the transition m → n be possible. If the power of the discharge or the gas pressure is sufficiently high generation will cease owing to equalization of the populations N_m, N_n of the levels m, n. In this case

$$\frac{N_m}{g_m} - \frac{N_n}{g_n} = \frac{1-r}{l} \sqrt{\frac{4\pi}{\ln 2}} \frac{\Delta\omega_D}{\lambda^2 A_{mn} g_m}, \tag{A.1}$$

where it is assumed that the line is of Gaussian form with width $\Delta\omega_D$ (at half height). The ratio of the fluxes Φ_{mj}, Φ_{nk} on the transitions m → j, n → k, will be

$$\frac{\Phi_{mj}}{\Phi_{nk}} = \frac{\omega_{mj}}{\omega_{nk}} \frac{A_{mj} N_m}{A_{nk} N_n} = \frac{\omega_{mj} A_{mj} g_m}{\omega_{nk} A_{nk} g_n} \left[1 + \frac{N_m/g_m - N_n/g_n}{N_n/g_n} \right]. \tag{A.2}$$

An evaluation of $N_m/g_m - N_n/g_n$ for typical conditions ($\lambda = 1\ \mu$, $l = 10^2$ cm, $1 - r = 10^{-2}$, $\Delta\omega_D = 5 \cdot 10^9$ sec^{-1}, $A_{mn} = 10^7$ sec^{-1}) from (A.1) gives 10^7 cm^{-3}. The population N_n/g_n in order of magnitude is 10^9 cm^{-3}. Hence, we can assume with great accuracy that at the instant when generation on the transition m → n ceases formula (A.2) means that

$$\frac{\Phi_{mj}}{\Phi_{nk}} = \frac{\omega_{mj} A_{mj} g_m}{\omega_{nk} A_{nk} g_n}. \tag{A.3}$$

Thus, by measuring Φ_{mj} / Φ_{nk}, we can determine the ratio A_{mj} / A_{nk} of the probabilities of transitions beginning from different levels.

Another criterion of the equality $N_m/g_m = N_n/g_n$, besides the cessation of generation, may be the reduction to zero of the gain on the transition m → n. In a number of cases this method may be preferable.

The described method in conjunction with measurement of the probabilities of transitions beginning from one level enables us to "combine" a large number of transitions in one system. Since gas lasers may generate on several hundred transitions, the field of application of the discussed method is very extensive.

This method was successfully applied in [33, 36] for the determination of the probabilities of 3s–2p Ne transitions.

3. The width and profile of the spontaneous emission and absorption lines of atoms in gas systems are usually determined by thermal motion. The radiation width of the line is very small ($\Gamma \sim 0.01\ k\bar{v}$) and is responsible for the line profile only in the far wing, which has been used in many investigations (see [116], for instance).

According to Chapter II, induced transitions in a strong monochromatic field lead to the appearance of a distinct structure, determined by the relaxation constants of the levels m, n, on the "Doppler background" of the spontaneous emission line. Measurements of this structure will obviously give the radiation widths of the lines (when the gas pressure and electron concentration are sufficiently low). Of particular interest is the case where the probabilities of decay of the upper and lower levels differ greatly. Here γ_m and γ_n can be determined separately from the spontaneous emission spectrum (see § 9, section **2**).

Unfortunately, at present, as far as we know, there are no experimental results relating to this question.

4. Absolute measurements of the output power also enable the determination of some combination of radiation constants. According to (21.10), the derivative of the flux with respect to the parameter Q/Q_0 or l/l_0 at the threshold point is

$$l_0 \frac{d\Pi}{dl}\bigg|_{l=l_0} = \sigma(1-r-a)\frac{4\pi\hbar\omega}{\lambda^2\gamma_{mn}}\frac{I_1}{I_2}. \tag{A.4}$$

From formula (17.39) for $\Omega = \Delta$ we obtain

$$\frac{I_1}{I_2} = \frac{\Gamma\Gamma_m\Gamma_n}{\Gamma_m+\Gamma_n}, \quad l_0 \frac{d\Pi}{dl}\bigg|_{l=l_0} = \sigma(1-r-a)\frac{8\pi\hbar\omega}{\lambda^2}\frac{\Gamma\Gamma_m\Gamma_n}{A_{mn}(\Gamma_m+\Gamma_n)}. \tag{A.5}$$

Thus, in this case the quantity $(A_{mn}/\Gamma)[(1/\Gamma_m) + (1/\Gamma_n)]$ is introduced.

We apply (A.5) to the transition Ne $\lambda = 6328$ Å. According to the results of [105], we have

$$\frac{\Gamma\Gamma_m\Gamma_n}{A_{mn}(\Gamma_m+\Gamma_n)} = 1.6\cdot 10^9 \text{ sec}^{-1}. \tag{A.6}$$

We assume that in the conditions of [105] the quantities Γ, Γ_j are determined by spontaneous transitions. Then

$$2\Gamma = \Gamma_m + \Gamma_n, \quad \Gamma_n = 0.84\cdot 10^8 \text{ sec}^{-1} \text{ [117]}. \tag{A.7}$$

Hence,

$$A_{mn}/\Gamma_m = 2.6\cdot 10^{-2}. \tag{A.8}$$

The treatment of the results of [105] in § 21 gave a value of 0.057 for the parameter ξ. Since $\gamma_n/k\bar{v} = 0.0074$, we can assume that the width of the level and ξ are determined by the upper level: $\xi = \gamma_m/k\bar{v}$. Then

$$\begin{aligned}\Gamma_m &= 2\gamma_m = 6.6\cdot 10^8 \text{ sec}^{-1},\\ A_{mn} &= 1.7\cdot 10^7 \text{ sec}^{-1}.\end{aligned} \tag{A.9}$$

According to [33, 36], the value of A_{mn}, determined by the method of Par. 2, is 10^{-7} sec^{-1}. The discrepancy may be due to inaccuracy in determining the parameter ξ or the invalidity of the assumption of the radiative nature of Γ, Γ_j. In the latter case we must assume that $\Gamma = \xi k\bar{v} = 6.6 \cdot 10^8$ sec^{-1} and

$$A_{mn} = \frac{3.4\cdot 10^7}{1+\Gamma_n/\Gamma_m} \text{ sec}^{-1}. \tag{A.10}$$

5. The quantity Γ can be determined by using the condition for generation on the following axial mode. Let initially generation occur on only one mode, the frequency of which coincides with ω_{mn}. Further increase in the level of excitation may give rise to generation on neighboring modes. According to (25.7), the excess over the threshold excitation at which this occurs will be

$$\frac{Q}{Q_0} - 1 = \frac{(2\Gamma)^2 + \Omega_\mu^2}{(k\bar{v})^2}, \quad \Omega_\mu = \frac{\pi c}{l}. \tag{A.11}$$

If $\Omega_\mu \sim 2\Gamma$, i.e.,

$$l \sim \frac{\pi c}{2\Gamma} = \frac{4.7\cdot 10^{10}}{\Gamma(\text{sec}^{-1})}(\text{cm}) \tag{A.12}$$

then $(Q/Q_0) - 1$ will be sensitive to the value of Γ. Hence, when $\Gamma \sim 4 \cdot 10^8$ sec^{-1} the measurements can be conducted at $l \sim 100$ cm.

6. The literature refers to several more effects which may be useful in the discussed problem. In [31] $\Gamma_m + \Gamma_n$ was measured for the transition $\lambda = 1.15\mu$, Ne $3p'[\frac{3}{2}]_2 - 4s'[\frac{1}{2}]_1^0$ from the relationship between the output power and the frequency $\Pi(\Omega)$. The obtained value ($\Gamma_m + \Gamma_n = 1.6 \cdot 10^8$ sec^{-1}) and the data for Γ_n ($\Gamma_n = 0.84 \cdot 10^8$ sec^{-1} [117]) give $\Gamma_m = 0.76 \cdot 10^8$ sec^{-1}.

An effect related to the dip on the curve of $\Pi(\Omega)$ is observed when the power of a gas laser is measured in relation to the external magnetic field H [118]. On the curve $\Pi(\Omega)$ there are also dips with a width which depends on Γ.

7. In the cases discussed in Pars. 3-6 the measured values are changed if the width of the line due to interactions becomes comparable with the radiative decay probabilities. We note that traditional spectroscopic techniques require the collision widths to be comparable with the Doppler width. Hence, it is essential here that the concentrations of perturbing particles be approximately two orders greater than in the investigation of the same collision processes by means of nonlinear effects. Thus, in the last case the experiment can be conducted in simpler and uniquely defined conditions. This applies, in particular, to the investigation of the statistical dependence of broadening due to interaction and the Doppler effect. The asymmetry of the line due to this statistical dependence is relatively slight if the collision frequency ν is less than $k\bar{v}$. At high collision frequencies, $\nu > k\bar{v}$, the line profile in ordinary conditions again becomes symmetric. In nonlinear effects the asymmetry is very pronounced also when $\Gamma \lessgtr \nu \ll k\bar{v}$ [curve $\Pi(\Omega)$, § 17, 21].

It was shown in § 17 that curve $\Pi(\Omega)$ is very sensitive to elastic collisions and a careful investigation at different pressures should presumably allow the separation of hard and soft collisions. Unfortunately, the experiments of [31] were interpreted by a different method and the material given in [31] is inadequate for any sound conclusions.

8. If the output is interrupted without disturbing the regime of excitation of the working levels (for instance, by introducing an opaque shutter into the resonator), the populations of the working levels are changed. It is found that the populations of other levels, not implicated in the output, are also changed [119]. This effect is due either to cascade transitions from the working levels or to inelastic processes. We can thus determine the spontaneous transition probabilities and the cross sections of atom-atom and atom-electron collisions involving highly excited atoms. The results of the first measurements conducted by this method are given in [32, 34].

This short review is not exhaustive, but it is clear from what has been said that gas lasers provide a whole set of new means for the direct determination of important atomic constants and the characteristics of collision processes.

Literature Cited

1. A. Einstein, Phys. Z., 18:121 (1917).
2. V. A. Fabrikant, F. A. Butaeva, and M. M. Vudynskii, Author's Certificate No. 576749/0-279/26, June 18, 1951. Byull. izobretenii, 20:29 (1959).
3. L. D. Landau and E. M. Lifshits, Electrodynamics of Continuous Media [in Russian], GITTL, Moscow (1957).
4. J. Weber, IRE Trans. Electr. Devices, No. 6 (1953).
5. N. G. Basov and A. M. Prokhorov, Zh. Éksp. Teor. Fiz. 27:431 (1954).
6. J. P. Gordon, H. J. Zeiger, and S. N. Townes, Phys. Rev., 95:282 (1954).
7. N. G. Basov and A. M. Prokhorov, Usp. Fiz. Nauk 57:485 (1955).
8. A. L. Schawlow and C. H. Townes, Phys. Rev. 112:1940 (1958).
9. N. G. Basov, O. N. Krokhin, and Yu. M. Popov, Usp. Fiz. Nauk 72:161 (1960) [Sov. Phys. – Usp., 12(5):1033 (1961)].

10. T. H. Maiman, Nature, 187 (4736):493 (1960).
11. A. Javan, W. R. Bennett, and D. R. Herriott, Phys. Rev. Lett., 6:106 (1961).
12. A. Javan, E. A. Balik, and W. L. Bond, J. Opt. Soc. Am., 52:96 (1962).
13. A. Javan, Proceedings of Third Conference on Quantum Electronics (1963).
14. C. Townes and A. Schawlow, Microwave Spectroscopy, McGraw-Hill, New York (1955).
15. V. I. Talanov, Izv. vuzov. Radiofizika, 7:564 (1964).
16. I. L. Fabelinskii, Molecular Scattering of Light, Plenum Press, New York (1968).
17. S. A. Akhmanov and R. V. Khokhlov, Problems of Nonlinear Optics [in Russian], Izd. VINITI, Moscow (1964).
18. L. V. Keldysh, Zh. Éksp. Teor. Fiz., 47:1945 (1964) [Sov. Phys. – JETP, 20:1307 (1965)].
19. Yu. P. Raizer, Usp. Fiz. Nauk, 87:29 (1965) [Sov. Phys. – Usp., 8:650 (1966)].
20. S. Panzer, Z. Angew. Math. Phys., 16:138 (1965).
21. C. K. N. Patel, Appl. Phys. Lett., 7:15 (1965).
22. S. G. Rautian and I. I. Sobel'man, Zh. Éksp. Teor. Fiz., 41:2018 (1961) [Sov. Phys. – JETP, 14(2):1433 (1962)].
23. I. Kasper and G. Pimentel, Appl. Phys. Lett., 5:231 (1964).
24. T. L. Andreeva, V. A. Dudkin, V. I. Malyshev, G. V. Mikhailov, V. N. Sorokin, and L. A. Novikova, Zh. Éksp. Teor. Fiz. 49:1408 (1965) [Sov. Phys. – JETP, 22:969 (1966)].
25. S. Jacobs, G. Gould, and P. Rabinowitz, Phys. Rev. Lett., 7:415 (1961).
26. W. R. Bennett, Phys. Rev., 126:580 (1962).
27. V. (W.) R. Bennett, Usp. Fiz. Nauk, 81:119 (1963).
28. W. W. Rigrod, Appl. Phys. Lett., 2:51 (1963).
29. T. Uchida, Appl. Opt., 4:129 (1965).
30. I. M. Belousova, O. B. Danilov, and B. A. Ermakov, Zh. Éksp. Teor. Fiz., 47:2013 (1964) [Sov. Phys. – JETP, 20:1351 (1965)].
31. A. Javan and A. Szoke, Phys. Rev. Lett., 10:521 (1963).
32. A. Javan and J. H. Parks, Phys. Rev., 139A:1351 (1965).
33. V. P. Chebotaev, Dissertation, IFP SO AN SSSR (1965).
34. A. S. Khaikin, Zh. Éksp. Teor. Fiz., 51:38 (1966) [Sov. Phys. – JETP, 24:25 (1967)].
35. G. G. Petrash and S. G. Rautian, Opt. Spektrosk., 18:336 (1965) [Opt. Spectrosc., 18:188 (1965)].
36. V. G. Kirpilenko, Dissertation, MFTI (1965).
37. S. G. Rautian and T. A. Germogenova, Opt. Spektrosk., 17:157 (1964) [Opt. Spetrosc., 17:85 (1964)].
38. A. Javan, Proceedings of International Conference on Quantum Electronics, San Juan (1965).
39. C. B. Moore, Appl. Opt., 4:252 (1965).
40. S. G. Rautian and I. I. Sobel'man, Zh. Éksp. Teor. Fiz., 39:217 (1960) [Sov. Phys. – JETP, 12:156 (1961)].
41. S. G. Rautian and I. I. Sobel'man, Opt. Spektrosk., 10:134 (1961) [Opt. Spectrosc., 10:65 (1961)].
42. S. G. Rautian and A. S. Selivanenko, Author's Declaration.
43. A. Lempicki and H. Samuelson, Phys. Lett., 4:133 (1963).
44. P. A. Bazhulin, L. D. Dergacheva, B. G. Distanov, G. V. Peregudov, A. M. Prokhorov, A. I. Sokolovskaya, and D. N. Shigorin, Opt. Spektrosk., 18:526 (1965) [Opt. Spectrosc., 18:298 (1965)].
45. S. G. Rautian and I. I. Sobel'man, Zh. Éksp. Teor. Fiz., 41:456 (1961) [Sov. Phys. – JETP, 14:328 (1962)].
46. T. I. Kuznetsova and S. G. Rautian, Zh. Éksp. Teor. Fiz., 43:1897 (1962) [Sov. Phys. – JETP, 16:1338 (1963)].
47. S. G. Rautian and I. I. Sobel'man, Zh. Éksp. Teor. Fiz., 44:834 (1963) [Sov. Phys. – JETP, 17:635 (1963)].

48. T. I. Kuznetsova and S. G. Rautian, Fiz. Tverd. Tela, 5:2105 (1963) [Sov. Phys. – Solid State, 5(8):1535 (1964)].
49. T. I. Kuznetsova and S. G. Rautian, Izv. vuzov. Radiofizika, 7:682 (1964).
50. S. G. Rautian, Fiz. Tverd. Tela, 6:1857 (1964) [Sov. Phys. – Solid State, 6(6):1462 (1964)].
51. T. A. Germogenova and S. G. Rautian, Otchet MIAN-FIAN (1963), Zh. Éksp. Teor. Fiz., 46:745 (1964) [Sov. Phys. – JETP, 19:507 (1964)].
52. T. I. Kuznetsova and S. G. Rautian, Zh. Éksp. Teor. Fiz., 49:1605 (1965) [Sov. Phys. – JETP, 22:1098 (1966)].
53. S. G. Rautian and I. I. Sobel'man, Preprint FIAN, A-145 (1965); Usp. Fiz. Nauk, 90:209 (1966) [Sov. Phys. – Usp., 9(5):701 (1967)].
54. S. G. Rautian, Zh. Éksp. Teor. Fiz., 52:1176 (1966) [Sov. Phys. – JETP, 24:788 (1967)]; Preprint FIAN, 133 (1966).
55. G. E. Notkin, S. G. Rautian, and A. A. Feoktistov, Zh. Éksp. Teor. Fiz., 52:1673 (1967) [Sov. Phys. – JETP, 25:1112 (1967)].
56. W. Heitler, The Quantum Theory of Radiation, University Press, Oxford (1936).
57. I. I. Sobel'man, Introduction to the Theory of Atomic Spectra [in Russian], Fizmatgiz (1963).
58. A. A. Vuylsteke, Elements of Maser Theory, D. Van Nostrand Co. (1960).
59. R. Karplus and L. Schwinger, Phys. Rev., 73:1020 (1948).
60. V. M. Fain and Ya. I. Khanin, Quantum Radiophysics, Izd. Sovetskoe Radio, Moscow (1965).
61. R. Cordover, A. Javan, J. Parks, and A. Szoke, Proceedings of International Conference on Quantum Electronics, San Juan, Puerto Rico (1965).
62. R. L. Fork and M. A. Pollack, Phys. Rev., 139A:1408 (1965).
63. I. I. Sobel'man, Usp. Fiz. Nauk, 54:552 (1954).
64. B. I. Stepanov, Dokl. Akad. Nauk SSSR, 148:74 (1963) [Sov. Phys. – Dokl., 8:37 (1963)].
65. L. S. Turovtseva (in press).
66. V. N. Fadeeva and N. M. Terent'ev, Tables of Values of Probability Integrals of Complex Argument [in Russian], Gostekhizdat (1954).
67. S. N. Bagaev, V. S. Kuznetsov, Yu. V. Troitskii, and B. I. Troshin, ZhÉTF Pis. Red., 1(4):21 (1965).
68. W. E. Lamb, Phys. Rev., 134A:1429 (1964); Proceedings of the Enrico Fermi International School of Physics (1963).
69. E. Fermi, Rev. Mod. Phys., 4:87 (1932).
70. E. A. Coddington and N. Levinson, Theory of Ordinary Differential Equations, McGraw-Hill, New York (1955).
71. B. Senitsky and G. Gould, Proceedings of Third Conference on Quantum Electronics, Paris (1963).
72. B. Senitsky, G. Gould, and S. Cutler, Phys. Rev., 130:1460 (1963).
73. B. Senitsky and S. Cutler, Microwave J., 7:62 (1964).
74. D. N. Klyshko, Yu. S. Konstantinov, and V. S. Tumanov, Izv. vuzov. Radiofizika, 8:513 (1965).
75. M. I. Podgoretskii and A. V. Stepanov, Preprint OIYaI (1960).
76. M. I. Podgoretskii and A. V. Stepanov, Zh. Éksp. Teor. Fiz., 40:561 (1961) [Sov. Phys. – JETP, 13(2):393 (1961)].
77. V. Weisskopf, Usp. Fiz. Nauk, 13:552 (1933).
78. S. Chandrasekhar, Stochastic Problems in Physics and Astronomy, Rev. Mod. Phys., 15:2 (1943).
79. I. S. Gradshtein and I. M. Ryzhik, Tables of Integrals, Sums, Series, and Products [in Russian], Fizmatgiz (1962).
80. L. A. Vainshtein and V. D. Zubakov, Extraction of Signals from Noise [in Russian], Izd. Sovetskoe Radio, Moscow (1960).

81. I. I. Privalov, Fourier Series [in Russian], Gostekhizdat, Moscow (1934).
82. V. L. Ginzburg, Dokl. Akad. Nauk SSSR, 30:397 (1941).
83. J. Keilson and J. E. Storer, Quart. Appl. Math., 10:243 (1952).
84. S. Chapman and T. Cowling, The Mathematical Theory of Non-Uniform Gases, Cambridge (1952).
85. E. P. Gross, Phys. Rev., 97:395 (1955).
86. I. V. Aleksandrov, Theory of Nuclear Magnetic Resonance [in Russian], Nauka, Moscow (1964).
87. W. R. Bennett, S. F. Jacobs, J. T. La Tourett, and P. Rabinowitz, Appl. Phys. Lett., 5:56 (1964).
88. L. A. Vainshtein and I. I. Sobel'man, Opt. Spektroskop., 6:440 (1959) [Opt. Spectrosc., 6:279 (1969)].
89. A. I. Burshtein, Zh. Éksp. Teor. Fiz., 49:1362 (1965) [Sov. Phys. – JETP, 22:939 (1966)].
90. S. G. Rautian, Candidate's Dissertation, FIAN (1957).
91. W. E. Lamb and B. L. Gyorffy, Proceedings of International Conference on Quantum Electronics, San Juan (1965).
92. V. L. Ginzburg, Propagation of Electromagnetic Waves in Plasma [in Russian], Fizmatgiz (1960).
93. L. M. Brekhovskikh, Waves in Layered Media [in Russian], Izd. AN SSSR, Moscow (1957).
94. L. A. Ostrovskii and E. I. Yakubovich, Zh. Éksp. Teor. Fiz., 46:963 (1964) [Sov. Phys. – JETP, 19:656 (1964)].
95. L. A. Ostrovskii and E. I. Yakubovich, Izv. vuzov. Radiofizika, 8:91 (1965).
96. A. N. Oraevskii, Masers [in Russian], Izd. AN SSSR, Moscow (1964).
97. A. P. Ivanov, B. I. Stepanov, B. M. Berkovskii, and I. L. Katsev, Dokl. Akad. Nauk BSSR, 6:147 (1962).
98. E. P. Zage, A. M. Samson, and B. I. Stepanov, Dokl. Akad. Nauk BSSR, 6:288 (1962).
99. A. Fox and T. Li, in the collection: Lasers [Russian translation], IL, Moscow (1963), p. 207.
100. T. I. Kuznetsova, Preprint FIAN, A-19 (1965).
101. L. A. Vainshtein, Zh. Éksp. Teor. Fiz., 44:1050 (1963) [Sov. Phys. – JETP, 17:709 (1963)].
102. J. Boyd and J. Gordon, in the collection: Lasers [Russian translation], IL, Moscow (1963), p. 363.
103. T. I. Kuznetsova, Zh. Tekh. Fiz., 34:419 (1964) [Sov. Phys. – Tech. Phys., 9:330 (1964)].
104. Yu. V. Troitskii and V. P. Chebotaev, Opt. Spektrosk., 20:362 (1965) [Opt. Spectrosc., 20:199 (1966)].
105. A. D. White, E. I. Gordon, and J. D. Rigden, Appl. Phys. Lett., 2:91 (1963).
106. I. M. Belousova, O. B. Danilov, and I. A. Erokhina, Zh. Éksp. Teor. Fiz., 44:1111 (1963) [Sov. Phys. – JETP, 17:748 (1963)].
107. N. G. Basov, E. P. Markin, and V. V. Nikitin, Opt. Spektrosk., 15:436 (1963) [Opt. Spectrosc., 15:235 (1963)].
108. A. Gurtovnik, Izv. vuzov. Radiofizika, 1:83 (1958).
109. V. N. Lugovoi, Radiotekhnika i élektronika, 6:1700 (1961).
110. C. L. Tang, H. Statz, and G. de Mars, Appl. Phys. Lett., 2:222 (1963).
111. A. A. Manenkov, R. M. Martirosyan, Yu. P. Pimenov, A. M. Prokhorov, and V. A. Sychugov, Zh. Éksp. Teor. Fiz., 47:2055 (1964) [Sov. Phys. – JETP, 20:1381 (1965)].
112. V. V. Korobkin, A. M. Leontovich, and M. N. Smirnova, Zh. Éksp. Teor. Fiz., 48:78 (1965) [Sov. Phys. – JETP, 21:53 (1965)].
113. W. Ewtuchow, Appl. Phys. Lett., 6:141 (1965).
114. C. K. N. Patel, W. R. Bennett, W. L. Faust, and R. A. McFarlane, Phys. Rev. Lett., 9:102 (1962).
115. W. L. Faust, R. A. McFarlane, C. K. N. Patel, and G. G. B. Garrett, Appl. Phys. Lett., 1:85 (1962).

116. A. Unsold, Physics of Stellar Atmospheres [Russian translation], Springer, Berlin (1938).
117. R. Ladenburg, Usp. Fiz. Nauk, 14:721 (1934).
118. V. I. Perel', Symposium on Nonlinear Optics [in Russian], Minsk (1965).
119. A. Javan, J. H. Parks, and A. Szoke, Bull. Am. Phys. Soc., 9:490 (1964).

INTERACTION OF ELECTROMAGNETIC FIELDS WITH ACTIVE MEDIA

T. I. Kuznetsova

Introduction

The electrodynamic problems which arise in the investigation of masers have several essential different features from the problems associated with an electromagnetic field in passive media. These features are due to the fact that masers contain substances capable of amplifying the field, and the radiative properties of the active substance are not prescribed, but depend on the characteristics of the field. In the case of sufficiently intense fields this dependence must be taken into consideration. Polarization of a substance with an inverse population due to a strong electromagnetic field was calculated in [1], where a quantitative formulation of the saturation effect was given. The first self-consistent problem for an active medium and a field producing saturation was formulated and solved in connection with the development of maser theory [2]. In this work values of the field strength and the dielectric constant averaged over the coordinates were introduced and the maser was thus replaced by an equivalent tank circuit. In subsequent works [3-5] on masers this approach was extended to the case of a monochromatic field, where the dielectric constant cannot be introduced; a system of equations for the field, polarization of the active medium, and inversion was obtained. The reduction of a maser to a tank circuit with concentrated parameters has been widely adopted in investigations of microwave masers. However. the potentialities of such a simplified approach are limited. For instance, this approximation is inadequate for the description of several effects in optical masers. Here, as is well known, the characteristic dimensions of the region of interaction of the field and substance greatly exceed the wavelength of the emission, i.e., the field cannot be regarded as constant relative to the coordinate. Hence, the understanding of the operation of lasers necessitates an investigation of the spatial distribution of the field and the polarization of the medium.

In several works [6-8] attempts have been made to solve the spatial problems for lasers by transport theory techniques. The authors of these papers assumed that there are noninterfering fluxes of radiation in the laser. They thereby unjustifiably ignored spatial structure of the order of a wavelength.

The next approach to the question of laser emission, based on the solution of the boundary-value electrodynamic problem, was first undertaken in our work [9]. In this paper we investigated the nonlinear spatial problem for a monochromatic field in a plane-parallel active

*Dissertation presented for the Degree of Candidate of Physicomathematical Sciences, 1966.

layer. We showed that in the case of a layer with highly reflecting boundaries the field varies almost periodically with the coordinate and resembles a standing wave. The properties of the medium also vary periodically and the inhomogeneity of the dielectric constant due to saturation is greater, the stronger the field. It was discovered that inhomogeneity over the extent of a wavelength has to be taken into account in the calculation of the output power. The inhomogeneity is manifested in the increase in emission power with excitation being slower than linear [2]. An even more significant role of spatial inhomogeneity was discovered in an examination of the possible conditions of operation of a laser.

The theoretical investigation of masers led to methods of investigating the stability of monochromatic generation [10-12]. These methods were subsequently extended to the case of resonators where the distance between the oscillation modes lay within the width of the spectral line. It was assumed in this case that the spatial structure of the two modes was the same. From the calculations made in [13] it was concluded that monochromatic generation in the infrared and visible regions is stable.

Questions of stable conditions of operation have also been examined within the framework of transport equations [7-8]. Since it was assumed in these papers that different modes have the same spatial structure, the obtained result did not differ from that of [13]. The mode closest in frequency to the center of the line of the active substance was found to be stable.

A similar viewpoint has been expressed by other authors, those of [14], for instance; it was assumed that nonmonochromatic generation is possible only with an inhomogeneously broadened line. These theoretical ideas contradicted the experimental data, which indicated multimode operation of lasers.

In our work [15] we obtained a qualitatively different result. We showed that the production of a strong field at one frequency is not the reason for the absence of generation of fields at other frequencies if the field and medium are inhomogeneous in space and the fields of different frequencies have a different spatial distribution. The cause of the instability of monochromatic generation in conditions typical for lasers was thus indicated.

The need to take the spatial distribution of the field and the inversion into account was subsequently recognized in the literature [16]. However, most of the theoretical works on this theme, apart from [17, 18], have been carried out on the assumption that the fields in operating conditions have the same coordinate dependence as in an empty resonator [19-21]. The correctness of this assumption was not obvious earlier, since the inhomogeneities of the medium due to saturation can tune the natural oscillations. Our work [22], devoted to the generation of a plane-parallel layer at arbitrarily high excitation levels showed that the field distribution is distorted when the excitation parameter changes. With high reflection coefficients and sufficiently high excitation the form of the field differs considerably from the natural oscillations of the empty plane-parallel layer.

In the case of nonplane laser models we can expect an even greater tuning of the natural oscillations by inhomogeneities of the medium.

There is an abundant literature [23-27] on the theory of two- and three-dimensional empty resonators at present, but the appearance of the first work on an open resonator confirmed the view that the eigenfunctions of the empty resonator are manifested in real lasers. Yet in many cases the laser medium is inhomogeneous in a direction perpendicular to the direction of generation. In our papers [28, 29] we considered prescribed (field-independent) transverse inhomogeneities of inversion. We showed that the smallest inhomogeneities (a change of $\sim 10^{-8}$ in the dielectric constant) can lead to focusing of the field, an increase in the natural frequencies, and a considerable change in the generation condition. The effect of transverse inhomogeneities of the medium on the properties of the natural oscillations narrows the region of applicability of the theory of empty resonators even in the case of gas lasers.

The formulation of electrodynamic problems for active media and the application of the results to some particular systems led to the need for a more thorough investigation of the properties of an active substance in a strong field. In particular, it became essential to investigate polarization of the medium in a field containing one strong spectral component and several weak ones. An investigation of the behavior of quantum-mechanical systems in a field of such form was made in [30-32]. In these papers, however, the weak field was assumed to be monochromatic and the gain for it was calculated (polarization was ignored). Later, in [33-37] calculations of the polarization of a quantum-mechanical system due to a weak field and a monochromatic strong one were begun. It was shown in [33, 35, 37] that if the weak field is monochromatic, then a component on a combination frequency appears in the polarization.

In our works [38, 39] we calculated the polarization of a two-level system of more general (in comparison with [33, 35, 37]) form and a three-level system. We found that with different longitudinal and transverse relaxation constants the intensity of the arising combination components depends significantly on the difference in frequencies of the weak and strong fields.

The appearance of combination components in the polarization must be taken into account in the analysis of the stability of monochromatic generation. In the general case, two interacting weak fields must be considered. But when there is a difference in the order of magnitudes of the longitudinal and transverse relaxation constants there is a broad region of frequencies and intensities of the strong field at which interaction of the weak fields can be neglected and questions of stability can be considered on the basis of the idea of a dielectric constant.

Coupling of weak fields through polarization must also occur in the propagation of a modulated signal in the amplifier. This means that the methods based on energy considerations and developed for cases of amplification of a monochromatic field [40-42] become inapplicable and it is necessary to consider the spatial problem for coupled fields. Such a problem was solved in our work [39].

Thus, the main content of the dissertation is an investigation of the spatial distribution of the field in systems containing an active medium and the influence of space effects on some characteristics of amplifiers and oscillators.

CHAPTER I

Calculation of Polarization of Medium

In this chapter we will calculate the mean dipole moment of the active substance due to the action of a prescribed field consisting of the sum of a monochromatic strong component and several weak components. The results obtained will be used in the subsequent chapters.

1. The dipole moment is determined by means of the density matrix. We assume that the interaction of the medium with the electromagnetic field is described by a two-level scheme.

The equations for the density matrix, averaged over the excitation moments, have the form [43]

$$\begin{gathered}\frac{d}{dt}\rho_{11}+2\gamma_1\rho_{11}=-i\frac{p}{\hbar}E(\rho_{12}-\overset{*}{\rho}_{12})+\Lambda_1,\\ \frac{d}{dt}\rho_{22}+2\gamma_2\rho_{22}=i\frac{p}{\hbar}E(\rho_{12}-\overset{*}{\rho}_{12})+\Lambda_2,\\ \frac{d}{dt}\rho_{12}+(\tilde{\Gamma}-i\omega_0)\rho_{12}=i\frac{p}{\hbar}E(\rho_{22}-\rho_{11}).\end{gathered}\tag{1.1}$$

Here E is the electric field strength, the subscript 2 denotes the upper level, and the subscript 1 the lower level; Λ_2 and Λ_1 are the rates of excitation of the considered levels, ω_0 is the natural frequency of the transition between them; the matrix elements of the dipole moment $p_{11} = p_{22} = 0$, $p_{12} = p$, and we can assume that $p = p^*$ and $p > 0$. The system is characterized by three relaxation constants: $\gamma_1, \gamma_2, \tilde{\Gamma}$.

We first consider the case where the strong and weak fields are monochromatic:

$$E = \frac{\hbar}{p} F \cos(\omega t + \varphi) + \frac{\hbar}{p} f \cos(\omega t + \Omega t + \Phi). \tag{1.2}$$

Here we introduce the two phases φ and Φ, since this notation will be more convenient later in the consideration of spatial problems. We will regard the first term in formula (1.2) as the strong field and the second term as the weak field. We assume that the frequencies ω and $\omega + \Omega$ are close to transition frequency ω_0: $|\omega - \omega_0| < \tilde{\Gamma}$, $|\omega + \Omega - \omega_0| < \tilde{\Gamma}$.

We convert to new unknown functions

$$\begin{aligned} \rho_{11} + \rho_{22} &= y_1, \\ \rho_{22} - \rho_{11} &= y_2, \\ \rho_{12} &= ie^{i\omega t + i\varphi} y_{12}. \end{aligned}$$

In addition, we introduce the symbols:

$$\begin{aligned} \Gamma &= \gamma_1 + \gamma_2, \\ \gamma &= \gamma_1 - \gamma_2, \\ \Lambda &= \Lambda_2 - \Lambda_1, \\ \lambda &= \Lambda_2 + \Lambda_1. \end{aligned} \tag{1.3}$$

From (1.1) using (1.2) and (1.3), we obtain a system of equations for the y functions

$$\begin{aligned} &\frac{d}{dt} y_1 + \Gamma y_1 = \gamma y_2 + \lambda, \\ &\frac{d}{dt} y_2 + \Gamma y_2 = \gamma y_1 + \Lambda - 2\left[F \cos(\omega t + \varphi) + f \cos(\omega t + \Omega t + \Phi)\right] \left(y_{12} e^{i\omega t + i\varphi} + \overset{*}{y}_{12} e^{-i\omega t - i\varphi}\right), \\ &\frac{d}{dt} y_{12} + (\tilde{\Gamma} + i\omega - i\omega_0) y_{12} = e^{-i\omega t - i\varphi} \left[F \cos(\omega t + \varphi) + f \cos(\omega t + \Omega t + \Phi)\right] y_2 . \end{aligned} \tag{1.4}$$

We now discard the terms containing the frequency 2ω on the right sides of (1.4) (as was done in [1, 30, 43]):

$$\begin{aligned} &\frac{d}{dt} y_1 + \Gamma y_1 = \gamma y_2 + \lambda, \\ &\frac{d}{dt} y_2 + \Gamma y_2 = \gamma y_1 + \Lambda - F(y_{12} + \overset{*}{y}_{12}) - f\left(y_{12} e^{-i\Omega t - i\Phi} + \overset{*}{y}_{12} e^{i\Omega t + i\Phi}\right), \\ &\frac{d}{dt} y_{12} + (\tilde{\Gamma} + i\omega - i\omega_0) y_{12} = {}^1\!/_2 F y_2 + {}^1\!/_2 f e^{i\Omega t + i\Phi} y_2. \end{aligned} \tag{1.5}$$

We will seek the solution of system (1.5) in the form of an exact solution of this system with $f = 0$ and a correction due to the weak field

$$y_n = X_n + x_n.$$

The zero-approximation equations have the form

$$\begin{aligned} &\frac{d}{dt} X_1 + \Gamma X_1 = \gamma X_2 + \lambda, \\ &\frac{d}{dt} X_2 + \Gamma X_2 = \gamma X_1 + \Lambda - F(X_{12} + \overset{*}{X}_{12}), \\ &\frac{d}{dt} X_{12} + (\tilde{\Gamma} + i\omega - i\omega_0) X_{12} = {}^1\!/_2 F X_2 . \end{aligned} \tag{1.6}$$

and are well known in the literature; for the corrections x_n we obtain, by discarding quantities of the form fx_n, the following system of equations:

$$\begin{gathered}\frac{d}{dt}x_1 + \Gamma x_1 - \gamma x_2 = 0,\\ \frac{d}{dt}x_2 + \Gamma x_2 - \gamma x_1 + F(x_{12} + \overset{*}{x}_{12}) = -f\left(X_{12}e^{-i\Omega t - i\Phi + i\varphi} + \overset{*}{X}_{12}e^{i\Omega t + i\Phi - i\varphi}\right),\\ \frac{d}{dt}x_{12} + (\tilde{\Gamma} + i\omega - i\omega_0)x_{12} - {}^1/_2 F x_2 = {}^1/_2 f e^{i\Omega t + i\Phi - i\varphi} X_2.\end{gathered} \tag{1.7}$$

We will consider the system of equations (1.6) and (1.7) in the range $-\infty < t < +\infty$ and require boundedness of the solutions for $t = \pm\infty$. In this case the conditions for the solution of (1.6), (1.7) are uniquely defined. In fact, the homogeneous system of equations corresponding to (1.6), (1.7) has the form

$$\begin{gathered}\frac{d}{dt}X_1 + \Gamma X_1 - \gamma X_2 = 0,\\ \frac{d}{dt}X_2 + \Gamma X_2 - \gamma X_1 + FX_{12} + F\overset{*}{X}_{12} = 0,\\ \frac{d}{dt}X_{12} + (\tilde{\Gamma} + i\omega - i\omega_0)X_{12} - {}^1/_2 F X_2 = 0,\\ \frac{d}{dt}\overset{*}{X}_{12} + (\tilde{\Gamma} - i\omega + i\omega_0)\overset{*}{X}_{12} - {}^1/_2 F X_2 = 0.\end{gathered} \tag{1.8}$$

The natural frequencies of this system are determined from the equation

$$\begin{vmatrix} i\nu + \Gamma & -\gamma & 0 & 0\\ -\gamma & i\nu + \Gamma & F & F\\ 0 & -\frac{1}{2}F & i\nu + \tilde{\Gamma} + i\omega - i\omega_0 & 0\\ 0 & -\frac{1}{2}F & 0 & i\nu + \tilde{\Gamma} - i\omega + i\omega_0\end{vmatrix} = 0. \tag{1.9}$$

It is easy to see that when ν is real equation (1.9) is not satisfied and (1.8) has solutions only with complex frequencies, i.e., unbounded solutions, when $t = \pm\infty$.

The solutions of equation (1.6) have the form

$$\begin{gathered}X_2 = \frac{\frac{\Lambda}{\Gamma} + \frac{\lambda\gamma}{\Gamma^2}}{1 - \frac{\gamma^2}{\Gamma^2} + \frac{F^2}{\Gamma\tilde{\Gamma}}\left[1 + \left(\frac{\omega - \omega_0}{\tilde{\Gamma}}\right)^2\right]^{-1}},\\ X_{12} = \frac{1}{2}\frac{F}{\tilde{\Gamma} + i(\omega - \omega_0)}X_2.\end{gathered} \tag{1.10}$$

System (1.7) can conveniently be solved, for instance, by means of the substitution

$$x_n = \bar{x}_n e^{i\Omega t + i\Phi - i\varphi} + \bar{\bar{x}}_n e^{-i\Omega t - i\Phi + i\varphi}.$$

From (1.7) we obtain a system of linear algebraic equations for $\bar{x}_n$, $\bar{\bar{x}}_n$:

$$\begin{gathered}(i\Omega + \Gamma)\bar{x}_1 - \gamma\bar{x}_2 = 0,\\ (i\Omega + \Gamma)\bar{x}_2 - \gamma\bar{x}_1 + F\bar{x}_{12} + F\overset{*}{\bar{\bar{x}}}_{12} = -f\overset{*}{X}_{12},\\ (i\Omega + \tilde{\Gamma} + i\omega - i\omega_0)\bar{x}_{12} - \frac{1}{2}F\bar{x}_2 = \frac{1}{2}fX_2,\\ (i\Omega + \tilde{\Gamma} - i\omega + i\omega_0)\overset{*}{\bar{\bar{x}}}_{12} - \frac{1}{2}F\bar{x}_2 = 0.\end{gathered} \tag{1.11}$$

Equations for $\bar{\bar{x}}_1 = \bar{x}_1^*$, $\bar{\bar{x}}_2 = \bar{x}_2^*$, $\bar{\bar{x}}_{12}$, $\overrightarrow{x}_{12}^*$ can easily be obtained by writing the expressions conjugate to (1.11)

From the solution of (1.11) we find

$$\bar{x}_2 = -\frac{1}{2} fFX_2 \frac{\dfrac{2\tilde{\Gamma} + i\Omega}{\tilde{\Gamma} + i\Omega} \dfrac{\tilde{\Gamma} + i\Omega - i\omega + i\omega_0}{\tilde{\Gamma} - i\omega + i\omega_0}}{\left(i\Omega + \Gamma - \dfrac{\gamma^2}{i\Omega + \Gamma}\right)\left(i\Omega + \tilde{\Gamma} + \dfrac{(\omega - \omega_0)^2}{i\Omega + \tilde{\Gamma}}\right) + F^2}, \tag{1.12}$$

$$\bar{x}_{12} = \frac{\frac{1}{2} fX_2 + \frac{1}{2} F\bar{x}_2}{\tilde{\Gamma} + i(\omega + \Omega - \omega_0)}, \tag{1.13}$$

$$\bar{\bar{x}}_{12} = \frac{\frac{1}{2} F\bar{\bar{x}}_2}{\tilde{\Gamma} + i(\omega - \Omega - \omega_0)}, \quad \bar{\bar{x}}_2 = \overset{*}{\bar{x}}_2. \tag{1.14}$$

The mean value of the dipole moment is expressed in terms of our calculated quantities in the following way:

$$p_m = 2p\,\mathrm{Re}\{ie^{i\omega t + i\varphi} y_{12}\} = 2p\,\mathrm{Re}\{iX_{12}e^{i\omega t + i\varphi} + i\bar{x}_{12}e^{i(\omega+\Omega)t + i\Phi} + i\bar{\bar{x}}_{12}e^{i(\omega-\Omega)t + i2\varphi - i\Phi}\}. \tag{1.15}$$

Thus, in the considered approximation, in addition to the components of the frequencies ω and $\omega + \Omega$, present in the initial electromagnetic field, there is also a component on the combination frequency $\omega - \Omega$.

We consider formula (1.15) more fully. The quantity X_{12} defines the polarization component on the strong-field frequency and is expressed [see (1.10)] by the product of the strong-field amplitude F, the population difference X_2, and the resonance factor $[\tilde{\Gamma} + i(\omega - \omega_0)]^{-1}$.

The polarization component on the weak-field frequency is given by $\bar{x}_{12}$. As formula (1.13) shows, $\bar{x}_{12}$ consists of two terms. The first term has a fairly simple structure and is the same as the quantity defining polarization on the strong-field frequency with F replaced by f and the resonance factor $[\tilde{\Gamma} + i(\omega - \omega_0)]^{-1}$ replaced by $[\tilde{\Gamma} + i(\omega + \Omega - \omega_0)]^{-1}$. If F tends to zero, the first term in $\bar{x}_{12}$ is finite.

The second term in $\bar{x}_{12}$ is of a more complex nature. The dependence of the second term on F does not reduce merely to the effect of the strong field F on the population difference. This is clear from expression (1.12) for $\bar{x}_2$, which contains, besides the population difference X_2, an additional factor dependent on F. When $F \to 0$, x_2 and the whole term become zero. The frequency dependence of the considered term is contained both in the factor $[\tilde{\Gamma} + i(\omega + \Omega - \omega_0)]^{-1}$ and in $\bar{x}_2$.

The quantity $\bar{\bar{x}}_{12}$ in (1.15) represents the polarization component on the frequency $\omega - \Omega$. This frequency is not contained in the initial field and its appearance indicates the nonlinear properties of the system in the strong radiation field. Formulas (1.14) and (1.13) show that the modulus of $\bar{\bar{x}}_{12}$ and the modulus of the second term in the expression $\bar{x}_{12}$ are of the same order. This means that the appearance of the combination frequency in the polarization occurs in all cases where the deviation of the frequency dependence of the gain from the usual dispersion curve becomes significant. These two effects are more pronounced, the greater the saturation. They disappear when the strong-field amplitude is sufficiently small. We determine at what values of F the polarization on the combination frequency becomes commensurable with the polarization on the frequency $\omega + \Omega$. This will obviously occur when $|F\bar{x}_2| \approx |fX_2|$, i.e., when

$$F^2 \geqslant \left| i\Omega + \widetilde{\Gamma} + \frac{(\omega-\omega_0)^2}{i\Omega+\widetilde{\Gamma}} \right| \left| i\Omega + \Gamma - \frac{\gamma^2}{i\Omega+\Gamma} \right|. \tag{1.16}$$

When $\Omega = 0$, condition (1.16) is the same as the condition that the field F causes saturation [see formula (1.10)]. At greater Ω greater values of F* are required for the satisfaction of (1.16).

The employed method of calculation can also be applied to quantum-mechanical systems with electromagnetic excitation (three-level systems). For such systems we obtained similar results for the excitation of combination frequencies and their dependence on the saturation [39].

Several substances used in lasers and masers are described by a system of second-order equations which differs from (1.1) and satisfies the condition of conservation of the sum of the populations on levels 1 and 2, i.e., a system of the form:

$$\begin{aligned} &\frac{d}{dt}\rho_{11} + \Gamma_1\rho_{11} - \Gamma_2\rho_{22} = -i\frac{p}{\hbar}E(\rho_{12} - \rho_{12}^*),\\ &\frac{d}{dt}\rho_{22} + \Gamma_2\rho_{22} - \Gamma_1\rho_{11} = i\frac{p}{\hbar}E(\rho_{12} - \rho_{12}^*),\\ &\frac{d}{dt}\rho_{12} + (\widetilde{\Gamma} - i\omega_0)\rho_{12} = i\frac{p}{\hbar}E(\rho_{22} - \rho_{11}). \end{aligned} \tag{1.17}$$

By the transformation

$$\rho_{22} - \rho_{11} = y_2,$$
$$\rho_{12} = ie^{i\omega t + i\varphi}y_{12}$$

and the condition

$$\rho_{11} + \rho_{22} = 1$$

this system takes the form

$$\begin{aligned} &\frac{d}{dt}y_2 + \Gamma y_2 = \Lambda - 2[F\cos(\omega t + \varphi) + f\cos(\omega t + \Omega t + \Phi)](y_{12}e^{i\omega t + i\varphi} + y_{12}^* e^{-i\omega t - i\varphi}),\\ &\frac{d}{dt}y_{12} + (\widetilde{\Gamma} + i\omega - i\omega_0)y_{12} = e^{-i\omega t - i\varphi}[F\cos(\omega t + \varphi) + f\cos(\omega t + \Omega t + \Phi)]y_2, \end{aligned} \tag{1.18}$$

where $\Gamma = \Gamma_1 + \Gamma_2$, $\Lambda = \Gamma_2 - \Gamma_1$. A comparison of (1.18) with (1.4) shows that the population difference and the polarization for case (1.17) can be obtained if we put $\gamma = 0$ in formulas (1.10) and (1.12). The final results which will be obtained in this chapter [(1.23), (1.28)-(1.31)] are also valid in this case.

The action of the sum of a monochromatic strong field and a weak field on a quantum-mechanical system was first calculated in [30]. The method of calculation used in [30] allowed a consideration only of a two-level system with two relaxation constants γ_1 and γ_2 [$\widetilde{\Gamma} = (\gamma_1 + \gamma_2)/2$]. It is even more significant that the gain of the weak field, and not the polarization, was calculated in (30). However, in the case of a nonmonochromatic weak field its spectral components affect one another (combination frequencies in polarization) and one cannot introduce the concept of gain. The fact of the appearance of combination frequencies in polarization was first reported in [33, 35]. In [33, 35] a two-level system entailing less general assumptions regarding the relaxation constants and a special form of weak field was considered. Hence, the results of [33, 35] cannot be applied to the case of a field consisting of a strong

*When $|\omega - \omega_0|/\widetilde{\Gamma} > \frac{1}{2}$ this statement is not always valid.

monochromatic component and many weak components* with an arbitrary relationship between amplitudes and phases.

2. We now consider the case of a nonmonochromatic weak field:

$$E=\frac{\hbar}{p}F\cos(\omega t+\varphi)+\frac{\hbar}{p}\sum_j f^{(j)}\cos(\omega t+\Omega_j t+\Phi_j).$$

It is easy to see that in the adopted approximation, linear with respect to the weak field, the contribution of each component $f^{(j)}$ to the polarization is independent of the other components, i.e., the polarization has the form

$$p_{\mathrm{m}}=2p\,\mathrm{Re}\{X_{12}e^{i\omega t+i\varphi}+\sum_j x_{12}^{(j)}e^{i(\omega t+\Omega_j t+\Phi_j)}+\sum_j x_{12}^{(j)}e^{i(\omega t-\Omega_j t+2\varphi-\Phi_j)}\}. \tag{1.19}$$

We turn our attention to the fact that the polarization $p_{\omega+\Omega}$ on some particular frequency $\omega+\Omega$ is due to the action of a field on frequency $\omega+\Omega$ and a field on frequency $\omega-\Omega$ on the system. It is essential that no other components of the weak field affect $p_{\omega+\Omega}$. Hence, a complete picture of the polarization on frequency $\omega+\Omega$ in the case of an arbitrary weak field can be obtained from an examination of two monochromatic weak fields on frequencies $\omega+\Omega$ and $\omega-\Omega$. We denote the amplitudes of these fields by $f^{(1)}$ and $f^{(2)}$, respectively. Then, using (1.12)-(1.14), (1.19), we find

$$p_{\omega+\Omega}=p\,\mathrm{Re}\left\{\left[i\frac{X_2}{\tilde{\Gamma}+i(\omega+\Omega-\omega_0)}f^{(1)}e^{i\Phi_1}-\right.\right.$$

$$\left.\left.-iX_2BF^2\left(f^{(1)}e^{i\Phi_1}\frac{1}{\tilde{\Gamma}+i(\omega+\Omega-\omega_0)}\frac{\tilde{\Gamma}-i(\omega-\Omega-\omega_0)}{\tilde{\Gamma}-i(\omega-\omega_0)}+f^{(2)}e^{-i\Phi_2+i2\varphi}\frac{1}{\tilde{\Gamma}+i(\omega-\omega_0)}\right)\right]e^{i(\omega t+\Omega t)}\right\}, \tag{1.20}$$

$$B=\frac{\dfrac{\tilde{\Gamma}+\frac{1}{2}i\Omega}{\tilde{\Gamma}+i\Omega}}{\left[\tilde{\Gamma}+i\Omega+\dfrac{(\omega-\omega_0)^2}{\tilde{\Gamma}+i\Omega}\right]\left[\Gamma+i\Omega-\dfrac{\gamma^2}{\Gamma+i\Omega}\right]+F^2}.$$

It is clear from (1.20) that the polarization depends on the phase relationshps between the fields $f^{(1)}$ and $f^{(2)}$. In the case of resonance of the strong-field frequency and the transition frequency $\omega=\omega_0$, the interference factor in (1.20) has the form

$$f^{(1)}e^{i\Phi_1}+f^{(2)}e^{i2\varphi-i\Phi_2}.$$

If in this case $\Phi_1+\Phi_2-2\varphi=\pi$ and $f^{(1)}=f^{(2)}$ the interference factor becomes zero. In this case polarization on frequency $\omega+\Omega$ is expressed by the product of the amplitude $f^{(1)}$, the population difference X_2, and the resonance factor $[\tilde{\Gamma}+i(\omega+\Omega-\omega_0)]^{-1}$. Conversely, when $\Phi_1+\Phi_2-2\varphi=0$, $f^{(1)}=f^{(2)}$, the interference factor has a maximum modulus, and the polarization is described by a function which differs most significantly from the simple dispersion curve. In the general case the effect of a weak field of frequency $\omega-\Omega$ depends on $\Phi_1+\Phi_2-2\varphi$, $f^{(2)}$, $f^{(1)}$ and the position of the strong-field frequency ω relative to the transition frequency ω_0. Hence, the manifestation of the considered effect can be very different, depending on the conditions of the particular physical problem.

3. We will obtain an expression for polarization in the case where the weak fields are increasing or decaying in time, i.e., have the form $fe^{\Omega''t}\cos(\omega t+\Omega' t+\Phi)$. For fields with complex frequencies we cannot find a solution of (1.7) bounded at $t=\pm\infty$, and, hence, we cannot

*The case of a field composed of two weak components was considered in [37], which was published a little before our paper [38].

use the previous principle of selecting the only solution. We will now select a solution of (1.7) based on the argument that when $\Omega'' = 0$ it is converted to the solution obtained at real frequencies. Repeating the above calculation, we obtain the following result for the polarization:

$$p_m = 2p\,\mathrm{Re}\{iX_{12}e^{i\omega t+i\varphi} + i\bar{x}_{12}e^{\Omega'' t+i(\omega+\Omega')t+i\Phi)} + i\bar{\bar{x}}_{12}e^{\Omega'' t+i(\omega-\Omega')t+i(2\varphi-\Phi)}\}, \tag{1.21}$$

$$\bar{x}_{12} = \frac{\frac{1}{2}fX_2 + \frac{1}{2}F\bar{x}_2}{\Gamma + \Omega'' + i(\omega + \Omega' - \omega_0)},$$

$$\bar{\bar{x}}_{12} = \frac{\frac{1}{2}F\bar{x}_2^*}{\tilde{\Gamma} + \Omega'' + i(\omega - \Omega' - \omega_0)},$$

$$\bar{x}_2 = -fFX_2\frac{\dfrac{\tilde{\Gamma} + \frac{1}{2}(\Omega'' + i\Omega')}{\tilde{\Gamma} + i\Omega' + \Omega''}\cdot\dfrac{\tilde{\Gamma} + \Omega'' - i(\omega - \Omega' - \omega_0)}{\tilde{\Gamma} - i(\omega - \omega_0)}}{\left(\Omega'' + i\Omega' + \Gamma - \dfrac{\gamma^2}{(\Omega'' + i\Omega' + \Gamma)}\right)\left(\Omega'' + i\Omega' + \tilde{\Gamma} + \dfrac{(\omega - \omega_0)^2}{\Omega'' + i\Omega' + \tilde{\Gamma}}\right) + F^2}.$$

It is clear from (1.21) that polarization couples fields with complex frequencies: $f^{(1)}e^{\Omega'' t} \times \cos(\omega t + \Omega' t + \Phi_1)$ and $f^{(2)}e^{\Omega'' t}\cos(\omega t - \Omega' t + \Phi_2)$.

We give the formula for the mean value of the dipole moment for weak fields with complex frequencies. We write the field acting on the system in the form

$$E = \frac{\hbar}{p}\,\mathrm{Re}\{Fe^{-i\varphi}e^{-i\omega t} + f^{(1)}e^{-i\Phi_1}e^{-i(\omega+\Omega)t} + f^{(2)}e^{i\Phi_2}e^{i(\omega-\Omega)t}\}, \tag{1.22}$$

where $\Omega = \Omega' + i\Omega''$.

For field (1.22), in view of (1.21), we obtain

$$\begin{aligned} p_m = p\,\mathrm{Re}\Big\{ & -i\frac{X_2}{\tilde{\Gamma} - i(\omega - \omega_0)}Fe^{-i\varphi}e^{-i\omega t} - i\frac{X_2}{\tilde{\Gamma} - i(\omega + \Omega - \omega_0)}f^{(1)}e^{-i\Phi_1}e^{-i(\omega+\Omega)t} + \\ & + iX_2AF^2\left[\frac{\tilde{\Gamma} + i(\omega - \Omega - \omega_0)}{\tilde{\Gamma} - i(\omega + \Omega - \omega_0)}\,\frac{1}{\tilde{\Gamma} + i(\omega - \omega_0)}f^{(1)}e^{-i\Phi_1} + \right. \\ & \left. + \frac{1}{\tilde{\Gamma} - i(\omega - \omega_0)}f^{(2)}e^{i\Phi_2 - i2\varphi}\right]e^{-i(\omega+\Omega)t} + i\frac{X_2}{\tilde{\Gamma} + i(\omega - \Omega - \omega_0)}f^{(2)}e^{i\Phi_2}e^{i(\omega-\Omega)t} - \\ & - iX_2AF^2\left[\frac{\tilde{\Gamma} - i(\omega + \Omega - \omega_0)}{\tilde{\Gamma} + i(\omega - \Omega - \omega_0)}\,\frac{1}{\tilde{\Gamma} - i(\omega - \omega_0)}f^{(2)}e^{i\Phi_2} + \frac{1}{\tilde{\Gamma} + i(\omega - \omega_0)}f^{(1)}e^{-i\Phi_1 + i2\varphi}\right]e^{i(\omega-\Omega)t}\Big\}, \end{aligned} \tag{1.23}$$

where

$$A = \frac{\tilde{\Gamma} - \frac{1}{2}i\Omega}{\tilde{\Gamma} - i\Omega}\left[\left(\Gamma - i\Omega - \frac{\gamma^2}{\Gamma - i\Omega}\right)\left(\tilde{\Gamma} - i\Omega + \frac{(\omega - \omega_0)^2}{\tilde{\Gamma} - i\Omega}\right) + F^2\right]^{-1}, \tag{1.24}$$

and X_2 is given by formula (1.10).

4. We give equations describing the behavior of strong and weak fields in a medium with the considered properties. The equations for the separate spectral components can be obtained by the known [44, 45] method by substituting in the equation for the electric field strength

$$\nabla^2 E - \frac{\varepsilon_0}{c^2}\frac{\partial^2 E}{\partial t^2} = \frac{4\pi}{c^2}\frac{\partial^2}{\partial t^2}(Np_m)$$

(where N is the density of active centers, ε_0 is the dielectric constant of the substance in the absence of excitation) the expressions for the field (1.22) and the polarization (1.23), (1.24).

We first introduce some new symbols, which simplify the writing of the equations for the fields.

We introduce complex field amplitudes and we will express the field in units defining the degree of saturation:

$$E = \frac{1}{\sigma}\,\mathrm{Re}\,\{E_0 e^{-i\omega t} + E_1 e^{-i(\omega+\Omega)t} + E_2 e^{i(\omega-\Omega)t}\}, \tag{1.25}$$

$$\frac{1}{\sigma^2} = \frac{\hbar^2}{p^2}\Gamma\tilde{\Gamma}\left(1-\frac{\gamma^2}{\Gamma^2}\right)\left[1+\frac{(\omega-\omega_0)^2}{\tilde{\Gamma}^2}\right]. \tag{1.26}$$

In addition, we put

$$4\pi N\frac{p^2}{\hbar}\frac{1}{\Gamma\tilde{\Gamma}}\frac{\Lambda+\lambda\frac{\gamma}{\Gamma}}{1-\frac{\gamma^2}{\Gamma^2}} = 4\pi N\frac{p^2}{\hbar\tilde{\Gamma}}\left(\frac{\Lambda_2}{2\gamma_2}-\frac{\Lambda_1}{2\gamma_1}\right) = \beta. \tag{1.27}$$

In these symbols the equations for the spectral components of the field have the form

$$\nabla^2 E_0 + \varepsilon_0\frac{\omega^2}{c^2}E_0 = i\beta\frac{\omega^2}{c^2}\frac{1}{1-i\frac{\omega-\omega_0}{\tilde{\Gamma}}}\frac{1}{1+E_0E_0^*}E_0, \tag{1.28}$$

$$\nabla^2 E_1 + \varepsilon_0\frac{(\omega+\Omega)^2}{c^2}E_1 = i\beta\frac{(\omega+\Omega)^2}{c^2}\frac{1}{1-i\frac{\omega+\Omega-\omega_0}{\tilde{\Gamma}}}\frac{1}{1+E_0E_0^*}\times$$
$$\times\left\{E_1 - D\left[\frac{1+i\frac{\omega-\Omega-\omega_0}{\tilde{\Gamma}}}{1+i\frac{\omega-\omega_0}{\tilde{\Gamma}}}E_0E_0^*E_1 + \frac{1-i\frac{\omega+\Omega-\omega_0}{\tilde{\Gamma}}}{1-i\frac{\omega-\omega_0}{\tilde{\Gamma}}}E_0^2E_2\right]\right\}, \tag{1.29}$$

$$\nabla^2 E_2 + \varepsilon_0\frac{(\omega-\Omega)^2}{c^2}E_2 = -i\beta\frac{(\omega-\Omega)^2}{c^2}\frac{1}{1+i\frac{\omega-\Omega-\omega_0}{\tilde{\Gamma}}}\frac{1}{1+E_0E_0^*}\times$$
$$\times\left\{E_2 - D\left[\frac{1-i\frac{\omega+\Omega-\omega_0}{\tilde{\Gamma}}}{1-i\frac{\omega-\omega_0}{\tilde{\Gamma}}}E_0E_0^*E_2 + \frac{1+i\frac{\omega-\Omega-\omega_0}{\tilde{\Gamma}}}{1+i\frac{\omega-\omega_0}{\tilde{\Gamma}}}E_0^{*2}E_1\right]\right\}, \tag{1.30}$$

where

$$D = \frac{1-\frac{1}{2}\frac{i\Omega}{\tilde{\Gamma}}}{1-\frac{i\Omega}{\tilde{\Gamma}}}\left[\frac{1}{1-\frac{i\Omega}{\Gamma}}\frac{1}{1-\frac{i\Omega}{\Gamma}}\frac{\left(1-\frac{i\Omega}{\Gamma}\right)^2-\frac{\gamma^2}{\Gamma^2}}{1-\frac{\gamma^2}{\Gamma^2}}\frac{\left(1-\frac{i\Omega}{\tilde{\Gamma}}\right)^2+\left(\frac{\omega-\omega_0}{\tilde{\Gamma}}\right)^2}{1+\left(\frac{\omega-\omega_0}{\tilde{\Gamma}}\right)^2}+E_0E_0^*\right]^{-1}. \tag{1.31}$$

We note that the quantity β, which is contained in (1.28)-(1.30), is small in all real cases:

$$\beta \ll 1 \tag{1.32}$$

(usually β is 10^{-6}-10^{-9}). This fact will be used later.

Equation (1.28) can be reduced to the form

$$\nabla^2 E_0 + \varepsilon\frac{\omega^2}{c^2}E_0 = 0, \tag{1.33}$$

if we put

$$\varepsilon = \varepsilon_0 + \Delta\varepsilon = \varepsilon_0 - \frac{i\beta}{1-i\frac{\omega-\omega_0}{\tilde{\Gamma}}}\frac{1}{1+E_0E_0^*}, \tag{1.34}$$

Expression (1.34) for the dielectric constant of an active medium in a strong monochromatic field is known in the literature [2]. Henceforth, for convenience, we will refer to formula (1.34).

Equations (1.29) and (1.30) for weak fields are coupled; we cannot introduce the dielectric constant at the weak-field frequencies. Equations (1.29), (1.30) reflect the interaction of weak fields through polarization in the presence of a strong field. This effect is analogous to the parametric optical effects described in [44]. The special features of the parametric interaction considered here are that it takes place within the width of the spectral line and depends in a fairly complex manner on the strong-field intensity.

Thus, from the calculation of the polarization we have obtained a system of equations (1.28)-(1.30) for a field containing one intense spectral component and two weak components of symmetric (relative to the strong-field frequency) frequencies.

The weak components are described by the coupled equations (1.29), (1.30). If we take into consideration several weak fields, and not just two, the system of equations for them can be broken up into pairs of equations of type (1.29), (1.30) for weak fields of symmetric frequencies. Hence, we need only consider equations (1.28)-(1.30). We will henceforth base our treatment on the system of equations (1.28)-(1.30).

CHAPTER II

One-Dimensional Laser Model. Monochromatic Generation

In this chapter we will solve the nonlinear spatial problem for a field in an active resonator.

The laser model which will be considered here represents the properties of real systems in a simplified way. In view of the relative simplicity, however, this model can be investigated fairly fully and can be used to determine the characteristic features of lasers due to spatial inhomogeneity of the field and inversion.

1. We will assume that between the planes z = 0 and z = b there is an infinite layer of active substance in the x and y directions, the polarization of which is related to the field strength by formulas (1.23). The equation for a monochromatic field

$$\sigma E = \mathrm{Re}\{E_0 e^{-i\omega t}\}$$

has the form

$$\frac{d^2E_0}{dz^2} + \varepsilon\frac{\omega^2}{c^2}E_0 = 0, \tag{2.1}$$

where

$$\varepsilon = \varepsilon_0 + \Delta\varepsilon, \tag{2.2}$$

$$\Delta\varepsilon = \frac{\mu}{1 + E_0E_0^*}, \quad \mu = \mu' - i\mu'' = \frac{-i\beta}{1 - i\dfrac{\omega - \omega_0}{\Gamma}}, \quad \beta \ll 1. \tag{2.3}$$

The quantities β and σ are defined in Chapter I [see (1.25) and (1.34)].

The solutions for the field at $z < 0$, $z > b$ must be chosen so that when $z = \pm\infty$ the fields have the form of waves leaving the layer. From the continuity of $E/(dE/dz)$ we obtain

$$E_0(0) = \frac{i}{\frac{\omega}{c}\sqrt{\varepsilon_0}} \frac{1 - R_1}{1 + R_1} \frac{dE_0}{dz}(0), \tag{2.4}$$

$$E_0(b) = -\frac{i}{\frac{\omega}{c}\sqrt{\varepsilon_0}} \frac{1 - R_2}{1 + R_2} \frac{dE_0}{dz}(b), \tag{2.5}$$

where R_2 is the reflection coefficient for a wave propagated along the z axis in a homogeneous medium with $\varepsilon = \varepsilon_0$, from the halfspace $z > b$, and R_1 is that for a wave propagated in a direction opposite to the z axis in a homogeneous medium with $\varepsilon = \varepsilon_0$, from the halfspace $z < 0$.

If the medium is linear outside the active layer and the dielectric constant changes slowly with the coordinate, R_1 and R_2 can be regarded as constants and the problem can be solved for the field in the region $0 < z < b$ with assigned boundary conditions (2.4), (2.5).

2. We will carry out some transformations of the initial equation (2.1). For brevity we will introduce a dimensionless independent variable $\xi = (\omega/c)\sqrt{\varepsilon_0}z$. Then equation (2.1) and the boundary conditions (2.4), (2.5) will take the form

$$\frac{d^2E_0}{d\xi^2} + \left(1 + \frac{\Delta\varepsilon}{\varepsilon_0}\right) E_0 = 0, \tag{2.6}$$

$$E_0(0) = i\frac{1 - R_1}{1 + R_1}\frac{dE_0}{d\xi}(0), \tag{2.7}$$

$$E_0(L) = -i\frac{1 - R_2}{1 + R_2}\frac{dE_0}{d\xi}(L), \quad L = \frac{\omega}{c}\sqrt{\varepsilon_0}b. \tag{2.8}$$

The general solution of equation (2.6) can be put in the form

$$E_0 = Ae^{-i\xi} + Be^{i\xi} - \int_0^\xi \sin(\xi - \xi') E_0(\xi') \frac{\Delta\varepsilon(\xi')}{\varepsilon_0} d\xi', \tag{2.9}$$

where A and B are constants. We put

$$Ae^{-i\xi} + \frac{1}{2i}e^{-i\xi}\int_0^\xi e^{i\xi'} E_0(\xi') \frac{\Delta\varepsilon(\xi')}{\varepsilon_0} d\xi' = p_1(\xi)e^{-i\xi},$$

$$Be^{i\xi} - \frac{1}{2i}e^{i\xi}\int_0^\xi e^{-i\xi'} E_0(\xi') \frac{\Delta\varepsilon(\xi')}{\varepsilon_0} d\xi' = p_2(\xi)e^{i\xi}.$$

Then E_0, as (2.9) shows, will take the following form:

$$E_0 = p_1(\xi)e^{-i\xi} + p_2(\xi)e^{i\xi}. \tag{2.10}$$

Putting (2.10) into (2.9) we obtain equations for p_1 and p_2.

$$\begin{aligned} p_1(\xi) - p_1(0) &= \frac{1}{2i}\frac{\mu}{\varepsilon_0}\int_0^\xi \frac{p_1(\xi') + p_2(\xi')e^{2i\xi'}}{1 + |p_1(\xi') + p_2(\xi')e^{2i\xi'}|^2} d\xi', \\ p_2(\xi) - p_2(0) &= -\frac{1}{2i}\frac{\mu}{\varepsilon_0}\int_0^\xi \frac{p_1(\xi')e^{-2i\xi'} + p_2(\xi')}{1 + |p_1(\xi') + p_2(\xi')e^{2i\xi'}|^2} d\xi'. \end{aligned} \tag{2.11}$$

Equations (2.11) are equivalent to (2.1). From (2.4), (2.5) we obtain boundary conditions for p_1 and p_2:

$$p_2(0) = -R_1 p_1(0), \tag{2.12}$$

$$p_1(L) = -R_2 p_2(L). \tag{2.13}$$

It can be seen from (2.11) that functions p_1 and p_2 vary slowly, viz.,

$$\left|\frac{dp_{1,2}}{d\xi}\right| = \left|\frac{\mu}{2i\varepsilon_0} e^{\mp i\xi} \frac{p_1 e^{-i\xi} + p_2 e^{i\xi}}{1+|p_1 e^{-i\xi} + p_2 e^{i\xi}|^2}\right| = \frac{|\mu|}{2\varepsilon_0} \frac{|p_1 e^{-i\xi} + p_2 e^{i\xi}|}{1+|p_1 e^{-i\xi} + p_2 e^{i\xi}|^2},$$

whence, using (2.3), we obtain

$$\left|\frac{dp_{1,2}}{d\xi}\right| \ll \frac{|p_1 e^{-i\xi} + p_2 e^{i\xi}|}{1+|p_1 e^{-i\xi} + p_2 e^{i\xi}|^2}. \tag{2.14}$$

The estimate (2.14) means that when ξ changes by an amount of the order of unity (or when the z coordinate changes by an amount of the order of the wavelength $\lambda = \frac{2\pi}{\frac{\omega}{c}\sqrt{\varepsilon_0}}$) functions $p_1(\xi)$ and $p_2(\xi)$ are not greatly altered in the absolute sense or in relation to the quantity $|p_1 e^{-i\xi} + p_2 e^{i\xi}| = |E_0|$. Hence, in calculating the integrals in formulas (2.11) we neglect the change in p_1 and p_2 on the segment $[\xi, \xi + \pi]$. In addition, we replace $[p_{1,2}(\xi + \pi) - p_{1,2}(\xi)]$ $1/\pi$ by $dp_{1,2}/d\xi$. Thus, from (2.11) we obtain the system of differential equations:

$$\begin{aligned} \frac{dp_1}{d\xi} &= \frac{\mu}{2\pi i\varepsilon_0} \int_0^\pi \frac{p_1(\xi) + p_2(\xi) e^{2i\xi'}}{1+|p_1(\xi) + p_2(\xi) e^{2i\xi'}|^2} d\xi', \\ \frac{dp_2}{d\xi} &= -\frac{\mu}{2\pi i\varepsilon_0} \int_0^\pi \frac{p_2(\xi) + p_1(\xi) e^{-2i\xi'}}{1+|p_1(\xi) e^{-2i\xi'} + p_2(\xi)|^2} d\xi'. \end{aligned} \tag{2.15}$$

System (2.15) can also be put in the form

$$\begin{aligned} \frac{dp_1}{d\xi} &= p_1 \frac{\overline{\Delta\varepsilon}}{2\varepsilon_0} + p_2 \frac{\overline{\overline{\Delta\varepsilon}}}{2\varepsilon_0}, \\ \frac{dp_2}{d\xi} &= -p_2 \frac{\overline{\Delta\varepsilon}}{2\varepsilon_0} - p_1 \frac{\overline{\overline{\Delta\varepsilon}}}{2\varepsilon_0}. \end{aligned} \tag{2.16}$$

Here we have introduced symbols for the mean value and for the coefficient of the expansion of $\Delta\varepsilon$ in a Fourier series on the segment $[\xi, \xi + \pi]$:

$$\begin{aligned} \overline{\Delta\varepsilon} &= \int_\xi^{\xi+\pi} \Delta\varepsilon \, d\xi', \\ \overline{\overline{\Delta\varepsilon}} &= \int_\xi^{\xi+\pi} \Delta\varepsilon e^{2i\xi'} d\xi'. \end{aligned} \tag{2.17}$$

It follows from the deduction of (2.16) that equations of this form can be obtained not only for $\Delta\varepsilon$ in the form (2.3), but also for other kinds of relationship between $\Delta\varepsilon$ and the saturating field, as well as for the case of field-independent inhomogeneity of $\Delta\varepsilon$.

We note that consideration of only the mean value of $\Delta\varepsilon$ would correspond to the geometrical-optics approximation. The terms with $\overline{\overline{\Delta\varepsilon}}$ on the right sides of (2.16) mean that reflected waves are taken into account: The wave $p_2(p_1)$, reflected on inhomogeneities of the medium, gives a secondary wave which interferes with $p_1(p_2)$.

After calculation of the integrals equations (2.15) take the form

$$\frac{dp_1}{d\xi}=\frac{\mu}{2i\varepsilon_0}\left\{p_1\frac{1}{\sqrt{(1+p_1p_1^*+p_2p_2^*)^2-4p_1p_1^*p_2p_2^*}}+p_2\frac{1}{2p_1^*p_2}\frac{\sqrt{(1+p_1p_1^*+p_2p_2^*)^2-4p_1p_1^*p_2p_2^*}-(1+p_1p_1^*+p_2p_2^*)}{\sqrt{(1+p_1p_1^*+p_2p_2^*)^2-4p_1p_1^*p_2p_2^*}}\right\},$$

$$\frac{dp_2}{d\xi}=-\frac{\mu}{2i\varepsilon_0}\left\{p_2\frac{1}{\sqrt{(1+p_1p_1^*+p_2p_2^*)^2-4p_1p_1^*p_2p_2^*}}+p_1\frac{1}{2p_2^*p_1}\frac{\sqrt{(1+p_1p_1^*+p_2p_2^*)^2-4p_1p_1^*p_2p_2^*}-(1+p_1p_1^*+p_1p_2^*)}{\sqrt{(1+p_1p_1^*+p_2p_2^*)^2-4p_1p_1^*p_2p_2^*}}\right\}. \quad (2.18)$$

For the solution of equations (2.18)* it is convenient to introduce the moduli and phases of the functions $p_{1,2}(\xi)$:

$$p_{1,2}(\xi)=P_{1,2}(\xi)\,e^{i\varphi_{1,2}(\xi)}.$$

From (2.18) we obtain a system of equations for $P_{1,2}$, $\varphi_{1,2}$:

$$\frac{dP_1}{d\xi}=-\frac{\mu''}{2\varepsilon_0}\frac{1}{\sqrt{1+(P_1+P_2)^2}+\sqrt{1+(P_2-P_1)^2}}\left[\frac{P_1+P_2}{\sqrt{1+(P_1+P_2)^2}}+\frac{P_1-P_2}{\sqrt{1+(P_2-P_1)^2}}\right], \quad (2.19)$$

$$\frac{dP_2}{d\xi}=\frac{\mu''}{2\varepsilon_0}\frac{1}{\sqrt{1+(P_1+P_2)^2}+\sqrt{1+(P_2-P_1)^2}}\left[\frac{P_1+P_2}{\sqrt{1+(P_1+P_2)^2}}+\frac{P_2-P_1}{\sqrt{1+(P_2-P_1)^2}}\right], \quad (2.20)$$

$$\frac{d\varphi_1}{d\xi}=\frac{\mu'}{\mu''}\frac{1}{P_1}\frac{dP_1}{d\xi},\quad \frac{d\varphi_2}{d\xi}=\frac{\mu'}{\mu''}\frac{1}{P_2}\frac{dP_2}{d\xi}.$$

It follows from the boundary conditions (2.4), (2.5) that

$$P_2(0)=|R_1|\,P_1(0), \quad (2.21)$$

$$P_1(L)=|R_2|\,P_2(L), \quad (2.22)$$

$$e^{i\varphi_2(0)}=-\frac{R_1}{|R_1|}e^{i\varphi_1(0)}, \quad (2.23)$$

$$e^{i\varphi_1(L)}=-\frac{R_2}{|R_2|}e^{i\varphi_2(L)+i2L}. \quad (2.24)$$

The general solution of equations (2.19), (2.20) has the form

$$\sqrt{1+(P_1+P_2)^2}-\sqrt{1+(P_2-P_1)^2}=\text{const}, \quad (2.25)$$

$$-\frac{\mu''}{\varepsilon_0}\xi+(P_2-P_1)(P_2+P_1)+\ln\frac{(P_2+P_1)\left[\sqrt{1+(P_2-P_1)^2}+1\right]+(P_2-P_1)\left[\sqrt{1+(P_2+P_1)^2}+1\right]}{(P_2+P_1)\left[\sqrt{1+(P_2-P_1)^2}+1\right]-(P_2-P_1)\left[\sqrt{1+(P_2+P_1)^2}+1\right]}=\text{const}, \quad (2.26)$$

$$\varphi_1-\frac{\mu'}{\mu''}\ln P_1=\text{const},\quad \varphi_2-\frac{\mu'}{\mu''}\ln P_2=\text{const}. \quad (2.27)$$

It follows from (2.25), (2.26) that if one of the quantities P_1, P_2, say P_2, becomes zero even at one point, it will be zero at any ξ, and the solution of (2.19) will have the form

$$P_2(\xi)\equiv 0,\quad \ln P_1+\frac{P_1^2}{2}+\frac{\mu''}{2\varepsilon_0}\xi=\text{const}.$$

---

*Equations similar to (2.18) were obtained in [17].

In view of this we can state that in the case $R_1 = 0$ or $R_2 = 0$ equations (2.19), (2.20) with boundary conditions (2.21)-(2.24) have no solutions. Let, for instance, $R_1 = 0$. It follows from (2.21) that $P_2(0) = 0$. But then $P_2(\xi) \equiv 0$, and when $P_2(L) = 0$, irrespective of R_2, and $P_1(L) \neq 0$, condition (2.22) cannot be satisfied. In relation to lasers this corresponds to the known fact that operation is impossible without reflection of the waves amplified by the active substance of the layer.

3. Solving equations (2.27) by using (2.21)-(2.24) we obtain the condition for the natural frequency

$$2L \equiv 2\frac{\omega}{c}\sqrt{\varepsilon_0}b = 2\pi m + \operatorname{Im}\ln\frac{1}{R_1R_2} - \frac{\mu'}{\mu''}\operatorname{Re}\ln\frac{1}{R_1R_2}, \tag{2.28}$$

where m is a whole number.

The first term on the right side of (2.28) is the natural frequency of a plane-parallel empty layer with ideal mirrors. The second term takes the phase change on reflection into account. The phase jump on reflection depends on the properties of the reflecting surface and is not a specific feature of active resonators. To simplify the writing we will henceforth assume that R_1 and R_2 are real and positive, i.e., $\operatorname{Im}\ln\frac{1}{R_1R_2} = 0$. The third term is a correction to the frequency due to the active substance filling the resonator. We note that this correction is independent of the amplitude of the oscillations. For another form of dependence of $\Delta\varepsilon$ on ω and $|E_0|^2$ this result may be invalid. In fact, since an increase in field amplitude alters the dielectric properties of the resonator and its medium becomes more homogeneous, then in the general case the natural frequencies may change. As can be derived from the deduction of (2.27), (2.28) the constancy of the natural frequencies in our case is ensured by the fact that the ratio $\operatorname{Re}\Delta\varepsilon/\operatorname{Im}\Delta\varepsilon$ is independent of the field.

When $P_{1,2} \to 0$ the model considered by us becomes a laser with constant spatial properties and, hence, condition (2.28) must be the same as the condition for the frequency which is obtained when the inhomogeneity of the laser medium is neglected [2]. We use the way of writing condition (2.28) adopted in the literature. We denote the natural frequency of the lossless empty resonator by ω_{res}, $\omega_{res} = (\pi m/b)(c/\sqrt{\varepsilon_0})$, and we also introduce the line width of the empty resonator with losses on the mirrors $\Delta\omega_{res} = (c/\sqrt{\varepsilon_{0b}})\ln[1/(R_1R_2)^{\frac{1}{2}}]$. Then (2.28) will take the form

$$\omega - \omega_0 = \frac{\omega_{res} - \omega_0}{1 + \frac{\Delta\omega_{res}}{\tilde{\Gamma}}}. \tag{2.29}$$

Formula (2.29) agrees completely with the results of [2].

4. The boundary values of the wave amplitudes are determined from the system of equations:

$$\sqrt{1+P_+^2(L)} - \sqrt{1+P_-^2(L)} = \sqrt{1+P_+^2(0)} - \sqrt{1+P_-^2(0)}, \tag{2.30}$$

$$\frac{\mu''}{\varepsilon_0}L = P_-(L)P_+(L) - P_-(0)P_+(0) + \ln\left[\frac{1+\frac{1-R_2}{1+R_2}\frac{\sqrt{1+P_+^2(L)}+1}{\sqrt{1+P_-^2(L)}+1}}{1-\frac{1-R_2}{1+R_2}\frac{\sqrt{1+P_+^2(L)}+1}{\sqrt{1+P_-^2(L)}+1}} \cdot \frac{1+\frac{1-R_1}{1+R_1}\frac{\sqrt{1+P_+^2(0)}+1}{\sqrt{1+P_-^2(0)}+1}}{1-\frac{1-R_1}{1+R_1}\frac{\sqrt{1+P_+^2(0)}+1}{\sqrt{1+P_-^2(0)}+1}}\right], \tag{2.31}$$

$$P_-(0) = -\frac{1-R_1}{1+R_1} P_+(0),$$
$$P_-(L) = \frac{1-R_2}{1+R_2} P_+(L), \tag{2.32}$$

where for brevity we put $P_+ = P_2 + P_1$, $P_- = P_2 - P_1$. From (2.31) it is easy to obtain the generation threshold by making the amplitudes P_+, P_- tend to zero:

$$\left(\frac{\mu}{\varepsilon_0} L\right)_{\text{thresh}} = \ln \frac{1}{R_1 R_2}. \tag{2.33}$$

The threshold condition has the same form as in the theory which ignores the inhomogeneity of the laser medium.

When the threshold value of the inversion in homogeneous lasers is exceeded the square of the field amplitude depends linearly on the excitation [2]. In our case, however, formulas (2.30)-(2.32) for P_-^2, P_+^2 indicate a more complex dependence of these quantities on μ'' (degree of excitation). Formulas (2.30)-(2.32) can be simplified for very small reflection coefficients $R_{1,2} \ll 1$, and for reflection coefficients close to unity, $1 - R_{1,2} \ll 1$. When $R_{1,2} \ll 1$ we find from (2.30) that the boundary values can be determined with a small relative error from the equations

$$\frac{\mu''}{\varepsilon_0} L = \ln \frac{1}{R_1 R_2} + P_+^2(0) + P_+^2(L) + \ln \sqrt{1 + P_+^2(0)}\sqrt{1 + P_+^2(L)},$$
$$R_1 \frac{P_+^2(0)}{\sqrt{1 + P_+^2(0)}} = R_2 \frac{P_+^2(L)}{\sqrt{1 + P_+^2(L)}}. \tag{2.34}$$

When $1 - R_{1,2} \ll 1$ the formula

$$\frac{\mu''}{\varepsilon_0} L = \left(1 - \frac{R_1 + R_2}{2}\right)\left(\sqrt{1 + P_+^2} + 1 + P_+^2\right), \tag{2.35}$$
$$P_+(0) = P_+(L) = P_+$$

gives a small relative error.

We will discuss in more detail a symmetrical resonator with $R_1 = R_2 = R$. We introduce a symbol for amplification on the half wave of the laser $\zeta = (\mu''/\varepsilon_0)(L/2)$; we put $(1-R)/(1+R) = n$ and $P_+^2(L) = y$ [we will show below that $P_+^2(L)$ is the maximum value of the square of the field amplitude along the laser]. Formulas (2.30)-(2.32) can then be written in the following way:

$$\zeta = ny + \ln \frac{1 + n \dfrac{\sqrt{1+y}+1}{\sqrt{1+n^2 y}+1}}{1 - n \dfrac{\sqrt{1+y}+1}{\sqrt{1+n^2 y}+1}}, \tag{2.36}$$

and the formulas corresponding to the limiting values of the reflection coefficients take the form

$$R \ll 1, \; \zeta = \ln \frac{1}{R} + y + \ln \sqrt{1+y}, \tag{2.37}$$

$$n \ll 1, \; \zeta = n\left(\sqrt{1+y} + 1 + y\right). \tag{2.38}$$

The threshold value of ζ is

$$\zeta_{\text{thresh}} = \ln \frac{1}{R}.$$

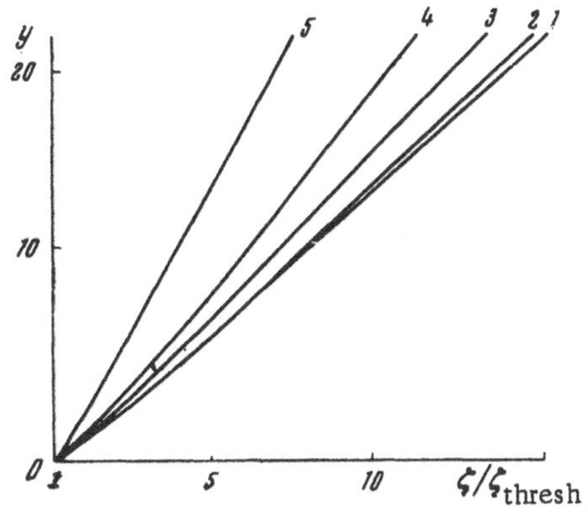

Fig. 1. Field intensity as a function of pumping. 1) R^2 = 96; 2) 50; 3) 11; 4) 3; 5) 0.07%.

It can be seen from formula (2.36), and from formulas (2.37), (2.38) for the limiting cases, that the derivative $d\zeta/dy$ at the threshold point is $^3/_2 n$. When ζ increases the relationship $y(\zeta)$ approaches a linear form (relatively, but not absolutely), and the derivative $d\zeta/dn$ tends to n.

Numerical calculations were made from formulas (2.36)-(2.38). Figure 1 shows the square of the field amplitude y (in saturation units) as a function of the excitation expressed in threshold units $\zeta/\zeta_{\text{thresh}}$. Curve 1 is drawn for the reflection coefficient R^2 = 96%. In the considered range of variation of the parameter $\zeta/\zeta_{\text{thresh}}$ this curve practically coincides with the limiting curve given by formula (2.35). The graphs constructed for smaller values of R lie above this and occupy the region between the limiting curve and the y axis. Curves 2, 3, 4, 5 correspond to values of the reflection coefficient R^2 = 50, 11, 3, and 0.07%.

The power output of the laser can be expressed in terms of the wave amplitudes $P_{1,2}$. For the radiation flux through unit area on the left boundary of the layer we obtain

$$\Pi_1 = \frac{c\sqrt{\varepsilon_0}}{8\pi\sigma^2}\frac{1-R_1}{1+R_1}P_+^2(0) \tag{2.39}$$

and for the radiation flux through unit area on the right boundary

$$\Pi_2 = \frac{c\sqrt{\varepsilon_0}}{8\pi\sigma^2}\frac{1-R^2}{1+R_2}P_+^2(L). \tag{2.40}$$

Figure 2 illustrates the relationship between the output power and the excitation. The figure shows the results of calculation in the case of a symmetrical resonator with $R_1 = R_2 = R$ [formula (2.36)] for the quantity $\frac{1-R}{1+R}P_+^2 = ny$, which is proportional to the energy flux from the laser. For convenience in comparing lasers with different reflection coefficients we have plotted on the x axis the quantity $\zeta - \zeta_{\text{thresh}}$ – the excess of excitation over the threshold, i.e., we have matched the threshold points of different lasers. The straight line 1 corresponds to the result of the theory of homogeneous lasers. Curve 6 is drawn from the limiting formula (2.37) for low reflection coefficients. All the graphs in the figure lie between the limiting curve 6 and the straight line 1. Calculation from formula (2.36) for R^2 = 3% gives a curve which practically coincides at the considered values of $\zeta - \zeta_{\text{thresh}}$ with the limiting curve 6. Curves 5, 4, 3, and 2 correspond to the values R^2 = 11, 50, 75, and 96%. The greatest differences in the value of emitted power from the results of calculation, from the theory which ignores the inhomogeneity of the laser medium, are near the threshold. The tangent of the angle of inclination for $\zeta - \zeta_{\text{thresh}} = 0$ is $^2/_3$ for curves 2, 3, 4, 5, 6 and is unity for the line 1. When $\zeta - \zeta_{\text{thresh}}$ increases, these differences decrease relatively. In the treatment of some physical questions it is important to take these differences into account.

5. We now consider the field distribution inside the laser. The relationship between the amplitudes P_1 and P_2 and the coordinates is given by formulas (2.25), (2.26), in which the constants of integration can be expressed by the values of P_1 and P_2 on the boundary. We note first of all that the functions P_1 and P_2 are monotonic. In fact, by converting the right sides of (2.19) and regarding P_1 and P_2 as positive (we exclude the case $P_1=0$ or $P_2=0$ from considera-

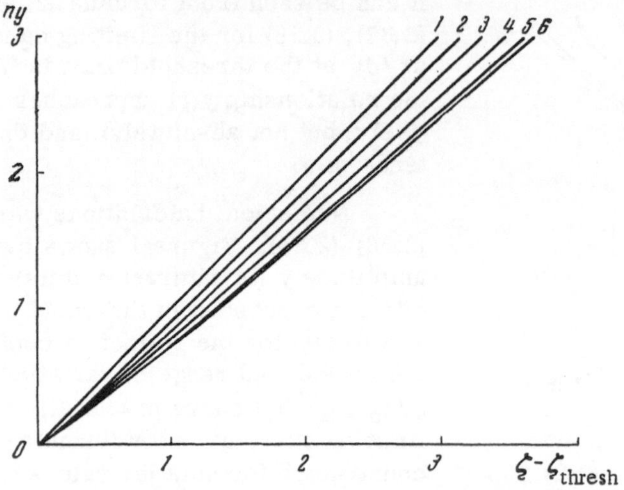

Fig. 2. Power emitted by laser as a function of pumping. 1) Inhomogeneity of laser medium neglected [6]; 2) R^2 = 96; 3) 75; 4) 50; 5) 11; 6) 3%.

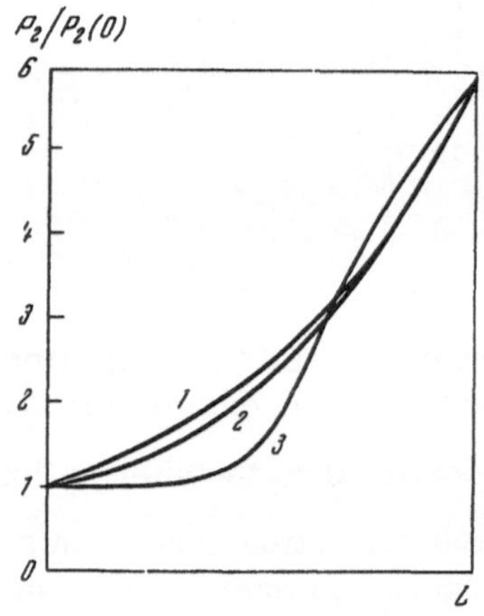

Fig. 3. Wave amplitude as function of coordinate of layer for case R^2 = 3%. 1) ζ = 1.75; 2) 2.5; 3) 17.5.

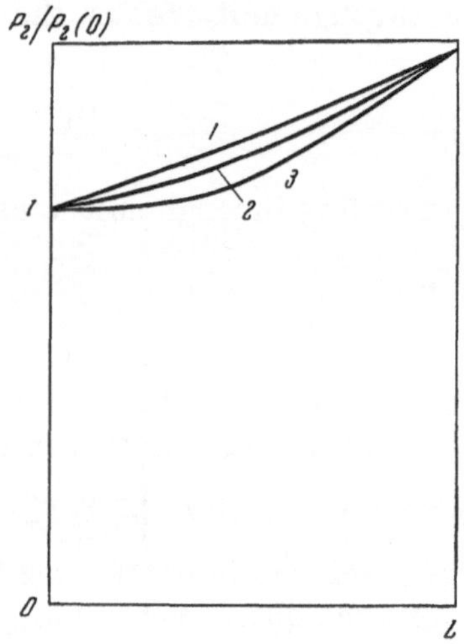

Fig. 4. Wave amplitude as a function of coordinate of layer for case R^2 = 50%. 1) ζ = 0.35; 2) 2.5; 3) 18.9.

tion), we can see that $dP_1/d\xi < 0$, and $dP_2/d\xi > 0$. Thus, P_1 decreases monotonically and P_2 increases monotonically on the segment [0, L]. It follows from this fact and from the boundary conditions (2.21) and (2.22) that

$$P_2(0) < P_2(\xi) < P_2(L) = \frac{1}{R_2} P_1(L) < \frac{1}{R_2} P_1(0) = \frac{1}{R_1 R_2} P_2(0),$$

i.e.,

$$P_2(0) < P_2(\xi) < \frac{1}{R_1 R_2} P_2(0), \tag{2.41}$$

and the analogous condition for P_1 is

$$P_1(L) < P_1(\xi) < \frac{1}{R_1 R_2} P_1(L). \tag{2.42}$$

It is clear from (2.41) and (2.42) that the maximum value of the relative change in amplitude depends only on the reflection coefficients. In particular, if the coefficients R_1 and R_2 are close to unity, the wave amplitudes do not vary much over the extent of the layer and do not differ much from one another.

The particular form of the relationship between the amplitude and the coordinate for some special cases is shown in the figures. A symmetrical resonator, for which $P_1(\xi) = P_2(L - \xi)$, was considered [the function $P_1(\xi)$ is symmetrical to the function $P_2(\xi)$ relative to the center of the resonator]. Figure 3 shows the relationship $P_2(\xi)$ for the case $R^2 = 3\%$. Curve 1 is plotted for $\zeta = \zeta_{\text{thresh}} = 1.75$. In this case the inhomogeneity of the dielectric constant is absent and the wave amplitudes vary exponentially in accordance with the form of the eigenfunctions of a resonator with homogeneous filling

$$E_0 \sim \sin\left[\left(\frac{m\pi}{b} - \frac{i}{2b}\ln\frac{1}{R_1R_2}\right)z + \frac{i}{2}\ln\frac{1}{R_1}\right].$$

Curve 2 is drawn for $\zeta = 2.5$, and curve 3 for $\zeta = 17.25$. Figure 4 shows graphs of $P_2(\xi)$ for the case $R^2 = 50\%$. Curve 1 is drawn for $\zeta = \zeta_{\text{thresh}} = 0.35$ and has an exponential form; curve 2 is drawn for $\zeta = 2.5$; curve 3 is for $\zeta = 18.9$. The figures show that with increase in excitation (ζ) the curvature of the graphs $P_2(\xi)$ increases. If the reflection coefficients are high, however, then at high values of ζ and close to the threshold the differences of $P_2(\xi)$ from a constant are insignificant [see (2.41)].

6. Since the wave amplitudes P_1 and P_2 in a laser with highly reflecting mirrors vary little along the layer, the variation of the field and dielectric constant with the coordinate in such a laser has a very simple form:

$$E_0 \approx e^{i\varphi}P_+ \sin\xi = e^{i\varphi}y^{1/2}\sin\xi, \quad P_+ = y^{1/2} = \text{const}, \quad \varphi = \text{const}, \tag{2.43}$$

$$\Delta\varepsilon = \frac{\mu}{1 + y\sin^2\xi}. \tag{2.44}$$

In this case the equality of the energy emitted by the active substance and the energy leaving the layer

$$-\frac{\omega}{8\pi}\frac{1}{\sigma^2}\int_0^l \operatorname{Im}(\Delta\varepsilon)\,E_0E_0^*\,dz = \Pi_1 + \Pi_2$$

[where Π_1 and Π_2 are given by formulas (2.39), (2.40)] can be written in the following form:

$$-\operatorname{Im}\left(\frac{\overline{\Delta\varepsilon}}{2\varepsilon_0} - \frac{\overline{\overline{\Delta\varepsilon}}}{2\varepsilon_0}\right) = \frac{1 - \dfrac{R_1 + R_2}{2}}{b\dfrac{\omega}{c}\sqrt{\varepsilon_0}}. \tag{2.45}$$

Here

$$\overline{\Delta\varepsilon} = \frac{1}{\pi}\int_{\xi}^{\xi+\pi} \Delta\varepsilon\,d\xi' = \mu\frac{1}{\sqrt{1+y}},$$
$$\overline{\overline{\Delta\varepsilon}} = \frac{1}{\pi}\int_{\xi}^{\xi+\pi} \Delta\varepsilon e^{-2i\xi'}\,d\xi' = \mu\left(\frac{1}{\sqrt{1+y}} - \frac{2}{y} + \frac{2}{y\sqrt{1+y}}\right). \tag{2.46}$$

Formulas (2.45) and (2.46) are equivalent, of course, to equation (2.35) given above.

We turn our attention to the fact that expression (2.45) clearly reveals the special features of an inhomogeneous laser: The emission of the laser depends not only on the averaged properties of its medium ($\overline{\Delta\varepsilon}$), but also on $\overline{\overline{\Delta\varepsilon}}$, which characterizes the inhomogeneity.

Thus, the slight inhomogeneity of the dielectric constant due to saturation is manifested in a one-dimensional active laser in the following way.

The natural frequencies generated for a selected form of $\Delta\varepsilon$ are independent of the inhomogeneity.

The eigenfunctions are the sum of two traveling waves with slowly varying amplitudes. The distribution of the amplitudes along the laser differs from the distribution in a homogeneous resonator. These differences may be considerable in a resonator with small reflection coefficients for excitation well above the threshold. When the mirrors have reflection coefficients close to unity the distortions of the eigenfunctions due to inhomogeneity are always small.

The inhomogeneity has a significant effect on the output power. As distinct from a laser with a uniform field and inversion the power of an inhomogeneous laser is a nonlinear function of the excitation.

CHAPTER III

Some Properties of Two-Dimensional Active Resonators

In the preceding chapter we solved the one-dimensional problem for a field in a medium with saturation.

In real devices used for lasers the dielectric properties of the volume which takes part in generation are not constant in a direction perpendicular to the generation axis. In particular, the imaginary part of the dielectric constant always has some inhomogeneity. In gas lasers, for instance, owing to the radial inhomogeneity of the gas discharge, the density of excited atoms decreases from the center of the discharge tube to the walls [46]. In solid-state lasers the condition for illumination leads to nonuniformity of inversion over the cross section of the specimen [47, 48].

In addition to these inhomogeneities, which are independent of the generation process, there is also a transverse inhomogeneity of the dielectric constant due to saturation. At present there are no known methods which allow an analytical investigation of two-dimensional, nonlinear electrodynamic problems. The difficulties in solving such problems are due to the fact that the smallest inhomogeneities of the dielectric constant in a direction perpendicular to the direction of generation lead to considerable changes in the properties of the emitting volume. To show how much the properties of the resonator are altered by the presence of transverse inhomogeneities, we consider the linear problem with an inhomogeneous dielectric constant. In addition, the solution of such a problem enables an assessment of the role of prescribed inversion inhomogeneities due to excitation processes.

1. We consider the following model of an inhomogeneous resonator. We assume that in the space between two infinite mirrors the dielectric constant depends only on one coordinate and varies according to the following law:

$$\varepsilon = 1 - i\Delta\varepsilon''(x) = 1 - i\delta \frac{1}{\mathrm{ch}^2 \frac{x}{h}} \tag{3.1}$$

(hypergeometric layer). The scheme of the resonator and a graph showing the variation of the imaginary part of ε are shown in Fig. 5. The variation of the field with time is chosen in the form

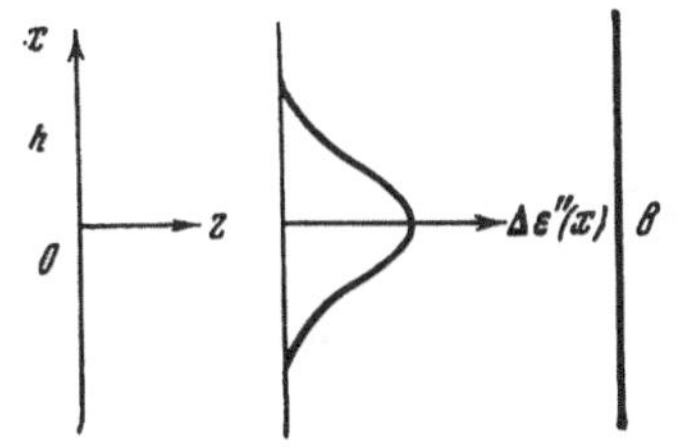

Fig. 5. Scheme of resonator and dielectric constant as a function of transverse coordinate.

$$E = \mathrm{Re}\{E_0 e^{-i\omega t}\}.$$

We consider the polarization $E_y = E$, $E_x = E_z = 0$.

We will assume that boundary conditions independent of x are assigned on the mirrors and we will seek the solution of the equation

$$\frac{\partial^2 E_0}{\partial x^2} + \frac{\partial^2 E_0}{\partial z^2} + \varepsilon \frac{\omega^2}{c^2} E_0 = 0$$

in the form of a product of functions of x and z, $E_0 = X(z)Z(z)$, where

$$\frac{d^2 Z}{dz^2} + (k_z + i\gamma_z)^2 Z = 0, \tag{3.2}$$

$$\frac{d^2 X}{dx^2} + \left\{(k_x + i\gamma_x)^2 - \frac{i\delta}{\mathrm{ch}^2 \frac{x}{h}} \left[(k_z + i\gamma_z)^2 + (k_x + i\gamma_x)^2\right]\right\} X = 0, \tag{3.3}$$

$$(k_x + i\gamma_x)^2 + (k_z + i\gamma_z)^2 = \frac{\omega^2}{c^2}. \tag{3.4}$$

The boundary conditions on the mirrors can conveniently be assigned in the form

$$\frac{E_0(x, 0)}{\frac{\partial E_0}{\partial z}(x, 0)} = \frac{i}{k_z + i\gamma_z} \frac{1 - R}{1 + R}, \quad \frac{E_0(x, b)}{\frac{\partial E_0}{\partial z}(x, b)} = -\frac{i}{k_z + i\gamma_z} \frac{1 - R}{1 + R}. \tag{3.5}$$

On the basis of (3.2), (3.5), we obtain

$$Z(z) = \sin\left[(k_z + i\gamma_z) z - i\gamma_z \frac{b}{2}\right], \tag{3.6}$$

$$k_z + i\gamma_z = \frac{m\pi}{b} - i\frac{1}{b}\ln\frac{1}{R} \quad (m \text{ is a whole number})$$

The solution of (3.3)* is determined from the condition that the function $X(\pm\infty)$ represents an emerging wave, i.e., has the form $e^{i\alpha|x|}$, $\mathrm{Re}\,\alpha > 0$. The solution of the boundary-value problem has the form

$$k_x + i\gamma_x = \frac{i}{h}\left(\sqrt{\frac{1}{4} - i\delta\frac{\omega^2}{c^2}h^2} - n - \frac{1}{2}\right) \quad (n = 0, 1, 2 \ldots), \tag{3.7}$$

$$X(x) = \left(\mathrm{ch}\frac{x}{h}\right)^{ih(k_x + i\gamma_x) - n} \cdot e^{-n\frac{x}{h}} \cdot F\left[ih(k_x + i\gamma_x) - n, -n, -ih(k_x + i\gamma_x), -e^{2\frac{x}{h}}\right]. \tag{3.8}$$

The hypergeometric function F with such parameters is a polynomial of degree n in $(-e^{2x/h})$. For illustration we write a few of the first eigenfunctions:

$$n = 0, \quad X = \left(\mathrm{ch}\frac{x}{h}\right)^{i(k_x + i\gamma_x)h},$$

$$n = 1, \quad X = \left(\mathrm{ch}\frac{x}{h}\right)^{i(k_x + i\gamma_x)h - 1} \mathrm{sh}\frac{x}{h},$$

*Equation (3.3) is well known in the literature [49-51] and its solutions have been used in the investigation of the reflection of radio waves from inhomogeneous transparent and absorbing layers of the ionosphere. The boundary-value problems for a hypergeometric layer (3.1), however, have been solved only for the case of a real potential [52].

$$n=2,\quad X=\left(\operatorname{ch}\frac{x}{h}\right)^{i(k_x+i\gamma_x)h-2}\left(\operatorname{ch}\frac{2x}{h}-\frac{i(k_x+i\gamma_x)h-2}{i(k_x+i\gamma_x)h-1}\right),$$

$$n=3,\quad X=\left(\operatorname{ch}\frac{x}{h}\right)^{i(k_x+i\gamma_x)h-3}\left(\operatorname{sh}\frac{3x}{h}-3\frac{i(k_x+i\gamma_x)h-3}{i(k_x+i\gamma_x)h-1}\operatorname{sh}\frac{x}{h}\right).$$

We will consider only the solutions (3.2), (3.3) for which the natural frequencies are approximately equal to ck_z (oscillations with almost axial directions):

$$\left|\frac{\omega-ck_z}{\omega}\right|\ll 1.$$

Then the natural frequency can conveniently be put in the form

$$\omega=\omega_0+\omega'+i\omega'',\quad \frac{\omega_0}{c}=k_z,\quad \left|\frac{\omega'+i\omega''}{\omega_0}\right|\ll 1. \tag{3.9}$$

2. When condition (3.9) holds, formula (3.7) gives an explicit expression for $k_x + i\gamma_x$:

$$k_x+i\gamma_x=\frac{i}{h}\left(\sqrt{\frac{1}{4}-i\delta\frac{\omega_0^2}{c^2}h^2}-n-\frac{1}{2}\right),$$

$$k_x=\frac{1}{h}\sqrt{-\frac{1}{8}+\sqrt{\left(\frac{1}{8}\right)^2+\frac{\delta^2}{4}\frac{\omega_0^4}{c^4}h^4}}, \tag{3.10}$$

$$\gamma_x=\frac{1}{h}\sqrt{\frac{1}{8}+\sqrt{\left(\frac{1}{8}\right)^2+\frac{\delta^2}{4}\frac{\omega_0^4}{c^4}h^4}}-n-\frac{1}{2}. \tag{3.11}$$

The behavior of the relationship between the dimensionless quantities k_xh and γ_xh and the dimensionless parameter $\frac{\delta}{2}\frac{\omega_0^2}{c^2}h^2$, which characterizes the inhomogeneity [(3.10), (3.11)] is shown in Figs. 6, 7. The graphs in Fig. 7 are numbered to correspond with the value of n in (3.11). We point out that $k_x + i\gamma_x$ is the x-component of the wave vector at high values of $|x|$, $|x|/h \gg 1$, where there is practically no inhomogeneity, $\Delta\varepsilon''(x) \ll \Delta\varepsilon''(0)$. The function X(x) (3.8) in this part of the layer has the form of a traveling wave, $X=\text{const}\, e^{i(k_x+i\gamma_x)|x|}$. When $\gamma_x > 0$ the fields decay at great distances from the resonator axis; when $\gamma_x < 0$ the fields increase with the coordinate (we will show below that such solutions must decay in time).

When there is no inhomogeneity ($\delta = 0$) all the solutions with $n \neq 0$ increase at infinity. The solution with $n = 0$ is bounded and represents the oscillation of a homogeneous layer, X(x) = const. With increase in $\frac{\delta}{2}\frac{\omega_0^2}{c^2}h^2$ there appear natural oscillations of higher transverse number bounded at infinity. The solution with number n becomes bounded when

$$\frac{\delta}{2}\frac{\omega_0^2}{c^2}h^2=\left(n+\frac{1}{2}\right)\sqrt{n(n+1)}. \tag{3.12}$$

With further increase in $\frac{\delta}{2}\frac{\omega_0^2}{c^2}h^2$ the field distribution is distorted. Greater inhomogeneities of ε ensure greater reflection and "contain" the field better. With increase in $\frac{\delta}{2}\frac{\omega_0^2}{c^2}h^2$ the field is completely localized in a narrower region close to the resonator axis. Figure 8 shows the distribution of the amplitude of the zero-point oscillation in relation to the coordinate for different values of the parameter $\frac{\delta}{2}\frac{\omega_0^2}{c^2}h^2$. Figure 9 shows the corresponding distribution for the first oscillation.

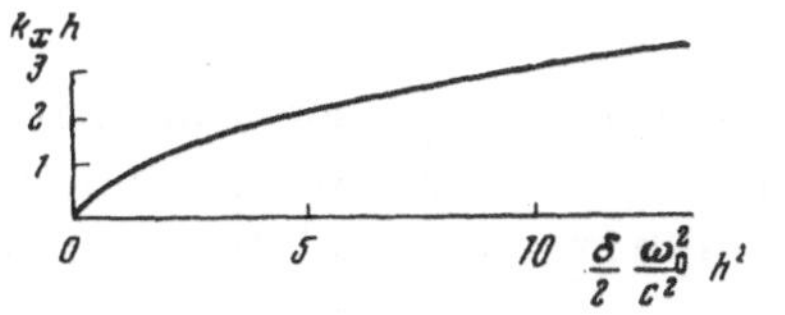

Fig. 6. Real part of the x-component of the wave vector (k_x) as a function of $\frac{\delta}{2}\frac{\omega_0^2}{c^2}h^2$. The relationship between k_x and $\frac{\delta}{2}\frac{\omega_0^2}{c^2}h^2$ is the same for all numbers of natural oscillations.

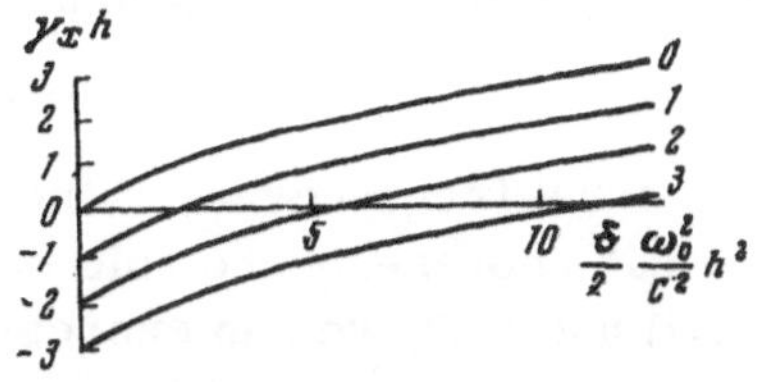

Fig. 7. Imaginary part of x-component of wave vector (γ_x) as a function of $\frac{\delta}{2}\frac{\omega_0^2}{c^2}h^2$. The figures indicate the number of the natural oscillations.

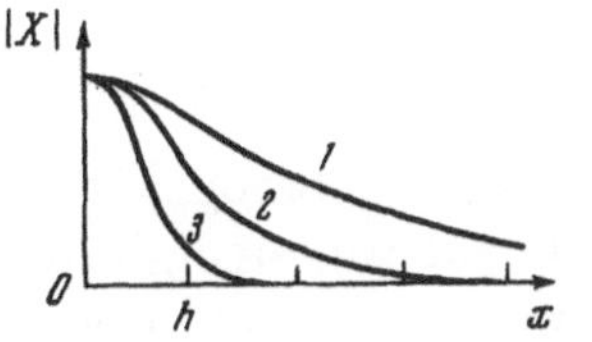

Fig. 8. Field amplitude of zero natural oscillation as a function of transverse coordinate. 1) $\frac{\delta}{2}\frac{\omega_0^2}{c^2}h^2 = 0.80$; 2) 4.05; 3) 20.

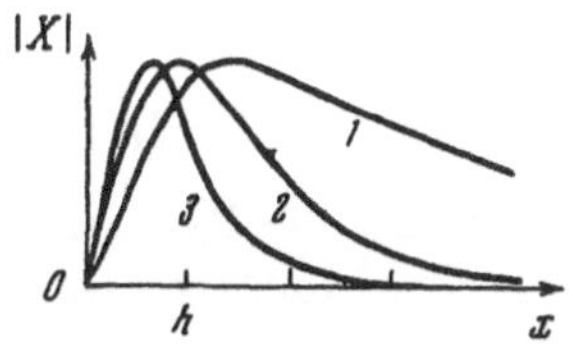

Fig. 9. Field amplitude of first natural oscillation as a function of transverse coordinate. 1) $\frac{\delta}{2}\frac{\omega_0^2}{c^2}h^2 = 2.9$; 2) 7.1; 3) 20.

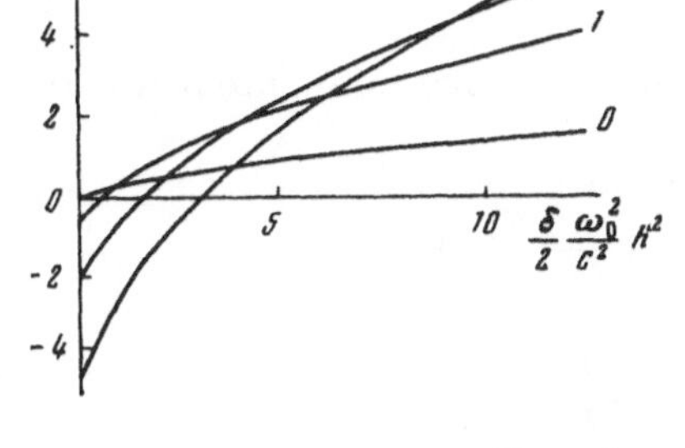

Fig. 10. The quantity $\frac{k_x^2 - \gamma_x^2}{2}h^2$, proportional to the real part of the correction to the frequency, as a function of $\frac{\delta}{2}\frac{\omega_0^2}{c^2}h^2$. The figures denote the numbers of the natural oscillations.

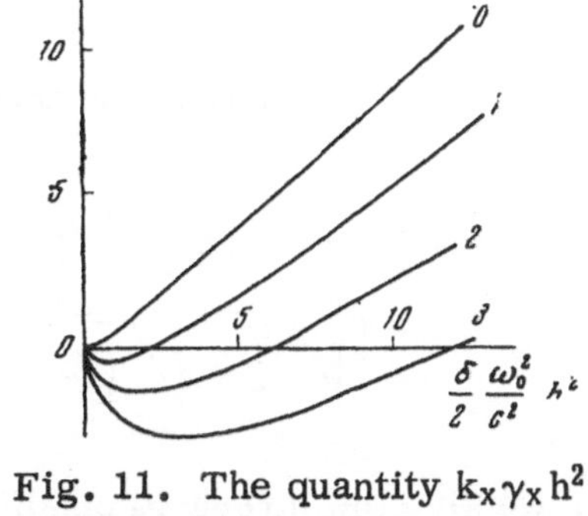

Fig. 11. The quantity $k_x\gamma_x h^2$, proportional to the imaginary part of the correction to the frequency, as a function of $\frac{\delta}{2}\frac{\omega_0^2}{c^2}h^2$. The figures denote the numbers of the natural oscillations.

3. Above we put the natural frequencies in the form

$$\omega = \omega_0 + \omega' + i\omega'',$$

separating the natural frequency ω_0 of the empty plane-parallel layer with ideal mirrors. The correction $\omega' + i\omega''$ represents the role of the transverse inhomogeneity and losses on the mirrors. Using (3.4) and (3.9) we can express the correction to the frequency by the real and imaginary parts of the components of the wave vector:

$$\frac{\omega_0\omega'}{c^2} = \frac{k_x^2 - \gamma_x^2}{2} = \frac{1}{h^2}\left[\left(n + \frac{1}{2}\right)\sqrt{\frac{1}{8} + \sqrt{\left(\frac{1}{8}\right)^2 + \frac{\delta^2}{4}\frac{\omega_0^4}{c^4}h^4}} - \frac{1}{2}\left(n + \frac{1}{2}\right)^2 - \frac{1}{8}\right], \tag{3.13}$$

$$\frac{\omega_0\omega''}{c^2} = k_x\gamma_x + k_z\gamma_z = \frac{1}{h^2}\sqrt{-\frac{1}{8} + \sqrt{\left(\frac{1}{8}\right)^2 + \frac{\delta^2}{4}\frac{\omega_0^4}{c^4}h^4}} \times$$
$$\times \left[\sqrt{\frac{1}{8} + \sqrt{\left(\frac{1}{8}\right)^2 + \frac{\delta^2}{4}\frac{\omega_0^4}{c^4}h^4}} - n - \frac{1}{2}\right] - \frac{1}{b}\frac{\omega_0}{c}\ln\frac{1}{R}. \tag{3.14}$$

Figure 10 shows the dimensionless quantity $\frac{k_x^2 - \gamma_x^2}{2}h^2$, proportional to the real part of the correction to the frequency [see (3.13)], as a function of the dimensionless quantity $\frac{\delta}{2}\frac{\omega_0^2}{c^2}h^2$, which characterizes the inhomogeneity. As was noted above, an increase in inhomogeneity leads to reduction of the region of localization of the field, i.e., the transverse wavelength is reduced. The real part of the frequency increases simultaneously, as formula (3.13) and Fig. 10 show.

Figure 11 shows the relationship between the dimensionless quantity $k_x\gamma_x h^2$ and $\frac{\delta}{2}\frac{\omega_0^2}{c^2}h^2$. This quantity is proportional to the imaginary part of the frequency in the case where the boundaries of the layer are ideal reflectors ($R = 1$). The solutions, which decay with the coordinate when $x = \pm\infty$ ($\gamma_x > 0$), are increasing in time. The solutions unbounded at $x = \pm\infty$ ($\gamma_x < 0$), decay in time. The threshold of generation of the n-th natural oscillation is attained if $\gamma_x = 0$ and has the form (3.12).

In the case where there are losses on the mirrors, $R \neq 1$, the threshold condition can be obtained from (3.14) by putting $\omega'' = 0$. It has the form

$$k_x\gamma_x = \frac{1}{b}\frac{\omega_0}{c}\ln\frac{1}{R}. \tag{3.15}$$

By transforming (3.15) we can obtain the threshold condition in a more convenient form:

$$2b\frac{c}{\omega_0}k_x\gamma_x = 2b\frac{c}{\omega_0}\frac{1}{h^2}\sqrt{-\frac{1}{8} + \sqrt{\left(\frac{1}{8}\right)^2 + \frac{\delta^2}{4}\frac{\omega_0^4}{c^4}h^4}} \times$$
$$\times\left[\sqrt{\frac{1}{8} + \sqrt{\left(\frac{1}{8}\right)^2 + \frac{\delta^2}{4}\frac{\omega_0^4}{c^4}h^4}} - n - \frac{1}{2}\right] = \ln\frac{1}{R^2}. \tag{3.16}$$

For comparison we give the threshold condition in a plane-parallel layer uniformly filled with active substance (2.33):

$$\delta\frac{\omega_0}{c}b = \ln\frac{1}{R^2}. \tag{3.17}$$

A comparison of (3.16) and (3.17) shows that in the case of an inhomogeneous layer larger values of δ are required to compensate for prescribed losses on the mirrors. This is due to the fact that the radiative properties of an inhomogeneous layer vary with the coordinate. Hence, the energy emitted close to the resonator axis ($x = 0$), where the imaginary part of the dielectric constant is greatest, not only compensates the losses on the mirrors in this region, but also maintains a steady regime in the part of the resonator where $\Delta\varepsilon''$ is less, and losses on the mirrors also occur.

The imaginary part of the dielectric constant, however, which is the radiation source, is also responsible for reflection of the electromagnetic field. Hence, the flux of electromagnetic energy in the direction of the x axis is not a constant characteristic of the resonator, but depends on δ.

The left sides of (3.17) and (3.16) differ by an amount

$$P = \delta \frac{\omega_0}{c} b - 2b \frac{c}{\omega_0} k_x \gamma_x = 2b \frac{c}{\omega_0} \frac{1}{h^2} \left(n + \frac{1}{2}\right) \sqrt{-\frac{1}{8} + \sqrt{\left(\frac{1}{8}\right)^2 + \frac{\delta^2}{4} \frac{\omega_0^4}{c^4} h^4}}. \quad (3.18)$$

This quantity represents the additional losses for the region adjoining the resonator axis and is related to the energy flux from the resonator axis in the region with lower excitation. We will call this quantity the transverse loss. As (3.18) shows, the transverse loss is different for oscillations of different number and increases with increase in the number n in proportion to $(n + \frac{1}{2})$.

4. We consider the following example of an inhomogeneous resonator. Let the distance between the mirrors be $b = 100$ cm, the width of the inhomogeneity $2h = 0.5$ cm. We assign an approximate value to the frequency $\omega_0/c \equiv 2\pi/\lambda = 2\pi \cdot 10^4\ \mathrm{cm}^{-1}$; the losses on the mirrors are $\ln(1/R^2) = 2\%$. From formula (3.16) we determine the threshold of generation of the zero and first oscillations, and then we can find the transverse loss and the correction to the frequency, and plot the field amplitude against the coordinate. We obtain

for $n = 0$:	for $n = 1$:
$\frac{\delta}{2} \frac{\omega_0^2}{c^2} h^2 = 0.8,$	$\frac{\delta}{2} \frac{\omega_0^2}{c^2} h^2 = 2.9,$
$\delta = 0.65 \cdot 10^{-8},$	$\delta = 2.35 \cdot 10^{-8},$
$\delta \frac{\omega_0}{c} b = 4\%,$	$\delta \frac{\omega_0}{c} b = 14.5\%,$
$P = 2\%,$	$P = 12.5\%,$
$\nu = \frac{\omega'}{2\pi} = 0.24$ MHz.	$\nu = \frac{\omega'}{2\pi} = 1.91$ MHz.

The field distribution is shown in Fig. 8, curve 1 ($n = 0$), and in Fig. 9, curve 1 ($n = 1$).

We consider the case of 16% losses on the mirrors. We obtain

for $n = 0$:	for $n = 1$:
$\frac{\delta}{2} \frac{\omega_0^2}{c^2} h^2 = 4.05,$	$\frac{\delta}{2} \frac{\omega_0^2}{c^2} h^2 = 7.1,$
$\delta = 3.3 \cdot 10^{-8},$	$\delta = 5.8 \cdot 10^{-8},$
$\delta \frac{\omega_0}{c} b = 20.6\%,$	$\delta \frac{\omega_0}{c} b = 36\%,$
$P = 4.6\%,$	$P = 20\%,$
$\nu = \frac{\omega'}{2\pi} = 0.88$ MHz.	$\nu = \frac{\omega'}{2\pi} = 3.56$ MHz.

The field distribution is shown in Fig. 8, curve 2 ($n = 0$), and in Fig. 9, curve 2 ($n = 1$).

In the considered examples the inhomogeneity is extremely small and is $\sim 10^{-8}$. The properties of the natural oscillations, however, differ significantly from the properties of the natural oscillations of the empty layer. The form of the fundamental natural oscillation (Fig. 8, curves 1 and 2) differs appreciably from a plane wave. We should draw attention to the transverse loss, the value of which is of the same order as the losses on the mirrors.

In the case of resonators of finite cross section in which there is a reflecting side boundary (or any mechanism concentrating the field), the properties of the natural oscillations depend on the inhomogeneity and the reflecting boundary. However, the consideration of the resonator as an infinite one is justified if the diameter of the resonator is much greater than the width of the inhomogeneity, and the quantity $\frac{\delta}{2}\frac{\omega_0^2}{c^2}h^2 \gtrsim 1$. Such a case may occur in semiconductor lasers. From the problem considered here we can assess qualitatively and quantitatively the effect of inhomogeneity on the properties of the natural oscillations.

A similar approach to the question of the modes in a semiconductor laser was made in [53], where a resonator of infinite cross section with a piecewise-constant inhomogeneity of ε was investigated. In this paper, however, the examination was restricted to the normal mode and to inhomogeneity values $\frac{|\Delta\varepsilon|}{2}\frac{\omega_0^2}{c^2}h^2 \ll 1$, for which the perturbation method can be used.

Much more convenient for calculations is a hypergeometric layer (3.1), where the problem of natural oscillations can be solved analytically.

Thus, a transverse inhomogeneity of the dielectric constant of the order of 10^{-8} can have a significant effect on the properties of the natural oscillations, and can lead to concentration of the field in the region of high excitation and to a change in the natural frequencies and the generation threshold.

We recall that in the preceding chapter we considered longitudinal inhomogeneities of the dielectric constant of the same order of magnitude and showed that the waves vary little over a length of $\lambda/2$ (the extent of the inhomogeneity). The different manifestation of small inhomogeneities in these two cases is due to the nature of the reflection of electromagnetic waves at small glancing angles.

CHAPTER IV

Weak Fields in a Resonator Generating Monochromatic Radiation

We turn now to a consideration of nonmonochromatic fields in active media.

We will assume that a strong nonmonochromatic field (causing saturation) in active media and weak fields (not causing saturation) are established in an active resonator. Owing to saturation the polarization on the weak-field frequency depends on the intensity of the strong field and, in addition, weak fields of symmetric (relative to the strong-field frequency) frequencies are coupled by polarization. Hence, a calculation of the weak fields which might be resonance oscillations of such a system is rather difficult. In this chapter we will find some solutions for weak fields on the basis of certain assumptions regarding the properties of the system and will determine in what conditions they are increasing in time.

The problem of weak fields in the presence of a strong field has a direct bearing on the question of the stability of monochromatic generation. If in an active resonator in which there is a monochromatic strong field, a field of any other frequency is increasing in time, the monochromatic regime is obviously unstable.

1. We take a one-dimensional laser model, as in Chapter II.

We will assume that the strong-field frequency is the same as the transition frequency of the active substance. To simplify the calculations we put $\gamma = 0$. In such conditions ($\omega - \omega_0 = 0$, $\gamma = 0$) the system of equations (1.28)-(1.31) for the field

$$\sigma E = \mathrm{Re}\{E_0 e^{-i\omega t} + E_1 e^{-i(\omega+\Omega)t} + E_2 e^{i(\omega-\Omega)t}\}$$

takes the form

$$\frac{d^2E_0}{dz^2} + \varepsilon_0\frac{\omega^2}{c^2}E_0 = i\frac{\omega^2}{c^2}\frac{\beta}{1+|E_0|^2}E_0,$$

$$\frac{d^2E_1}{dz^2} + \varepsilon_0\frac{(\omega+\Omega)^2}{c^2}E_1 = i\frac{(\omega+\Omega)^2}{c^2}\frac{\beta}{1+|E_0|^2}\frac{1}{1-i\Omega T_2}\left[E_1 - \frac{1-\frac{1}{2}i\Omega T_2}{(1-i\Omega T_1)(1-i\Omega T_2)+|E_0|^2}(|E_0|^2E_1 + E_0^2E_2)\right], \tag{4.1}$$

$$\frac{d^2E_2}{dz^2} + \varepsilon_0\frac{(\omega-\Omega)^2}{c^2}E_2 = -i\frac{(\omega-\Omega)^2}{c^2}\frac{\beta}{1+|E_0|^2}\frac{1}{1-i\Omega T_2}\left[E_2 - \frac{1-\frac{1}{2}i\Omega T_2}{(1-i\Omega T_1)(1-i\Omega T_2)+|E_0|^2}(|E_0|^2E_2 + E_0^{*2}E_1)\right].$$

Here $T_1 = \Gamma^{-1}$, $T_2 = \tilde{\Gamma}^{-1}$. The system (4.1) can be written in the more convenient form:

$$\frac{d^2E_0}{dz^2} + \varepsilon_0\frac{\omega^2}{c^2}E_0 = -\frac{\omega^2}{c^2}\Delta\varepsilon E_0, \tag{4.2}$$

$$\frac{d^2E_1}{dz^2} + \varepsilon_0\frac{(\omega+\Omega)^2}{c^2}E_1 = \frac{(\omega+\Omega)^2}{c^2}[-\delta\varepsilon E_1 + \alpha E_1 + \alpha_{12}E_2], \tag{4.3}$$

$$\frac{d^2E_2}{dz^2} + \varepsilon_0\frac{(\omega-\Omega)^2}{c^2}E_2 = -\frac{(\omega-\Omega)^2}{c^2}[-\delta\varepsilon E_2 + \alpha E_2 + \alpha_{21}E_1], \tag{4.4}$$

if we put

$$\Delta\varepsilon = \frac{-i\beta}{1+|E_0|^2},$$

$$\delta\varepsilon = \frac{\Delta\varepsilon}{1-i\Omega T_2} = \frac{1}{1-i\Omega T_2}\frac{-i\beta}{1+|E_0|^2},$$

$$\alpha = \frac{-i\beta}{1+|E_0|^2}\frac{1-\frac{1}{2}i\Omega T_2}{1-i\Omega T_2}\frac{|E_0|^2}{(1-i\Omega T_1)(1-i\Omega T_2)+|E_0|^2}, \tag{4.5}$$

$$\alpha_{12} = \frac{-i\beta}{1+|E_0|^2}\frac{1-\frac{1}{2}i\Omega T_2}{1-i\Omega T_2}\frac{E_0^2}{(1-i\Omega T_1)(1-i\Omega T_2)+|E_0|^2},$$

$$\alpha_{21} = \frac{-i\beta}{1+|E_0|^2}\frac{1-\frac{1}{2}i\Omega T_2}{1-i\Omega T_2}\frac{E_0^{*2}}{(1-i\Omega T_1)(1-i\Omega T_2)+|E_0|^2}.$$

We require the spectral components of the field to satisfy the following boundary conditions:

$$E_0(0) = \frac{i}{\frac{\omega}{c}\sqrt{\varepsilon_0}}\frac{1-R_1}{1+R_1}\frac{dE_0}{dz}(0), \quad E_0(b) = \frac{-i}{\frac{\omega}{c}\sqrt{\varepsilon_0}}\frac{1-R_2}{1+R_2}\frac{dE_0}{dz}(b),$$

$$E_1(0) = \frac{i}{\frac{\omega}{c}\sqrt{\varepsilon_0}}\frac{1-R_1}{1+R_1}\frac{dE_1}{dz}(0), \quad E_1(b) = \frac{-i}{\frac{\omega}{c}\sqrt{\varepsilon_0}}\frac{1-R_2}{1+R_2}\frac{dE_1}{dz}(b), \tag{4.6}$$

$$E_2(0) = \frac{-i}{\frac{\omega}{c}\sqrt{\varepsilon_0}}\frac{1-R_1^*}{1+R_1}\frac{dE_2}{dz}(0), \quad E_2(b) = \frac{i}{\frac{\omega}{c}\sqrt{\varepsilon_0}}\frac{1-R_2}{1+R_2}\frac{dE_2}{dz}(b).$$

We assume that the reflection coefficients on the boundaries are close to unity, $1 - R_{1,2} \ll 1$. In this case, as was shown in Chapter II [see (2.43)], the strong field E_0 is an almost periodic function

$$E_0 \approx e^{i\varphi} y^{1/2} \sin \frac{\omega}{c} \sqrt{\varepsilon_0}\, z, \quad y = \text{const}, \quad \varphi = \text{const}. \tag{4.7}$$

We introduce characteristics, averaged over the length $\pi c/\omega\sqrt{\varepsilon_0} = \lambda/2$, of the active medium

$$\begin{aligned}
\delta\bar{\varepsilon} &= \frac{1}{\lambda/2} \int\limits_z^{z+\lambda/2} \delta\varepsilon(z')\, dz', \\
\delta\bar{\bar{\varepsilon}} &= \frac{1}{\lambda/2} \int\limits_z^{z+\lambda/2} \delta\varepsilon(z')\, e^{2i\frac{\omega}{c}\sqrt{\varepsilon_0} z'}\, dz', \\
\bar{\alpha} &= \frac{1}{\lambda/2} \int\limits_z^{z+\lambda/2} \alpha(z')\, dz', \\
\bar{\bar{\alpha}} &= \frac{1}{\lambda/2} \int\limits_z^{z+\lambda/2} \alpha(z')\, e^{2i\frac{\omega}{c}\sqrt{\varepsilon_0} z'}\, dz'.
\end{aligned} \tag{4.8}$$

A similar averaging was carried out in Chapter II [see (2.17), (2.46)].

If we assume that the strong field is given exactly by formula (4.7), then $\delta\bar{\varepsilon}$, $\delta\bar{\bar{\varepsilon}}$, $\bar{\alpha}$, $\bar{\bar{\alpha}}$ are independent of z. In addition, if the phase of the strong field is constant (4.7), we can put $E_0 = E_0^*$; then, as can be seen from (4.5)

$$\alpha_{12} = \alpha_{21} = \alpha. \tag{4.9}$$

Calculations from formulas (4.8) give

$$\begin{aligned}
\delta\bar{\varepsilon} &= \frac{1}{1 - i\Omega T_2} \overline{\Delta\varepsilon} = \frac{-i\beta}{1 - i\Omega T_2} \frac{1}{\sqrt{1+y}}, \\
\delta\bar{\bar{\varepsilon}} &= \frac{1}{1 - i\Omega T_2} \overline{\overline{\Delta\varepsilon}} = \frac{-i\beta}{1 - i\Omega T_2} \left[\frac{1}{\sqrt{1+y}} - \frac{2}{y}\left(1 - \frac{1}{\sqrt{1+y}}\right)\right],
\end{aligned} \tag{4.10}$$

$$\begin{aligned}
\bar{\alpha} - \bar{\bar{\alpha}} &= -i\beta \frac{1 - \frac{1}{2} i\Omega T_2}{1 - i\Omega T_2} \frac{2}{y} \left\{1 + \frac{1}{(1 - i\Omega T_1)(1 - i\Omega T_2) - 1} \left[\frac{1}{\sqrt{1+y}} - \frac{(1 - i\Omega T_1)^{3/2} (1 - i\Omega T_2)^{3/2}}{\sqrt{(1 - i\Omega T_1)(1 - i\Omega T_2) + y}}\right]\right\}, \\
\bar{\alpha} + \bar{\bar{\alpha}} &= -i\beta \frac{1 - \frac{1}{2} i\Omega T_2}{1 - i\Omega T_2} \frac{2}{y} \left\{\frac{(1 - i\Omega T_1)(1 - i\Omega T_2) + 1 + y}{\sqrt{(1 - i\Omega T_1)(1 - i\Omega T_2)}\sqrt{(1 - i\Omega T_1)(1 - i\Omega T_2) + y} + \sqrt{1+y}} - 1\right\}.
\end{aligned} \tag{4.11}$$

In view of the smallness of the parameter β and, hence, of α, $\delta\varepsilon$, we will seek weak fields in the form

$$\begin{aligned}
E_1 &= A_1 e^{i\frac{\omega}{c}\sqrt{\varepsilon_0} z} + B_1 e^{-i\frac{\omega}{c}\sqrt{\varepsilon_0} z}, \\
E_2 &= A_2 e^{i\frac{\omega}{c}\sqrt{\varepsilon_0} z} + B_2 e^{-i\frac{\omega}{c}\sqrt{\varepsilon_0} z},
\end{aligned}$$

where $A_{1,2}$, $B_{1,2}$ are "slow" amplitudes. Equations for $A_{1,2}$, $B_{1,2}$ are derived in the same way as the equations for the strong-field amplitudes in Chapter II. Denoting the dimensionless coordinate

$$\frac{\omega}{c}\sqrt{\varepsilon_0}\, z = \xi$$

and using (4.8) and (4.9), we obtain from (4.3) and (4.4)

$$\begin{aligned}
2i\frac{dA_1}{d\xi} &= -2\frac{\Omega}{\omega}A_1 - \delta\bar{\varepsilon}/\varepsilon_0 A_1 - \delta\bar{\bar{\varepsilon}}/\varepsilon_0 B_1 + \bar{\alpha}/\varepsilon_0(A_1+A_2) + \bar{\bar{\alpha}}/\varepsilon_0(B_1+B_2),\\
2i\frac{dB_1}{d\xi} &= 2\frac{\Omega}{\omega}B_1 + \delta\bar{\varepsilon}/\varepsilon_0 B_1 + \delta\bar{\bar{\varepsilon}}/\varepsilon_0 A_1 - \bar{\alpha}/\varepsilon_0(B_1+B_2) - \bar{\bar{\alpha}}/\varepsilon_0(A_1+A_2),\\
2i\frac{dA_2}{d\xi} &= 2\frac{\Omega}{\omega}A_2 + \delta\bar{\varepsilon}/\varepsilon_0 A_2 + \delta\bar{\bar{\varepsilon}}/\varepsilon_0 B_2 - \bar{\alpha}/\varepsilon_0(A_1+A_2) - \bar{\bar{\alpha}}/\varepsilon_0(B_1+B_2),\\
2i\frac{dB_2}{d\xi} &= -2\frac{\Omega}{\omega}B_2 - \delta\bar{\varepsilon}/\varepsilon_0 B_2 - \delta\bar{\bar{\varepsilon}}/\varepsilon_0 A_2 + \bar{\alpha}/\varepsilon_0(B_1+B_2) + \bar{\bar{\alpha}}/\varepsilon_0(A_1+A_2).
\end{aligned} \tag{4.12}$$

The boundary conditions for $A_{1,2}$, $B_{1,2}$ follow from (4.6) and have the form

$$\begin{aligned}
A_1(0) &= -R_1B_1(0),\\
B_1(L) &= -R_2A_1(L), \qquad L = \frac{\omega}{c}\sqrt{\varepsilon_0}B\\
B_2(0) &= -R_1A_2(0),\\
A_2(L) &= -R_2B_2(L).
\end{aligned} \tag{4.13}$$

The system of equations (4.1) illustrates the fact that in the considered conditions there interact, firstly, waves of the same frequency and opposite direction and, secondly, waves of different frequencies. The coupling of waves of opposite direction is due to the inhomogeneity of the medium. In a homogeneous medium $\delta\bar{\bar{\varepsilon}} = 0$, $\bar{\bar{\alpha}} = 0$, and the wave A_1 (A_2) is not coupled with B_1 (B_2). The interaction of fields of different frequencies can be attributed, as already mentioned in Chapter I, to the special nature of the behavior of a quantum-mechanical system in a strong field; this interaction depends on α.

The system of equations (4.12) with boundary conditions (4.13) on conversion to functions $A_1 \pm B_2$, $B_1 \pm A_2$ is broken up into two independent systems:

$$\begin{aligned}
2i\frac{d}{d\xi}(A_1+B_2) &= -\left(2\frac{\Omega}{\omega} + \delta\bar{\varepsilon}/\varepsilon_0 - \frac{\bar{\alpha}+\bar{\bar{\alpha}}}{\varepsilon_0}\right)(A_1+B_2) - \left(\delta\bar{\bar{\varepsilon}}/\varepsilon_0 - \frac{\alpha+\bar{\bar{\alpha}}}{\varepsilon_0}\right)(B_1+A_2),\\
2i\frac{d}{d\xi}(B_1+A_2) &= \left(2\frac{\Omega}{\omega} + \delta\bar{\varepsilon}/\varepsilon_0 - \frac{\bar{\alpha}+\bar{\bar{\alpha}}}{\varepsilon_0}\right)(B_1+A_2) - \left(\delta\bar{\bar{\varepsilon}}/\varepsilon_0 - \frac{\bar{\alpha}+\bar{\bar{\alpha}}}{\varepsilon_0}\right)(A_1+B_2),
\end{aligned} \tag{4.14}$$

$$\begin{aligned}
2i\frac{d}{d\xi}(A_1-B_2) &= -\left(2\frac{\Omega}{\omega} + \delta\bar{\varepsilon}/\varepsilon_0 - \frac{\bar{\alpha}-\bar{\bar{\alpha}}}{\varepsilon_0}\right)(A_1-B_2) - \left(\delta\bar{\bar{\varepsilon}}/\varepsilon_0 + \frac{\bar{\alpha}-\bar{\bar{\alpha}}}{\varepsilon_0}\right)(B_1-A_2),\\
2i\frac{d}{d\xi}(B_1-A_2) &= \left(2\frac{\Omega}{\omega} + \delta\bar{\varepsilon}/\varepsilon_0 - \frac{\bar{\alpha}-\alpha}{\varepsilon_0}\right)(B_1-A_2) + \left(\delta\bar{\bar{\varepsilon}}/\varepsilon_0 + \frac{\bar{\alpha}-\bar{\bar{\alpha}}}{\varepsilon_0}\right)(A_1-B_2)
\end{aligned} \tag{4.15}$$

with boundary conditions

$$\begin{cases}
(A_1+B_2)_{\xi=0} = -R_1(B_1+A_2)_{\xi=0},\\
(B_1+A_2)_{\xi=L} = -R_2(A_1+B_2)_{\xi=L},
\end{cases} \tag{4.16}$$

$$\begin{cases}
(A_1-B_2)_{\xi=0} = -R_1(B_1-A_2)_{\xi=0},\\
(B_1-A_2)_{\xi=L} = -R_2(A_1-B_2)_{\xi=L}.
\end{cases} \tag{4.17}$$

The linearly independent solutions of system (4.14) have the form

$$\begin{cases}
A_1+B_2 = -\dfrac{\delta\bar{\bar{\varepsilon}}-\bar{\alpha}-\bar{\bar{\alpha}}}{2\varepsilon_0}e^{i\eta\xi},\\
B_1+A_2 = \left(\dfrac{\Omega}{\omega} + \dfrac{\delta\bar{\varepsilon}-\bar{\alpha}-\bar{\bar{\alpha}}}{2\varepsilon_0} - \eta\right)e^{i\eta\xi},
\end{cases} \tag{4.18}$$

$$\begin{cases} A_1 + B_2 = \left(\frac{\Omega}{\omega} + \frac{\delta\bar{\varepsilon} - \bar{\alpha} - \bar{\bar{\alpha}}}{2\varepsilon_0} - \eta\right) e^{-i\eta\xi}, \\ B_1 + A_2 = -\frac{\delta\bar{\bar{\varepsilon}} - \bar{\alpha} - \bar{\bar{\alpha}}}{2\varepsilon_0} e^{-i\eta\xi}, \end{cases} \tag{4.19}$$

$$\eta = \sqrt{\left(\frac{\Omega}{\omega} + \frac{\delta\bar{\varepsilon} - \bar{\alpha} - \bar{\bar{\alpha}}}{2\varepsilon_0}\right)^2 - \left(\frac{\delta\bar{\bar{\varepsilon}} - \bar{\alpha} - \bar{\bar{\alpha}}}{2\varepsilon_0}\right)^2}. \tag{4.20}$$

Similarly for (4.15):

$$\begin{cases} A_1 - B_2 = -\frac{\delta\bar{\bar{\varepsilon}} + \bar{\alpha} - \bar{\bar{\alpha}}}{2\varepsilon_0} e^{i\eta\xi}, \\ B_1 - A_2 = \left(\frac{\Omega}{\omega} + \frac{\delta\bar{\varepsilon} - \bar{\alpha} + \bar{\bar{\alpha}}}{2\varepsilon_0} - \eta\right) e^{i\eta\xi}, \end{cases} \tag{4.21}$$

$$\begin{cases} A_1 - B_2 = \left(\frac{\Omega}{\omega} + \frac{\delta\bar{\varepsilon} - \bar{\alpha} + \alpha}{2\varepsilon_0} - \eta\right) e^{-i\eta\xi}, \\ B_1 - A_2 = -\frac{\delta\bar{\bar{\varepsilon}} + \bar{\alpha} - \bar{\bar{\alpha}}}{2\varepsilon_0} e^{-i\eta\xi}, \end{cases} \tag{4.22}$$

$$\eta = \sqrt{\left(\frac{\Omega}{\omega} + \frac{\delta\bar{\varepsilon} - \bar{\alpha} + \bar{\bar{\alpha}}}{2\varepsilon_0}\right)^2 - \left(\frac{\delta\bar{\bar{\varepsilon}} + \bar{\alpha} - \bar{\bar{\alpha}}}{2\varepsilon_0}\right)^2}. \tag{4.23}$$

It follows from the requirement that the linear combination of (4.18) and (4.19) satisfy the boundary conditions (4.16) that

$$e^{2i\eta L} = \frac{R_1\left(\frac{\Omega}{\omega} + \frac{\delta\bar{\varepsilon} - \bar{\alpha} - \bar{\bar{\alpha}}}{2\varepsilon_0} - \eta\right) - \frac{\delta\bar{\bar{\varepsilon}} - \bar{\alpha} - \bar{\bar{\alpha}}}{2\varepsilon_0}}{\frac{\Omega}{\omega} + \frac{\delta\bar{\varepsilon} - \bar{\alpha} - \bar{\bar{\alpha}}}{2\varepsilon_0} - \eta - R_1 \frac{\delta\bar{\bar{\varepsilon}} - \bar{\alpha} - \bar{\bar{\alpha}}}{2\varepsilon_0}} \, \frac{R_2\left(\frac{\Omega}{\omega} + \frac{\delta\bar{\varepsilon} - \bar{\alpha} - \bar{\bar{\alpha}}}{2\varepsilon_0} - \eta\right) - \frac{\delta\bar{\bar{\varepsilon}} - \bar{\alpha} - \bar{\bar{\alpha}}}{2\varepsilon_0}}{\frac{\Omega}{\omega} + \frac{\delta\bar{\varepsilon} - \bar{\alpha} - \bar{\bar{\alpha}}}{2\varepsilon_0} - \eta - R_2 \frac{\delta\bar{\bar{\varepsilon}} - \bar{\alpha} - \bar{\bar{\alpha}}}{2\varepsilon_0}}; \tag{4.24}$$

(4.24) and (4.20) are the system of equations which determine the natural frequency (more precisely, the frequency correction Ω).

In exactly the same way it follows from (4.21), (4.22), (4.17) that

$$e^{2i\eta L} = \frac{R_1\left(\frac{\Omega}{\omega} + \frac{\delta\bar{\varepsilon} - \bar{\alpha} + \bar{\bar{\alpha}}}{2\varepsilon_0} - \eta\right) - \frac{\delta\bar{\bar{\varepsilon}} + \bar{\alpha} - \bar{\bar{\alpha}}}{2\varepsilon_0}}{\frac{\Omega}{\omega} + \frac{\delta\bar{\varepsilon} - \bar{\alpha} + \bar{\bar{\alpha}}}{2\varepsilon_0} - \eta - R_1 \frac{\delta\bar{\bar{\varepsilon}} + \bar{\alpha} - \bar{\bar{\alpha}}}{2\varepsilon_0}} \, \frac{R_2\left(\frac{\Omega}{\omega} + \frac{\delta\bar{\varepsilon} - \bar{\alpha} + \bar{\bar{\alpha}}}{2\varepsilon_0} - \eta\right) - \frac{\delta\bar{\bar{\varepsilon}} + \bar{\alpha} - \bar{\bar{\alpha}}}{2\varepsilon_0}}{\frac{\Omega}{\omega} + \frac{\delta\bar{\varepsilon} - \bar{\alpha} + \bar{\bar{\alpha}}}{2\varepsilon_0} - \eta - R_2 \frac{\delta\bar{\bar{\varepsilon}} + \bar{\alpha} - \bar{\bar{\alpha}}}{2\varepsilon_0}}; \tag{4.25}$$

(4.25) and (4.23) are also a system of equations which determine the natural frequency. Equations (4.25), (4.23) differ from (4.24), (4.20) in that the weak fields of symmetric frequencies in these two cases have different phase relationships. We transform (4.24) and (4.25) and introduce the symbol Ω_{res} for the difference in natural frequencies of the empty resonator: $\Omega_{\text{res}} = (\pi/L)\omega$. Then from (4.24), (4.25), (4.20), (4.23) we obtain equations for the frequency in the form

$$\begin{aligned} \eta - m\frac{\Omega_{\text{pes}}}{\omega} = &-\frac{i}{2L}\ln \frac{R_1\left(\frac{\Omega}{\omega} + \frac{\delta\bar{\varepsilon} - \bar{\alpha} \mp \bar{\bar{\alpha}}}{2\varepsilon_0} - \eta\right) - \frac{\delta\bar{\bar{\varepsilon}} \mp \bar{\alpha} - \bar{\bar{\alpha}}}{2\varepsilon_0}}{\frac{\Omega}{\omega} + \frac{\delta\bar{\varepsilon} - \bar{\alpha} \mp \bar{\bar{\alpha}}}{2\varepsilon_0} - \eta - R_1 \frac{\delta\bar{\bar{\varepsilon}} \mp \bar{\alpha} - \bar{\bar{\alpha}}}{2\varepsilon_0}} - \\ &-\frac{i}{2L}\ln \frac{R_2\left(\frac{\Omega}{\omega} + \frac{\delta\bar{\varepsilon} - \bar{\alpha} \pm \bar{\bar{\alpha}}}{2\varepsilon_0} - \eta\right) - \frac{\delta\bar{\bar{\varepsilon}} \pm \bar{\alpha} - \bar{\bar{\alpha}}}{2\varepsilon_0}}{\frac{\Omega}{\omega} + \frac{\delta\bar{\varepsilon} - \bar{\alpha} \pm \bar{\bar{\alpha}}}{2\varepsilon_0} - \eta - R_2 \frac{\delta\bar{\bar{\varepsilon}} \pm \bar{\alpha} - \bar{\bar{\alpha}}}{2\varepsilon_0}} \quad (m = 0, \pm 1, \pm 2 \ldots), \end{aligned} \tag{4.26}$$

where

$$\eta = \sqrt{\left(\frac{\Omega}{\omega} + \frac{\delta\bar{\varepsilon} - \bar{\alpha} \mp \alpha}{2\varepsilon_0}\right)^2 - \left(\frac{\delta\bar{\bar{\varepsilon}} \mp \bar{\alpha} - \bar{\bar{\alpha}}}{2\varepsilon_0}\right)^2}. \tag{4.27}$$

Equations (4.26), (4.27) contain the difference in natural frequencies of the empty resonator Ω_{res} and also the quantities $\delta\bar{\varepsilon}$, $\delta\bar{\bar{\varepsilon}}$, $\bar{\alpha}$, $\bar{\bar{\alpha}}$, which characterize the active medium and depend on the strong-field intensity. The weak-field frequency, generally speaking, can differ significantly from the resonant frequency of the empty resonator. For instance, when $T_1^{-1} \ll \Omega_{res} \ll T_2^{-1}$, $|\bar{\alpha}| \ll |\Delta\bar{\varepsilon}|$, $|\bar{\bar{\alpha}}| \ll |\Delta\bar{\varepsilon}|$, $\bar{\bar{\Delta\varepsilon}} \approx \bar{\Delta\varepsilon}$, $\bar{\Delta\varepsilon}/2\varepsilon_0 = -i^3/_4\,(\Omega_{res}/\omega)$ we can obtain from (4.26), (4.27)

$$\operatorname{Re}\Omega \approx \pm\Omega_{res}\sqrt{m^2-\left(\frac{3}{4}\right)^2} \quad (m=\pm 1, \pm 2, \pm 3\ldots),$$

i.e., when excitation greatly exceeds the threshold $\left(\left|\frac{\Delta\varepsilon}{2\varepsilon_0}\right| = \frac{3}{4}\frac{\Omega_{res}}{\omega} \gg \left(1-\frac{R_1+R_2}{2}\right)\frac{1}{L} = \left|\frac{\Delta\varepsilon}{2\varepsilon_0}\right|_{thresh}\right)$ the natural frequencies become nonequidistant, and their real parts are shifted towards the strong-field frequency $\left(\Omega_{res}\sqrt{m^2-\left(\frac{3}{4}\right)^2} < \Omega_{res}|m|\right)$.

2. We give some solutions of (4.26) and (4.27) for some specially chosen parameters.

We will consider the solutions of (4.26) when $m \neq 0$, and, in addition, we will assume that the right side of (4.26) can be expanded in terms of $1-R_1$, $1-R_2$. Then from (4.26), (4.27) we obtain

$$\frac{\Omega}{\omega}+\frac{\delta\bar{\varepsilon}-\bar{\alpha}\pm\bar{\bar{\alpha}}}{2\varepsilon_0} = -i\frac{1-\frac{R_1+R_2}{2}}{L} + \sqrt{\left(\frac{\Omega_{res}}{\omega}\right)^2+\left(\frac{\delta\bar{\bar{\varepsilon}}\mp\bar{\alpha}-\bar{\bar{\alpha}}}{2\varepsilon_0}\right)^2} \times$$
$$\times\left[1-i\frac{1-\frac{R_1+R_2}{2}}{L}\,\frac{\frac{\delta\bar{\bar{\varepsilon}}\mp\bar{\alpha}-\bar{\bar{\alpha}}}{2\varepsilon_0}}{\left(\frac{\Omega_{res}}{\omega}\right)^2+\left(\frac{\delta\bar{\bar{\varepsilon}}\mp\bar{\alpha}-\bar{\bar{\alpha}}}{2\varepsilon_0}\right)^2}\right],$$

$$\tag{14.28}$$

$$\frac{\Omega}{\omega}+\frac{\delta\bar{\varepsilon}-\bar{\alpha}\pm\bar{\bar{\alpha}}}{2\varepsilon_0} = -i\frac{1-\frac{R_1+R_2}{2}}{L} - \sqrt{\left(\frac{\Omega_{res}}{\omega}\right)^2+\left(\frac{\delta\bar{\bar{\varepsilon}}\mp\bar{\alpha}-\bar{\bar{\alpha}}}{2\varepsilon_0}\right)^2} \times$$
$$\times\left[1-i\frac{1-\frac{R_1+R_2}{2}}{L}\,\frac{\frac{\delta\bar{\bar{\varepsilon}}\mp\bar{\alpha}-\bar{\bar{\alpha}}}{2\varepsilon_0}}{\left(\frac{\Omega_{res}}{\omega}\right)^2+\left(\frac{\delta\bar{\bar{\varepsilon}}\mp\bar{\alpha}-\bar{\bar{\alpha}}}{2\varepsilon_0}\right)^2}\right].$$

If the excess of inversion over the threshold value is not too high, the frequency determined from (4.28) is close to the frequency of the empty resonator. Evaluations from formulas (4.10), (4.11), and (2.35) show that if the inversion in threshold units does not exceed $\frac{1}{\left(1-\frac{R_1+R_2}{2}\right)^2}$,

$$\frac{\beta}{\beta_{thresh}} < \frac{1}{\left(1-\frac{R_1+R_2}{2}\right)^2}, \tag{4.29}$$

then the quantities $\delta\bar{\varepsilon}$, $\delta\bar{\bar{\varepsilon}}$, $\bar{\alpha}$, $\bar{\bar{\alpha}} \ll \Omega_{res}/\omega$ and (4.28) reduce to

$$\frac{\Omega}{\omega} = \frac{\Omega_{res}}{\omega}m - i\frac{1-\frac{R_1+R_2}{2}}{L} - \frac{\delta\bar{\varepsilon}-\bar{\alpha}\mp\bar{\bar{\alpha}}}{2\varepsilon_0}. \tag{4.30}$$

In addition, in certain conditions we can ignore the interaction of weak fields of different frequencies.

We consider more fully the case where the constants of the substance satisfy the relationship

$$T_2^{-1} \gg T_1^{-1},$$

and the distance between the natural frequencies of the empty resonator Ω_{res} is such that

$$T_2^{-1} \gg \Omega_{res} \gg T_1^{-1}. \tag{4.31}$$

We note that for a ruby laser, for instance

$$T_2^{-1} = 10^{10} - 10^{11}\ \text{sec}^{-1},$$
$$T_1^{-1} = 10^{3} - 10^{4}\ \text{sec}^{-1},$$

and the distance between the natural frequencies of the empty resonator in typical cases is

$$\Omega_{res} = 10^{8} - 10^{9}\ \text{sec}^{-1}.$$

As noted in Chapter I, the interaction of weak fields of different frequencies, which always takes place in the presence of a strong field, is manifested most effectively close to the strong-field frequency ($|\Omega| \lessdot T_1^{-1}$). In the case of frequency differences $\Omega \approx \Omega_{res} \gg T_1^{-1}$ this interaction can be neglected if the saturation does not exceed the value of $\Omega_{res}T_1$:

$$\beta/\beta_{thresh} < \Omega_{res} T_1, \quad \Omega_{res} T_1 \gg 1. \tag{4.32}$$

When conditions (4.31), (4.32) are fulfilled, the quantities $\bar{\alpha}$, $\bar{\bar{\alpha}}$, determined from formulas (4.11), are much less than $\delta\bar{\varepsilon}$ and can be dropped from expression (4.30) for the natural frequency. Then (4.30) will take the form

$$\frac{\Omega}{\omega} = \frac{\Omega_{res}}{\omega} m - i \frac{1 - \frac{R_1 + R_2}{2}}{L} - \frac{\delta\bar{\varepsilon}}{2\varepsilon_0}$$

or, in view of the condition $\Omega_{res}T_2 \ll 1$

$$\operatorname{Re}\Omega = \Omega' \approx m\Omega_{res}$$

$$\operatorname{Im}\Omega = -\left(1 - \frac{R_1 + R_2}{2}\right)\frac{\omega}{L} - \frac{\omega}{2\varepsilon_0}\operatorname{Im}\delta\bar{\varepsilon} \approx \frac{\omega}{2\varepsilon_0}\frac{-\operatorname{Im}\overline{\Delta\varepsilon}}{1 + (\Omega' T_2)^2} - \frac{\omega}{L}\left(1 - \frac{R_1 + R_2}{2}\right). \tag{4.33}$$

3. The sign of the imaginary part of the frequency in (4.33) decides whether the solution increases or decays in time. The generation threshold of the weak field is obtained from the requirement that $\operatorname{Im}\Omega = 0$:

$$\frac{1}{2\varepsilon_0}\frac{-\operatorname{Im}\overline{\Delta\varepsilon}}{1 + (\Omega' T_2)^2} - \frac{1}{L}\left(1 - \frac{R_1 + R_2}{2}\right) = 0. \tag{4.34}$$

We recall that the energy balance equation for the strong field, as shown in Chapter II (2.45), has the form

$$\frac{-\operatorname{Im}(\overline{\Delta\varepsilon} - \overline{\overline{\Delta\varepsilon}})}{2\varepsilon_0} - \frac{1}{L}\left(1 - \frac{R_1 + R_2}{2}\right) = 0. \tag{4.35}$$

Expression (4.34) contains the mean value of the imaginary part of the dielectric constant $\operatorname{Im}\overline{\Delta\varepsilon}$; the factor $1/[1 + (\Omega' T_2)^2]$ denotes the reduction in amplification with increasing distance from the center of the line. In addition to the mean value $\operatorname{Im}\overline{\Delta\varepsilon}$, (4.35) contains the Fourier coefficient $\operatorname{Im}\overline{\overline{\Delta\varepsilon}}$, which denotes the reduction in amplification of the strong field due to inhomogeneity of the medium.

The difference in (4.34) and (4.35) is clearly illustrated as follows. In the considered conditions both the strong and weak fields are close to standing waves with different wave vectors. At the antinodes of the strong field the imaginary part of the dielectric constant is a minimum, and at the nodes it is a maximum. However, near the nodes the strong field itself is small and these parts of the volume make practically no contribution to the total emission of the laser. In the region of the antinodes the field is a maximum, but here $-\mathrm{Im}\,\overline{\Delta\varepsilon}$ has a small value. Hence, radiation on the strong-field frequency is characterized by a smaller value than $-\mathrm{Im}\,\overline{\Delta\varepsilon}$. The imaginary part of the dielectric constant on the weak field frequency $-\mathrm{Im}\,\overline{\Delta\varepsilon}/[1+(\Omega' T_2)^2]$, has the same period as the strong field. The weak field, however, has a different wavelength and its nodes do not coincide with the maxima of $-\mathrm{Im}\,\Delta\varepsilon$. Hence, the emission of the weak field is characterized by the mean value $-\mathrm{Im}\,\Delta\varepsilon/[1+(\Omega' T_2)^2]$.

Consequently, despite the fact that at each individual point of the laser the dielectric constant for the weak field is smaller than for the strong field, the inhomogeneity of the medium can result in the total amplification for the weak field becoming greater than for the strong field.

A comparison of (4.34) and (4.35) shows that the generation threshold for the weak field is attained if

$$1+(\Omega' T_2)^2 < \frac{-\mathrm{Im}\,\overline{\Delta\varepsilon}}{-\mathrm{Im}\,(\overline{\Delta\varepsilon}-\overline{\overline{\Delta\varepsilon}})} = \frac{\frac{1}{2}y}{\sqrt{1+y}-1}. \tag{4.36}$$

Small values of saturation are sufficient for the satisfaction of (4.36). For instance, if the difference frequency Ω' is 1/10 of the line width, $\Omega' T_2 = 1/10$, then (4.36) is satisfied at a saturation $y \geq 0.04$.

The fulfillment of the condition for weak field generation means that if a monochromatic field is established in any interval of time in the laser, then the fields on other frequencies, due to spontaneous emission, will increase in time, which leads to a change in the generation conditions.

Thus, resonance oscillations of a plane-parallel active layer may be considerably altered by the presence of the monochromatic strong field, which alters the properties of the active medium. The resonance oscillations of such a layer are fields containing two spectral components. The frequencies and spatial distribution of weak fields, generally speaking, depend on the inhomogeneity of the medium due to the strong field. However, when certain limitations are imposed on the strong-field intensity the weak fields are almost standing waves. For such fields the wave vector and the real part of the frequency depend only on the distance between the mirrors, and the imaginary part of the frequency depends only on the mean value of the inversion.

In conditions typical for a ruby laser there are weak fields which increase in time. This means instability of monochromatic generation. This form of instability is due entirely to the difference in the spatial distribution of the strong and weak fields.

The relationship between the spectral composition of the emission and the spatial distribution of the field and inversion has been clearly confirmed by several experimental investigations [54, 55].

CHAPTER V

Nonlinear Interaction of Weak Fields in the Presence of a Strong Monochromatic Field in a Traveling-Wave Amplifier

We consider the propagation of an electromagnetic wave consisting of a monochromatic strong component and several weak components in an active medium in a traveling-wave system. In such a system the coupling of weak fields by polarization is particularly pronounced, since it is not complicated by the interaction of waves of different direction [see the discussion of (4.12)]. In addition, this formulation of the problem is of great practical interest. It is connected with the question of propagation of the traveling wave of a modulated signal with a low modulation index in a maser and the signal distortions which occur in this case.

As before, we will assume that the properties of the active medium are such that we can use the equations given in Chapter I. Masers employing paramagnetic crystals composed of material which can be described by a three-level scheme are most widely used in practice at present. In our paper [39] we calculated the magnetic moment of a three-level quantum-mechanical system in the case of saturation of the pumping transition (this condition is always fulfilled in paramagnetic masers). We showed that in this case valid equations for the field were (1.28)-(1.31) in which the constants Γ, β were expressed differently by the microcharacteristics of the substance, and the constant γ was zero.

1. We will assume that the strong field is in resonance with the transition frequency and the weak field contains two components with symmetric frequencies. To simplify the problem and also having in mind the example of a paramagnetic maser, we put the constant γ equal to zero. In the one-dimensional case in such conditions the system of equations (1.28)-(1.31) for the field

$$\sigma E = \mathrm{Re}\,\{E_0 e^{-i\omega t} + E_1 e^{-i(\omega+\Omega)t} + E_2 e^{i(\omega-\Omega)t}\}$$

takes the form

$$\frac{d^2E_0}{dz^2} + \varepsilon_0 \frac{\omega^2}{c^2} E_0 = i\frac{\omega^2}{c^2}\frac{\beta}{1+|E_0|^2} E_0,$$

$$\frac{d^2E_1}{dz^2} + \varepsilon_0 \frac{(\omega+\Omega)^2}{c^2} E_1 = i\frac{(\omega+\Omega)^2}{c^2}\frac{\beta}{1+|E_0|^2}\frac{1}{1-i\Omega T_2}\left[E_1 - \frac{1-\frac{1}{2}i\Omega T_2}{(1-i\Omega T_1)(1-i\Omega T_2)+|E_0|^2}(|E_0|^2E_1 + E_0^2 E_2)\right],$$

$$\frac{d^2E_2}{dz^2} + \varepsilon_0 \frac{(\omega-\Omega)^2}{c^2} E_2 = -i\frac{(\omega-\Omega)^2}{c^2}\frac{\beta}{1+|E_0|^2}\frac{1}{1-i\Omega T_2}\left[E_2 - \frac{1-\frac{1}{2}i\Omega T_2}{(1-i\Omega T_1)(1-i\Omega T_2)+|E_0|^2}(|E_0|^2E_2 + E_0^{*2} E_1)\right]. \tag{5.1}$$

Here $T_1 = \Gamma^{-1}$, $T_2 = \tilde{\Gamma}^{-1}$.

To solve system (5.1) we put the unknown functions in the form of a product of rapidly oscillating phase factors and complex amplitudes

$$E_0 = v_0(z)\, e^{i\frac{\omega}{c}\sqrt{\varepsilon_0}z},\quad E_1 = v_1(z)\, e^{i\frac{\omega+\Omega}{c}\sqrt{\varepsilon_0}z},\quad E_2 = v_2(z)\, e^{-i\frac{\omega-\Omega}{c}\sqrt{\varepsilon_0}z}$$

(waves propagating along the z axis are considered).

Using the smallness of the parameter β we can show that the functions v_0, v_1, and v_2 do not change much over the length $c/\omega\sqrt{\varepsilon_0}$, and obtain the following equations for these functions:

$$\frac{dv_0}{dz} = \beta\frac{\omega}{c}\sqrt{\varepsilon_0}\frac{v_0}{1+v_0v_0^*},$$

$$\frac{dv_1}{dz} = \beta\frac{\omega}{c}\sqrt{\varepsilon_0}\frac{1}{1-i\Omega T_2}\left[\frac{v_1}{1+v_0v_0^*} - \frac{1-\frac{1}{2}i\Omega T_2}{1+v_0v_0^*}\frac{v_1v_0v_0^* + v_2v_0^2}{(1-i\Omega T_1)(1-i\Omega T_2)+v_0v_0^*}\right], \tag{5.2}$$

$$\frac{dv_2}{dz} = \beta\frac{\omega}{c}\sqrt{\varepsilon_0}\frac{1}{1-i\Omega T_2}\left[\frac{v_2}{1+v_0v_0^*} - \frac{1-\frac{1}{2}i\Omega T_2}{1+v_0v_0^*}\frac{v_2v_0v_0^* + v_1v_0^{*2}}{(1-i\Omega T_1)(1-i\Omega T_2)+v_0v_0^*}\right].$$

The first equation shows that amplification alters only the modulus of the complex amplitude v_0. Since by an appropriate choice of time origin we can choose any input value of the phase v_0, we will assume that the function v_0 is real. Using $v_0 = v_0^*$ we can perform further transformations and obtain uncoupled equations for $v_1 + v_2$ and $v_1 - v_2$:

$$\frac{dv_0}{dz} = \alpha \frac{v_0}{1+v_0^2},$$

$$\frac{d}{dz}(v_1 + v_2) = \frac{\alpha}{1 - i\Omega T_2} \frac{v_1 + v_2}{1+v_0^2} - 2\alpha \frac{1 - \frac{1}{2} i\Omega T_2}{1 - i\Omega T_2} \frac{v_0^2}{(1 - i\Omega T_1)(1 - i\Omega T_2) + v_0^2} \frac{v_1 + v_2}{1+v_0^2}, \tag{5.3}$$

$$\frac{d}{dz}(v_1 - v_2) = \frac{\alpha}{1 - i\Omega T_2} \frac{v_1 - v_2}{1+v_0^2}.$$

Here we put $\alpha = (\omega/c)\sqrt{\varepsilon_0}\beta$.

Equations (5.3) are easily integrated

$$\frac{v_0^2(z) - v_0^2(0)}{2} + \ln \frac{v_0(z)}{v_0(0)} = \alpha z, \tag{5.4}$$

$$v_1(z) = \left[\frac{v_0(z)}{v_0(0)}\right]^{\frac{1}{1-i\Omega T_2}} \left\{\frac{v_1(0) + v_2(0)}{2} \left[\frac{(1 - i\Omega T_1)(1 - i\Omega T_2) + v_0^2(0)}{(1 - i\Omega T_1)(1 - i\Omega T_2) + v_0^2(z)}\right]^{1/2 + \frac{1/2}{1-i\Omega T_2}} + \frac{v_1(0) - v_2(0)}{2}\right\}, \tag{5.5}$$

$$v_2(z) = \left[\frac{v_0(z)}{v_0(0)}\right]^{\frac{1}{1-i\Omega T_2}} \left\{\frac{v_1(0) + v_2(0)}{2} \left[\frac{(1 - i\Omega T_1)(1 - i\Omega T_2) + v_0^2(0)}{(1 - i\Omega T_1)(1 - i\Omega T_2) + v_0^2(z)}\right]^{1/2 + \frac{1/2}{1-i\Omega T_2}} - \frac{v_1(0) - v_2(0)}{2}\right\}. \tag{5.6}$$

Formula (5.4) for a strong field corresponds to the result of paper [42], which was devoted to amplification of a monochromatic field. Formulas (5.5) and (5.6) describe the propagation of interacting weak fields.

Formulas (5.4)-(5.6) are valid not only for inversion of the medium, but also for the absence of inversion ($\alpha < 0$).

In addition, if the weak field contains more than two spectral components the obtained formulas are also applicable and by v_1 and v_2 we mean any two weak components of symmetric frequencies.

2. It follows from formulas (5.5), (5.6) that if condition $v_1(0) + v_2(0) = 0$, is fulfilled at $z = 0$, then $v_1(z) + v_2(z) = 0$ for any z. It is also clear from (5.5), (5.6) that if $v_1(0) - v_2(0) = 0$, then $v_1(z) - v_2(z) = 0$. This means that in the case of frequency ($v_1 + v_2 = 0$) and amplitude ($v_1 - v_2 = 0$) modulation the form of the modulation is preserved.

If, however, $v_1(0) + v_2(0) \neq 0$ and $v_1(0) - v_2(0) \neq 0$, the relationship between v_1 and v_2 is altered. In particular, if the weak field is monochromatic at the input of the system, $v_1(0) \neq 0$, $v_2(0) = 0$, then a field of combination frequency $v_2(z) \neq 0$ will appear in the system.

We compare cases of frequency and amplitude modulation. If $v_1(0) = -v_2(0)$, it follows from (5.5), (5.6) that

$$v_1(z) = -v_2(z) = \left[\frac{v_0(z)}{v_0(0)}\right]^{\frac{1}{1-i\Omega T_2}} v_1(0). \tag{5.7}$$

If $v_1(0) = v_2(0)$, then

$$v_1(z) = v_2(z) = \left[\frac{v_0(z)}{v_0(0)}\right]^{\frac{1}{1-i\Omega T_2}} \left[\frac{(1-i\Omega T_1)(1-i\Omega T_2)+v_0^2(0)}{(1-i\Omega T_1)(1-i\Omega T_2)+v_0^2(z)}\right]^{1/2+\frac{1/2}{1-i\Omega T_2}} v_1(0). \tag{5.8}$$

In the case of frequency modulation (5.7) the amplitude gain of the side components differs from that of the central frequency only by the index $1/(1 - i\Omega T_2)$, which characterizes the departure from resonance.

In the case of amplitude modulation (5.8) the amplitude gain of the side components differs from the corresponding gain in (5.7) by the factor $\left[\frac{(1-i\Omega T_1)(1-i\Omega T_2)+v_0^2(0)}{(1-i\Omega T_1)(1-i\Omega T_2)+v_0^2(z)}\right]^{1/2+\frac{1/2}{1-i\Omega T_2}}$. This factor is due to the nonlinear interaction of weak fields of symmetric frequencies in an active medium in the presence of a strong field. In the case where the strong-field intensity at the input is sufficiently great, viz.

$$v_0^2(0) \geqslant v_m^2,$$

$$v_m^2 = \frac{1}{2}\left[\Omega^2 T_1 T_2 + \sqrt{\Omega^4 T_1^2 T_2^2 + 4(1+\Omega^2 T_1^2)}\right], \tag{5.9}$$

the side components of the amplitude-modulated signal are not amplified in propagation in the amplifying medium, but are attenuated. When $v_0^2(0) < v_m^2$ the side components are amplified until $v_0^2(z)$ becomes equal to v_m^2, and then they begin to decay.

The special features of propagation of amplitude-modulated signals are illustrated in the figures. We consider the relationship between $g = G^{\frac{1}{1-i\Omega T_2}}\left[\frac{(1-i\Omega T_1)(1-i\Omega T_2)+v_0^2(0)}{(1-i\Omega T_1)(1-i\Omega T_2)+Gv_0^2(0)}\right]^{1+\frac{1}{1-i\Omega T_2}}$ (gain for squares of amplitudes of side components) and $G = v_0^2(z)/v_0^2(0)$ (gain of intensity of weak field). Figure 12 shows curves for the modulus of g and Fig. 13 shows curves for $\arg(\sqrt{g})$. The behavior of the graphs depends greatly on the ratio $v_0^2(0)/v_m^2$, where v_m^2 is given by formula (5.9). The figures show the dependence of g on G for values of the frequency difference Ω which satisfy the condition $T_1^{-1} \ll \Omega \ll T_2^{-1}$. When this condition holds, the graphs drawn for different Ω but for the same values of $v_0^2(0)/v_m^2$ are practically the same. All the graphs for $|g|$ with $G > 1$ lie below the straight line $|g| = G$; this means that the gain for the side components is always less than that for the strong field. The curves for $|g|$ attain their maxima when $G = v_m^2/v_0^2(0)$. When $G \ll v_m^2/v_0^2(0)$ the differences between the gain for the side components and G are insignificant; when $G \gtrsim v_m^2/v_0^2(0)$ the differences become significant. The graphs also show cases of attenuation of the side components in the active medium if $G > v_m^2/v_0^2(0) > 1$.

Since formulas (5.4)-(5.6) are valid not only for an inverse medium, but also for the case of absence of inversion, the figures show the course of graphs in the region $G > 1$ and $G < 1$. The graphs, in particular, show cases of amplification of the side components in an absorbing medium.

The special features of the phase variation of the amplitude-modulated signal consist in the fact that there is an advance in phase in addition to the geometric advance. This also depends on the strong-field amplification and the saturation at the input, and is different for different modulation frequencies, as Fig. 13 illustrates.

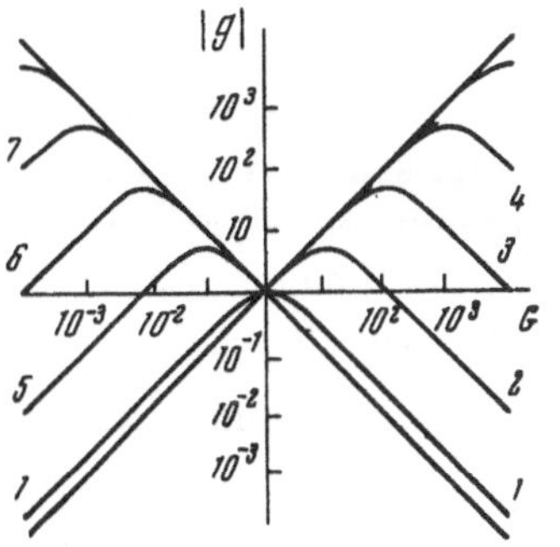

Fig. 12. Dependence of $|g|$ on G. 1) $v_0^2(0) = v_m^2 \simeq \Omega T_1$; 2) $v_0^2(0)/\Omega T_1 = 10^{-1}$; 3) 10^{-2}; 4) 10^{-3}; 5) 10; 6) 10^2; 7) 10^3.

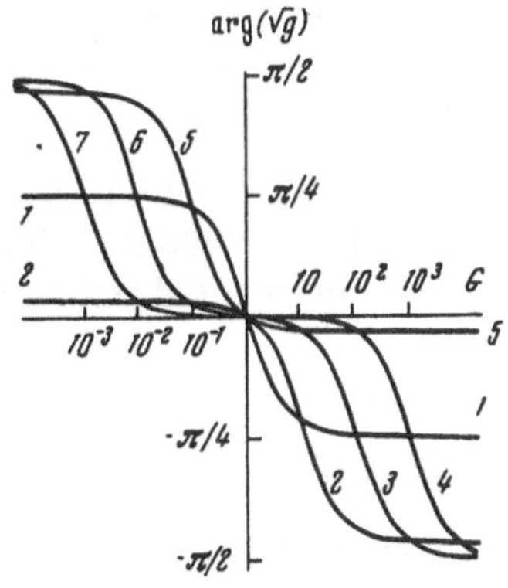

Fig. 13. Dependence of arg $(\sqrt{g})$ on G. The figures correspond with those in Fig. 12.

3. The differences in the gains G, $G^{1/(1-i\Omega T_2)}$ and g mean that the modulated signals propagated in amplifiers are distorted. Since in the case of amplitude modulation the gain of the side components $|g|$ in all the considered cases is less than the gain of the central component G, the relationship between the side components and the central one at the output of the amplifier differ from the relationship at the input: the modulation index is reduced. In addition, when there are many side components in the signal spectrum the relationships between the different side components are altered owing to the dependence of g on the frequency modulation.

If the input signal contains only one side component $\omega + \Omega$, a field of combination frequency $\omega - \Omega$ will appear at the output owing to the difference in the values of G and $G^{1/(1-i\Omega T_2)}$.

These distortions are due to the nonlinear interactions of weak fields in the presence of a strong field.

At modulation frequencies $\Omega \sim T_2^{-1}$ coupling of the weak fields occurs only when the saturation $v_0^2 \sim T_1/T_2$ [see (1.16) and (5.9)]. For instance, in paramagnetic masers $T_1/T_2 \sim 10^5$ and intensities of the order of T_1/T_2 are not attained in real systems. In such a case, when $\Omega \sim T_2^{-1}$ the gains of the side components of the amplitude-modulated and frequency-modulated signals are the same, $g = G^{1/(1-i\Omega T_2)}$. The distortions in this case are similar to the distortions in a linear amplifier with a gain G, which depends, however, on the saturation (5.4).

At modulation frequencies $\Omega \ll T_2^{-1}$ and real values of saturation the special features of the gain of the amplitude-modulated signal may be manifested, but frequency-modulated signals are not distorted when $\Omega \ll T_2^{-1}$.

The differences in the gains of amplitude- and frequency-modulated signals are greatest at low frequencies, $\Omega \lessdot T_1^{-1}$; here these differences are manifested when the saturation factor $v_0^2 \approx 1$.

As an example we consider an amplifier for which $T_1 = 10^{-3}$ sec and $T_2 = 10^{-8}$ sec, and the gain in linear operation $G(v_0^2(0) \to 0) = G_0 = 100$. These figures are typical of paramagnetic masers. We assume that there is an amplitude-modulated signal at the input and the modulation frequency is 1 kHz, $\Omega = 2\pi \cdot 10^3$ sec^{-1}. If saturation is such that the gain G at the central frequency is 2, the side components in these conditions are not amplified and are not absorbed, and $|g| = 1$. The modulation index of the signal is halved in this case. If in the case of the same saturation at the input there is only one side component at a distance of 1 kHz from the central one, then at the output there will be a component symmetric to it and with an intensity 4% of that of the initial weak component.

The question of nonlinear distortions of signals in masers has been considered in the literature [56, 57]. In [57], however, the case of low carrier intensities $v_0^2 \ll 1$ was considered. The calculation carried out in [57] showed that the intensities of the arising combination components were 9-10 orders lower than the intensities of the fundamental frequency components. Specially conducted experiments [58, 59] confirmed this result.

As was shown above, saturation of masers gives rise to stronger nonlinear effects.

In [56] a saturated amplifier was considered and the reduction in the modulation index of the amplitude-modulated signal was calculated. The calculation was made on the assumption that the gain is quasi-statically determined by the signal amplitude at the input. This assumption is valid only at low modulation frequencies (small in comparison with the population relaxation constant). Hence, the results of [56] are applicable only when $\Omega \ll T_1^{-1}$.

As already observed, the manifestation of nonlinear effects depends significantly on the modulation frequency and at low Ω the effects are greatest. Hence, the results given in [56] are overestimates of the nonlinear distortions for modulation frequencies $\Omega > T_1^{-1}$.

Thus, nonlinear interaction of weak fields in the presence of a strong field in an active medium leads to distortion of the signal spectrum. The results obtained here can be used to determine signal distortions in amplifiers in various specific cases.

Summary

1. The question of calculating the polarization of an active medium in a strong electromagnetic field was considered. The polarization of a two-level quantum-mechanical system with three different relaxation constants in a field containing a monochromatic strong component (ω) and several weak components ($\omega + \Omega_j$) was calculated. The polarization components $p_{\omega-\Omega_j}$ on the combination frequencies ($\omega - \Omega_j$) were determined.

We showed that if the components of the longitudinal and transverse relaxations have a different order of magnitude, the ratio $p_{\omega-\Omega}/p_{\omega+\Omega}$ depends significantly on the frequency difference Ω of the weak and strong fields. At small Ω saturation of the order of unity is sufficient for $|p_{\omega-\Omega}|$ to be comparable with $|p_{\omega+\Omega}|$. At larger frequency differences greater values of saturation are required for this.

2. Owing to the appearance of combination components in the polarization, weak fields of symmetric frequencies must be considered together. Equations for weak fields interacting through polarization in the presence of a strong field were obtained.

3. Boundary-value problems for systems with an active medium were posed. A method of solution of one-dimensional boundary-value problems based on the smallness of the nonlinear part of the polarization was developed. We showed that the field can be represented in the form of two waves with amplitudes which vary little over a wavelength. A system of equations was obtained for amplitudes averaged over a wavelength.

4. Monochromatic generation in a one-dimensional laser model in the case of homogeneous line broadening was calculated. We showed that the natural frequencies are independent of the excitation level owing to the fact that in the adopted model of the active medium the ratio $\Delta\varepsilon''/\Delta\varepsilon'$ is independent of the field. We found that the output power depends nonlinearly on the excitation level and is less than the value given by calculations in which the inhomogeneity of the field and medium is ignored.

5. We showed that owing to inhomogeneities of the medium due to saturation the field distribution in a plane active resonator may differ from the distribution in an empty plane-parallel layer. These changes are small when the reflection coefficients on the boundaries of the layer are high (for any value of the excitation parameter the field is almost a standing wave). When the reflection coefficients on the boundaries are low the field is considerably distorted with change in the excitation parameter and in the case of high excitation may differ significantly from the distribution in an empty plane-parallel layer.

6. Using a simple model we investigated the effect of inhomogeneity of inversion in a direction perpendicular to the direction of generation on the properties of the natural oscillations. We calculated the natural oscillations of a layer in which a hypergeometric dependence of the imaginary part of the dielectric constant on the transverse coordinate was assigned. We showed that for oscillations of low transverse numbers the role of the inhomogeneity is given by the parameter $\frac{\delta}{2}\frac{\omega_0^2}{c^2}h^2$, where δ is the maximum inhomogeneity, and h is the width of the inhomogeneity. When $\frac{\delta}{2}\frac{\omega_0^2}{c^2}h^2 \gtrsim 1$ the inhomogeneity has an appreciable focusing effect on the field and considerably alters the generation condition.

For a width of inhomogeneity of the order of millimeters and a frequency in the optical range the parameter $\frac{\delta}{2}\frac{\omega_0^2}{c^2}h^2$ becomes unity when $\delta \sim 10^{-8}$, i.e., the slightest inhomogeneities of inversion can lead to considerable alteration of the natural oscillations of active resonators.

7. The boundary value problem for weak fields in a one-dimensional active resonator containing a strong monochromatic field was posed. An equation was obtained for the natural frequencies with due allowance for the interaction of weak fields through polarization and inhomogeneity of inversion due to saturation. We showed that in the general case the natural frequencies can differ significantly from the natural frequencies of an empty plane-parallel layer.

8. Weak fields increasing in time were found in conditions typical for a ruby laser. We showed that in such conditions the instability of monochromatic generation is entirely due to the difference in the spatial distribution of fields of different frequencies.

9. We investigated the propagation of a signal containing a strong spectral component of resonate frequency and several weak components in an active medium in a traveling wave system. We showed that the frequency (amplitude)-modulated signal remains frequency (amplitude)-modulated.

10. We showed that the interaction of weak fields through polarization in the presence of a strong field leads to distortion of the signal spectrum in saturation conditions. In the case of frequency modulation the signal distortions are similar to those in a linear amplifier with the gain reduced due to saturation. In the case of amplitude modulation specific nonlinear distortions are produced.

I express my deep thanks to S. G. Rautian for discussion and to M. M. Sushchinskii for interest in this work.

Literature Cited

1. R. Karplus and J. Schwinger, Phys. Rev., 73:1020 (1948).
2. N. G. Basov and A. M. Prokhorov, Dokl. Akad. Nauk SSSR, 101:47 (1955).
3. Yu. L. Klimontovich and R. V. Khokhlov, Zh. Éksp. Teor. Fiz., 32:1151 (1957) [Sov. Phys. – JETP, 5:937 (1957)].

4. V. M. Fain, Zh. Éksp. Teor. Fiz., 33:945 (1957) [Sov. Phys. – JETP, 6:726 (1958)].
5. A. N. Oraevskii, Radiotekhnika i élektronika, 4:719 (1959).
6. A. P. Ivanov, B. I. Stepanov, B. M. Berkovskii, and I. L. Katsev, Dokl. AN BSSR, 6:147 (1962).
7. B. I. Stepanov, Dokl. AN BSSR, 6:629 (1962).
8. B. I. Stepanov, Dokl. AN SSSR, 148:74 (1963).
9. T. I. Kuznetsova and S. G. Rautian, Zh. Éksp. Teor. Fiz., 43:1897 (1962) [Sov. Phys. – JETP, 16:1338 (1963)].
10. Kh. Yu. Khaldre and R. V. Khokhlov, Izv. vuzov, Radiofizika, 1(5-6):60 (1958).
11. A. S. Gurtovnik, Izv. vuzov. Radiofizika, 1(5-6):83 (1958).
12. A. N. Oraevskii, Masers [in Russian], Nauka, Moscow (1964).
13. V. N. Lugovoi, Radiotekhnika i élektronika, 6:1700 (1961).
14. W. R. Bennett, Phys. Rev., 126:580 (1962).
15. T. I. Kuznetsova and S. G. Rautian, Fiz. Tverd. Tela, 5:2105 (1963) [Sov. Phys. – Solid State, 5(8):1535 (1964)].
16. C. L. Tang, H. Statz, and G. A. de Mars, Appl. Phys. Lett., 2:222 (1963).
17. L. A. Ostrovskii and E. I. Yakubovich, Zh. Éksp. Teor. Fiz., 46:963 (1964) [Sov. Phys. – JETP, 19:656 (1964)].
18. L. A. Ostrovskii and E. I. Yakubovich, Izv. vuzov. Radiofizika, 8:91 (1965).
19. C. L. Tang, H. Statz, and G. A. de Mars, J. Appl. Phys., 34:2289 (1963).
20. Yu. A. Anan'ev and B. M. Sedov, Zh. Éksp. Teor. Fiz., 48:782 (1965) [Sov. Phys. – JETP, 21:517 (1965)].
21. B. L. Lifshits and V. N. Tsikunov, Zh. Éksp. Teor. Fiz., 49:1843 (1965) [Sov. Phys. – JETP, 22:1260 (1966)].
22. T. I. Kuznetsova and S. G. Rautian, Izv. vuzov. Radiofizika, 7:682 (1964).
23. A. G. Fox and T. Li, Bell Syst. Tech. J., 40:453 (1961).
24. G. D. Boyd and J. P. Gordon, Bell Syst. Tech. J., 40:489 (1961).
25. L. A. Vainshtein, Zh. Éksp. Teor. Fiz., 44:1050 (1963) [Sov. Phys. – JETP, 17:709 (1963)].
26. L. A. Vainshtein, Zh. Éksp. Teor. Fiz., 45:684 (1963) [Sov. Phys. – JETP, 18:471 (1964)].
27. L. A. Vainshtein, Zh. Tekh. Fiz., 34:193 (1964) [Sov. Phys. – Tech. Phys., 9:157 (1964)].
28. T. I. Kuznetsova, Zh. Tekh. Fiz., 34:419 (1964) [Sov. Phys. – Tech. Phys., 9:330 (1964)].
29. T. I. Kuznetsova, Zh. Tekh. Fiz., 36:58 (1966) [Sov. Phys. – Tech. Phys., 11:40 (1966)]; Preprint FIAN, A-19 (1965).
30. S. G. Rautian and I. I. Sobel'man, Zh. Éksp. Teor. Fiz., 41:456 (1961) [Sov. Phys. – JETP, 14:328 (1962)].
31. T. A. Germogenova and S. G. Rautian, Zh. Éksp. Teor. Fiz., 46:745 (1964) [Sov. Phys. – JETP, 19:507 (1964)].
32. S. G. Rautian and T. A. Germogenova, Opt. Spectrosk., 17:157 (1964) [Opt. Spectrosc., 17:85 (1964)].
33. J. R. Fontana, R. H. Pantell, and R. G. Smith, J. Appl. Phys., 33:2085 (1962).
34. B. Senitsky and G. Gould, Proceedings of Third International Conference on Quantum Electronics (1964), p. 1751.
35. B. Senitzky, G. Gould, and S. Cutler, Phys. Rev., 130:1460 (1963).
36. B. Senitzky and S. Cutler, Microwave J., 11(1):62 (1964).
37. D. N. Klyshko, Yu. S. Konstantinov, and V. S. Tumanov, Izv. vuzov. Radiofizika, 8:513 (1965).
38. T. I. Kuznetsova and S. G. Rautian, Zh. Éksp. Teor. Fiz., 49:1605 (1965) [Sov. Phys. – JETP, 22:1098 (1966)].
39. N. V. Karlov and T. I. Kuznetsova, Radiotekhnika i élektronika, 12:284 (1967).
40. N. G. Basov, O. N. Krokhin, and Yu. M. Popov, Usp. Fiz. Nauk, 72:161 (1960) [Sov. Phys. – Usp., 12(5):1033 (1961)].

41. B. I. Stepanov, Dokl. AN BSSR, 5:489 (1961).
42. N. V. Karlov, Yu. P. Pimenov, and A. M. Prokhorov, Radiotekhnika i élektronika, 6:410 (1961).
43. W. E. Lamb, Phys. Rev., 134A:1429 (1964).
44. S. A. Akhmanov and R. V. Khokhlov, Problems of Nonlinear Optics [in Russian], Izd. VINITI, Moscow (1964).
45. V. M. Fain and Ya. I. Khanin, Quantum Radiophysics [in Russian], Izd. Sovetskoe Radio, Moscow (1965).
46. W. R. Bennett and J. W. Knutson, Bull. Am. Phys. Soc., 9:500 (1964).
47. C. H. Cooke, J. McKenna, and J. G. Skinner, Appl. Optics, 3:957 (1964).
48. J. G. Skinner, Appl. Optics, 3:963 (1964).
49. K. Rawer, Ann. Phys., 35:385 (1939); 42:294 (1942).
50. L. M. Brekhovskikh, Waves in Layered Media [in Russian], Izd. AN SSSR, Moscow (1957).
51. V. L. Ginzburg, Propagation of Electromagnetic Waves in a Plasma [in Russian], Fizmatgiz, Moscow (1960).
52. L. D. Landau and E. M. Lifshits, Quantum Mechanics [in Russian], Fizmatgiz, Moscow (1963).
53. J. McWarter, Solid-State Electronics, 6:417 (1963).
54. C. L. Tang, H. Statz, G. A. de Mars, and D. T. Wilson, Phys. Rev., 136:A1 (1964).
55. V. Evtuchov, Appl. Phys. Lett., 6:141 (1965).
56. L. L. Moskvitin and Yu. E. Naumov, Radiotekhnika i élektronika, 9:2105 (1964).
57. E. O. Schultz-DuBois, Proc. IEEE, 52:644 (1964).
58. W. J. Tabor, F. S. Chen, and E. O. Schultz-DuBois, Proc. IEEE, 52:656 (1964).
59. F. Bosch, Z. angew. Phys., 18:254 (1965).

ELECTRODYNAMICS OF LASERS WITH AN OPEN RESONATOR AND INHOMOGENEOUS FILLING

A. F. Suchkov

The field produced inside a solid-state laser is usually represented as a superposition of the normal modes of the empty resonator. It is assumed in this case that the complex dielectric constant of the active medium filling the resonator is homogeneous in space. However, the inhomogeneity of the crystals used in real lasers, the unsteady output, and pumping inhomogeneity have the result that the real and imaginary parts of the complex dielectric constant are, generally speaking, functions of the coordinates and time. Spatial inhomogeneities significantly affect the operation of solid-state lasers.

An investigation of the effect of inhomogeneities of the active medium in a direction transverse to the laser axis is of most interest for the study of the transverse structure of the field, the angular divergence of the emission, the coherence of the emission, and the steadiness of the oscillations.

An effective method of calculating the spatial and temporal structure of the generation field in the near- and far-field regions in the presence of inhomogeneities of the complex dielectric constant in a direction perpendicular to the laser axis was developed in [1].

In [1] there was obtained a system of two equations describing the dynamics of the field within a single axial mode and the inverse population in an infinite plane-parallel layer with ideally reflecting boundaries:

$$\frac{\partial F}{\partial t}=\frac{ic^2}{2\varepsilon'(0)\omega_0}\frac{\partial^2 F}{\partial x^2}-\frac{i\omega_0}{2\varepsilon'(0)}(\delta\varepsilon'(x,t)-i\varepsilon''(x,t))F, \tag{1a}$$

$$\frac{\partial N}{\partial t}=n\left(\rho(x)-\frac{1}{\tau}\right)-N\left(\rho(x)+\frac{1}{\tau}\right)-2W_iN|\overline{F^2(x,t)}|. \tag{1b}$$

Here F = F(x, t) is the slowly varying complex field amplitude. The x axis is directed along the layer. The dielectric constant is assumed to be homogeneous in the direction of the z axis (in a direction perpendicular to the boundaries of the layer) and has the form

$$\varepsilon(x,t)=\varepsilon'(0)+\delta\varepsilon'(x)+i\varepsilon''(x,t). \tag{2}$$

In the equation for the active medium N = N(x, t) is the inverse population, n is the concentration of active atoms, τ is the lifetime of the atom on the upper level for spontaneous emission, W_i is the induced emission probability, and $\rho(x, t)$ is the probability of transition of an atom to the excited state by the pumping field. Radiation losses on reflection at the mirrors, which occur in real lasers, are assumed to be uniformly distributed over the thickness of the layer and are included in the imaginary part of the dielectric constant. The deduction of the equation of the field dynamics entailed the use of the usual properties of laser emission: high monochromaticity and sharp directionality. We can thus seek the solution of the Maxwell equations in the form of a superposition of standing waves, the amplitudes and phases of which along the layer are slowly varying functions of the coordinates and time, so that the field inside the resonator has the form

$$E(x, z, t) = F(x, t) \sin \frac{\pi n z}{L} e^{-i\omega_0 t}. \quad (3)$$

Equations (1) can easily be extended to the case of a bounded layer (crystal) with totally reflecting side surfaces. If the active medium is a strip bounded by the planes $|x| = R$, then equation (1a) must be solved with boundary conditions corresponding to totally reflecting side surfaces:

$$F(R) = F(-R) = 0. \quad (4)$$

In this case the solution of equation (1a) can be sought in the form of a superposition of normal modes of the empty resonator with complex amplitudes $a_k(t)$, which depend on the time. In [1] an infinite system of equations for the amplitudes $a_k(t)$ was obtained:

$$\frac{da_k(t)}{dt} = \frac{\omega_0}{2\varepsilon'(0)} \sum \varepsilon''_{km}(t) e^{i(\Omega_k - \Omega_m)t} a_m(t), \quad (5)$$

$$\varepsilon''_{km} = \int_{-R}^{R} u_k(x) \varepsilon''(x, t) u_m(x) dx.$$

Here $u_k(x)$, and Ω_k are the eigenfunctions and the eigenvalues of equation (1a) with $\varepsilon''(x, t) \equiv 0$ and boundary conditions (4).

This system of equations can be used to calculate the characteristics of the emission in closed resonant cavities when diffraction effects on the edges of the mirrors can be neglected. This can be done if the transverse dimensions of the field are limited by the spatial distribution of the active medium, and not by diffraction effects on the edges of the mirrors (see [2], [3], for instance). However, the method developed in [1] cannot be used to calculate the operating conditions of a laser if diffraction reflections at the edges of the mirrors play an important role. The consideration of these diffraction effects provides the basis for the well-developed theory of open resonators. This theory enables us to find the complex frequencies Ω_k and eigenfunctions $F_k(x)$ (field-amplitude and phase distributions on the surface of the mirrors) of the modes of the empty resonator. However, the eigenfunctions $F_k(x)$ do not have the property of orthogonality and completeness. Hence, the generation field in an open resonator with inhomogeneous filling cannot be represented in the form of a superposition of normal modes of the open resonator. The edge effects and inhomogeneities must be considered simultaneously.

In this paper, which is based on L. A. Vainshtein's rigorous solution of the problem of reflection of an electromagnetic wave at the open end of a waveguide, we consider the question of the relative influence of inhomogeneity of the dielectric constant and edge effects and extend the previously developed method of calculating the spatial and temporal characteristics of the laser to the case where these effects must be taken into account.

We will consider edge effects within the framework of the following model. In an infinite plane-parallel layer we introduce a discontinuity of the dielectric constant such that the reflection coefficient for it is as close as possible to that of the open end of the waveguide. A region of an infinite plane-parallel layer bounded on both sides by a discontinuity of dielectric constant we will regard as an open resonator with inhomogeneous filling.

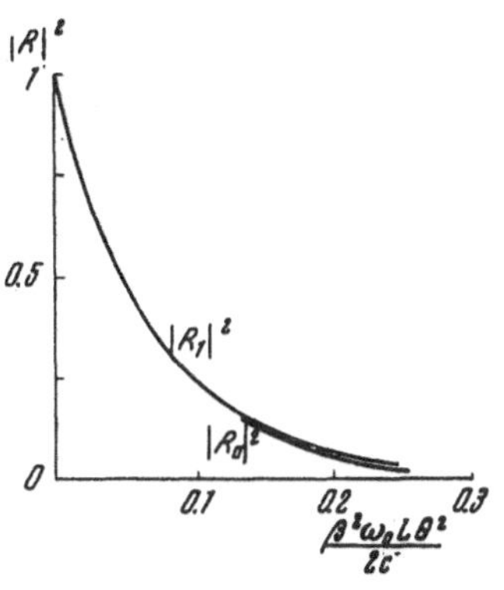

Fig. 1. Reflection coefficients $|R_0|^2$ and $|R_1|^2$ as functions of angle of incidence.

Let the wave number $k = 2\pi/\lambda$ be connected with the distance L between the mirrors by the relationship

$$k = \frac{\pi}{L}(n + 2p), \tag{6}$$

where n is a large number equal to the number of half-waves between the plates of the waveguide, $|p| < 1/2$, and is connected with the wave parameters by the relationships

$$k_x^2 = \frac{4\pi^2 np}{L^2} = \frac{2\omega_0\omega_h}{c^2} = \left(\frac{2\pi}{\Lambda}\right)^2, \tag{7}$$

where k_x and Λ are the component of the wave vector and the wavelength, respectively, along the waveguide, ω_0 is the critical frequency, and $(\omega_0 + \omega_h)$ is the frequency of the incident wave. At small $|p|$ the reflection coefficient for such a wave on the open end of the waveguide has the form [4]

$$R_0 = -e^{i\beta(i+1)\sqrt{4\pi p}}, \tag{8}$$

where $\beta = 0.824$.

If the infinite layer has a discontinuity of dielectric constant $\Delta\varepsilon \ll 1$, a wave propagating at a small angle $\alpha = k_x/k_z$ to the interface (glancing incidence), is reflected with a reflection coefficient R_1, determined by the Fresnel formulas [5]. Assuming that $\varepsilon_1, \varepsilon_2 \sim 1$; $|\Delta\varepsilon| = |\varepsilon_2 - \varepsilon_1| \ll 1$, $\alpha \ll 1$, we have, irrespective of the polarization of the incident wave,

$$R_1 = \frac{\alpha - \sqrt{\Delta\varepsilon + \alpha^2}}{\alpha + \sqrt{\Delta\varepsilon + \alpha^2}}. \tag{9}$$

Expanding R_0 and R_1 in a series to the first linear term in $\alpha = (4p/n)^{1/2}$ and equating the two expressions, we obtain

$$\Delta\varepsilon = \frac{2ic}{\beta^2\omega_0 L}. \tag{10}$$

Thus, for the value of $\Delta\varepsilon$ given by expression (10) the reflection coefficients for the incident wave at the open end of the waveguide and on the discontinuity $\Delta\varepsilon$ of the dielectric constant in the infinite layer are the same in a linear approximation.

Let the dielectric constant in the infinite plane-parallel layer have the form

$$\varepsilon(x) = \begin{cases} 1 + \delta\varepsilon'(x) + i\varepsilon_0'' - i\varepsilon''(x, t), & |x| < a, \\ 1 + \delta\varepsilon'(-a) + i\varepsilon_0'' + \dfrac{2ic}{\beta^2\omega_0 L}, & x < a, \\ 1 + \delta\varepsilon'(a) + i\varepsilon_0'' + \dfrac{2ic}{\beta^2\omega_0 L}, & x > a. \end{cases} \tag{11}$$

Here ε_0'' is responsible for the losses on reflection at the mirrors, and $\varepsilon''(x, t)$ is determined by the inverse population. Then the system of equations (1) will correctly describe the spatial and temporal picture of the field in an open resonator with inhomogeneous filling so long as the diffraction losses are small. However, as can be seen from Fig. 1, where $|R_0|^2$ and $|R_1|^2$ are shown as functions of the angle of incidence $\theta = k_x/k_z$, there is an appreciable difference in the reflection coefficients R_0 and R_1 only in the region where their absolute value is small and the reflected wave can be neglected. Hence, the system of equations (1a) and (1b) will give a true qualitative description of the generation field in the case of large losses too.

The selected form of $\varepsilon(x)$ means strong absorption when $|x| > a$ ($|\Delta\varepsilon| = 2c/\beta^2\omega_0 L \sim 10^{-7}$) and the transmitted wave rapidly decays. Hence there is no need to consider an infinite layer.

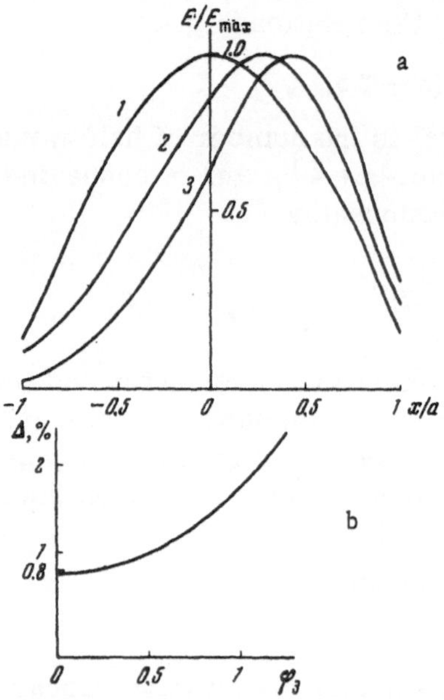

Fig. 2. Dominant mode of resonator with inclined mirrors. a) Field amplitude distribution for different angles of inclination (φ_3) of the mirrors: 1) $\varphi_3 = 0$; 2) 0.5"; 3) 1": diffraction losses Δ as a function of angle of inclination of mirrors (for one transit).

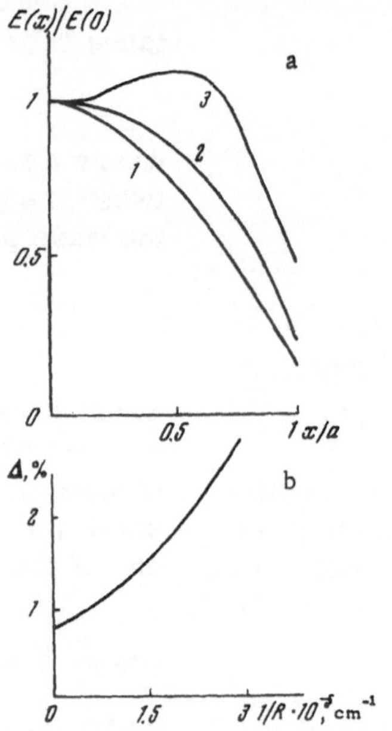

Fig. 3. Dominant mode of resonator with convex mirrors. a) Field amplitude distribution on mirror with radius of curvature: 1) $1/R = 0$; 2) $1.5 \cdot 10^{-5}$; 3) $3 \cdot 10^{-5}$; b) diffraction losses Δ as function of radius of curvature of mirrors (for one transit).

Equation (1a) can be solved in the bounded case $|x| = R > a$ with zero boundary conditions $F(\pm R) = 0$. The value of R depends on the specific conditions of the problem. It is essential, of course, that the transmitted wave actually decays.

We now seek the field in the form of a superposition of modes of the empty resonator possessing the property of orthogonality and completeness and with complex amplitudes, which depend on the time

$$F(x, t) = \sum a_k(t)\, u_k(x)\, e^{-i\Omega_k t}, \tag{12}$$

where

$$u_k(x) = \begin{cases} \frac{1}{\sqrt{R}} \cos \frac{\pi k x}{2R} & (k = 1, 3, 5 \ldots), \\ \frac{1}{\sqrt{R}} \sin \frac{\pi k x}{2R} & (k = 2, 4, 6 \ldots), \end{cases}$$

$$\Omega_k = \frac{1}{2\omega_0}\left[\frac{\pi c k}{2R}\right]^2,$$

and obtain a system of equations for the amplitudes $a_k(t)$, similar to (5) $[a_k(t) = A'_k(t) + iA''_k(t)]$:

$$\begin{aligned} \frac{dA'_k(t)}{dt} &= \frac{\omega_0}{2}\sum A'_m(t)\, \varepsilon''_{km}(t) + \frac{\omega_0}{2}\sum A''_m(t)\, \varepsilon'_{km} + (\Omega_k - \Omega_1)\, A''_k(t), \\ \frac{dA''_k(t)}{dt} &= \frac{\omega_0}{2}\sum A''_m(t)\, \varepsilon''_{km}(t) - \frac{\omega_0}{2}\sum A'_m(t)\, \varepsilon'_{km} - (\Omega_k - \Omega_1)\, A'_k(t). \end{aligned} \tag{13}$$

The system of equations (13) in conjunction with the equation for the inverse population (1b) can be used for specific calculations of the operation of lasers with an open resonator of any almost plane-parallel form with any kind of pumping.

We demonstrate this method by some specific examples.

1. When ε_0'' and $\varepsilon''(x, t)$ are zero [see (1.1)], the steady-state solutions of system (13) will describe the normal modes of an open resonator, the geometry of which depends on the inhomogeneity of the refractive index $\delta\varepsilon'(x)$. The system of equations (13) was integrated numerically on a M-20 electronic digital computer with $k_{max} = 21$. Figure 2a shows the obtained field amplitude distributions on the mirror in a resonator with inclined mirrors for the dominant mode, while Fig. 2b shows the diffraction losses as a function of the angle of inclination of the mirrors. The calculations were made with the following parameters: mirror width $2a = 1/3$ cm, distance between mirrors L = 75 cm, $\omega_0 = 3 \cdot 10^{15}$ sec^{-1}. This corresponds to a number of Fresnel zones $N = a^2/L\lambda \approx 5.6$.

As was to be expected, the results of the calculations agree completely with those of Fox and Li [6, 7] and are also in complete correspondence with Vainshtein's theory of open resonators. Figure 3a shows a more interesting case, where the inhomogeneity of the refractive index corresponds to a negative lens (or a resonator with cylindrical mirrors with their convex sides facing one another). The calculation was made with the same values of parameters as in Fig. 2. The relationship between the diffraction losses and the curvature 1/R of the mirrors is shown in Fig. 3b. A characteristic feature of the field amplitude distribution on the mirror is the dip in the center of the resonator at sufficiently high radii of curvature. As in the case of a resonator with inclined mirrors, the field is concentrated in the "broader" part of the resonator.

2. The developed method of numerical integration of the system of equations for the amplitudes $a_k(t)$ was used to calculate the spatial and temporal characteristics of the emission of a laser with plane mirrors inclined at a small angle to one another. The calculation was made with the following values of laser parameters, which are close to the parameters of the ruby laser used by Korobkin and Leontovich to investigate the fine structure of the peak and beating of angular modes: mirror width $2a = 0.25$ cm, distance between mirrors L = 75 cm, $\varepsilon_0'' = 1/3 \cdot 10^{-7}$ (corresponding lifetime of photon in resonator $T_0 = 10^{-8}$ sec), excess of pumping over threshold 50%; the initial distribution of the inverse population and distribution of the pumping light intensity are almost homogeneous and are shown in Fig. 4e.

The angle of inclination of the mirrors is $\theta = 1.5''$, $\varepsilon''(x, t)$ in the center of the amplification line is determined from the inverse population by the relationship

$$\varepsilon''(x, t) = \frac{4\pi e^2 f N(x, t)}{m\gamma\omega_0}. \quad (14)$$

The calculations were made with the following parameters, which are typical for ruby: $f = 10^{-5}$, $\omega_0 = 3 \cdot 10^{15}$ sec^{-1}, $\gamma = 3 \cdot 10^{12}$ sec^{-1}, $n = 2 \cdot 10^{19}$ cm^3, $\tau = 3 \cdot 10^{-3}$ sec. Random initial values of the amplitudes $a_k(t)$ were chosen to correspond with the spontaneous emission intensity. Figure 4a shows the total field energy W in the resonator as a function of time. In the first three peaks up to the time t_1 the structure of the field on the mirror corresponds to the simplest mode of the laser. The field amplitude distribution $|E(x)|$ at instant t_1 is shown in Fig. 4b. As the figure shows, the centroid of the field distribution is displaced relative to the laser axis. The displacement is in the direction of the broad edge. The directionality graph of the emission is also displaced, as can be seen from Fig. 4g, which shows the field phase distribution $\alpha(x)$ at instant t_1. The directionality graph is also shifted in the direction of the "broad" edge of the resonator.

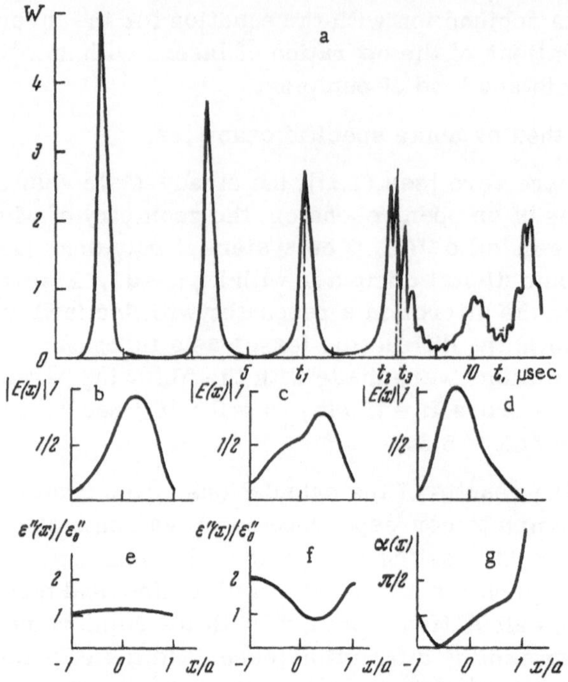

Fig. 4. Temporal and spatial characteristics of generation field in laser with inclined mirrors.

Physically the displacement of the field and directionality graph is due to the presence of an energy flux towards the broad edge. These results agree qualitatively with the results of experimental investigations (see [8], for instance).

Owing to nonuniform burnup of the active substance the distribution of the inverse population takes such a form (Fig. 4f) that generation of the next "mode" becomes possible. Two-frequency generation sets in and the subsequent peaks are modulated with a frequency close to the beat frequency of the modes of the empty resonator. The nature of the solution inside the peak can be determined by the following simple procedure. The system of equations (5) for a generator with two modes has the form

$$\begin{gathered}\frac{da_1}{dt} = a_2\varepsilon_{11} + a_2\varepsilon_{12}e^{-i\Omega t},\\ \frac{da_2}{dt} = a_1\varepsilon_{11}{}^{i\Omega t} + a_2\varepsilon_{22},\\ \varepsilon_{km} = -\frac{\omega_0}{2\,\varepsilon'(0)}\int_{-R}^{R} u_k^0\,\varepsilon''(x,\,t)\,u_m^0\,dx \qquad (m,\,k = 1,\,2),\\ \Omega = \Omega_2 - \Omega_1.\end{gathered} \tag{15}$$

We will find the linear generation regime, i.e., we assume that the active substance is not burned up, and let $\varepsilon_{11} = \varepsilon_{22} = 0$, $\varepsilon_{12} \neq 0$. In other words, the conditions for excitation of the two modes are fulfilled, but the modes interact through the substance. The system of equations (15) has two linearly independent solutions:

$$\begin{gathered}u_1(x,\,t) = e^{-\frac{i\Omega t}{2}+i\omega t}\left[u_1^0 - u_2^0\frac{i}{\varepsilon_{12}}\left(\frac{\Omega}{2} - \omega\right)\right],\\ u_2(x,\,t) = e^{\frac{i\Omega t}{2}-i\omega t}\left[u_1^0 < u_2^0\frac{i}{\varepsilon_{12}}\left(\frac{\Omega}{2} + \omega\right)\right],\end{gathered} \tag{16}$$

where $\omega = [(\Omega^2/4) - \varepsilon_{12}^2]^{\frac{1}{2}}$. The newly obtained modes are propagated into one another by $2\omega < \Omega$. Thus, we have attraction of the mode frequencies. When $\varepsilon_{12} = \Omega/2$ there is one frequency, and when $\varepsilon_{12} > \Omega/2$ one solution becomes increasing, and the other decaying. The general solution has the form

$$F(x, t) = A_1 u_1 + A_2 u_2.$$

Hence, for the value of $\int_{-R}^{R} F^* F dx$, which is proportional to the total field energy W in the resonator, we have

$$\int_{-R}^{R} F^* F dx \sim (A_1^2 + A_2^2) + \frac{4\varepsilon_{12}}{\omega} A_1 A_2 \cos 2\omega t. \tag{17}$$

In this very simple example we have nondecaying oscillations of total field energy with frequency 2ω in the resonator. Expression (17) has a simple physical meaning: Let $u_1^0 \sim \cos x$, $u_2^0 \sim \sin 2x$, $\varepsilon''(x) = \beta x$. The change in field energy in the resonator is determined by the sign of the integral $I = \int_{-R}^{R} E^2 \beta x dx$. When u_1^0 and u_2^0 are in phase, the major part of the field energy is in the region $x > 0$. Hence, integral I is positive and the total field energy increases. When u_1^0 and u_2^0 are in opposite phase, the field lies mainly in the region $x < 0$. Integral I is negative and the field energy decreases.

The fact that the numerical solution of the system of equations acquires a two-frequency nature follows from Fig. 4c and 4d, which show the field amplitude distributions at instant t_2, when the energy W is a maximum, and at t_3, when the total energy is a minimum. Thus, the nature of the solution within the peak is qualitatively similar to that obtained for a laser with two modes. It should be noted that the modulation frequency is not exactly equal to the beat frequency of the modes of the empty resonator, but also depends on the nonuniformity of the distribution of the inverse population (17). A fuller analysis of the results of integration showed that the modulation frequency decreases with the passage of time. For instance, in the eighth peak (not shown in Fig. 4) the modulation frequency is 20% less than in the fourth. This is due to the fact that interaction of the modes is increased with the passage of time owing to the increase in inhomogeneity of the distribution of inverse population. The obtained order of excitation of the laser modes, the field concentration in the broad part of the resonator, the displacement of the emission directionality graph towards the broad edge and the modulation of the total emission power within one intensity peak (fine structure of peak) are in good qualitative agreement with experimental results [8, 9].

Literature Cited

1. A. F. Suchkov, Zh. Éksp. Teor. Fiz., 49:1495 (1965) [Sov. Phys. – JETP, 22:1026 (1966)].
2. V. S. Letokhov and A. F. Suchkov, Zh. Éksp. Teor. Fiz., 50:1148 (1966) [Sov. Phys. – JETP, 23:764 (1966)].
3. V. S. Letokhov and A. F. Suchkov, Zh. Éksp. Teor. Fiz., 52:282 (1966) [Sov. Phys. – JETP, 25:182 (1967)].
4. L. A. Vainshtein, Zh. Éksp. Teor. Fiz., 44:1050 (1963) [Sov. Phys. – JETP, 17:709 (1963)].
5. L. D. Landau and E. M. Lifshits, Electrodynamics of Continuous Media [in Russian], Fizmatgiz, Moscow (1964).
6. A. G. Fox and T. Li, Proc. IRE, 48:1904 (1960).

7. A. G. Fox and T. Li, Bell Syst. Tech. J., 40:453 (1961).
8. V.L. Broude, V. V. Zaika, V. I. Kravchenko, and M. S. Soskin, Zh. prikl. spektroskopii, 3:3 (1962).
9. V. V. Korobkin and A. M. Leontovich, Zh. Éksp. Teor. Fiz., 49:7 (1965) [Sov. Phys. – JETP, 22:6 (1966)].

DYNAMICS OF A Q-SWITCHED LASER

V. S. Letokhov and A. F. Suchkov

Introduction

The most effective method of generating giant pulses of coherent light – "Q-modulation" ("Q-switching") of a laser – was proposed by Hellwarth [1]. The first theoretical investigations [1-6] of the energy, maximum power, and the rate of rise and decay of the giant pulse were based on equations balancing the total energy of the radiation field and the total number of active particles in the resonator. Such equations represent a simple model in which the density of the inverse population and the energy density of the radiation field are assumed homogeneous over the volume of the resonator. Hence, such a model ignores the electrodynamic effects responsible for the spatial and temporal development of the generation field, and the effect of the initial spatial distribution of the inverse population and the resonator geometry on the dynamics of pulse generation. The delay time, length, and shape of the giant pulse calculated for such a simplified model can differ considerably (even qualitatively [7]) from the actual values. As will be shown below, this is due to transverse development of the generation region during a time of the order of the length of the giant pulse. The need to consider the spatial development of the generation region in a Q-switched laser was pointed out in [8].

A systematic theoretical consideration of the spatial and temporal generation of a giant light pulse will entail the solution of the essentially nonlinear unsteady interaction of the many oscillation modes of the field in a resonator with an inhomogeneous inversely populated medium, the dimensions of which are much greater than the wavelength of the generated emission. The spatial and temporal development of the giant pulse depends significantly on the inhomogeneities of the inverse population density and the refractive index across the laser axis. An effective method of investigating the nonlinear unsteady interaction of many modes in the presence of inhomogeneity of the inverse population and refractive index was developed and used to examine the self-modulating regimes of a solid-state laser in [9]. In the present paper the investigation of the dynamics of a Q-switched laser is based on the same method. The main results are published in [10, 11].

The generation of the giant pulse can be divided into two phases. In the first phase after Q-switching the amplitude of the electromagnetic field in the oscillation modes increases exponentially from the spontaneous noise level to a level almost sufficient to saturate the gain of the medium, but no appreciable saturation occurs. Hence, the first phase of pulse development, the duration of which is called the delay time [2], can be called the linear generation region. During this phase a generation "jet" is established in the central region of thecrystal. During the second phase saturation of the gain of the medium occurs, i.e., the majority of active particles emit. During this time the giant pulse proper is generated. The second phase can be called the region of nonlinear development. During this phase the initial "jet" develops in a transverse direction and generation involves the whole crystal.

§ 1. Formulation of the Problem. Basic Equations

We will consider a model of a Q-switched laser which provides a complete representation of the electrodynamic effects in the development of a giant pulse. On the z axis (laser axis) the active medium is completely bounded by reflecting plane-parallel mirrors at a distance L from one another.

We regard the losses actually occurring in the laser (absorption and removal of radiation) as uniformly distributed throughout the volume. We also assume that the active medium is characterized by a complex dielectric constant which is homogeneous along the y axis, but varies in the direction of the x axis, perpendicular to the laser axis:

$$\varepsilon(x, t) = \varepsilon_0 + \delta\varepsilon'(x) + i\varepsilon_0'' - i\varepsilon''(x, t). \tag{1}$$

Here ε_0'' describes the radiation losses and is connected with the lifetime T_0 of the photon in the resonator by the relationship

$$\varepsilon_0'' = \frac{1}{\omega_0 T_0}, \tag{2}$$

while $\varepsilon''(x, t)$ describes the amplification of the radiation and is connected with the inverse population density N(x, t) by the relationship

$$\varepsilon''(x, t) = \frac{\lambda}{2\pi}\sigma N(x, t), \tag{3}$$

where ω_0 and λ are the frequency and wavelength of the generated emission, $\sigma = \sigma(\omega_0)$ is the cross section of the radiative transition.

In Q-switched lasers many longitudinal (axial) oscillation modes are excited and, hence, the periodic change in the inverse population in the longitudinal direction is largely smoothed out. As a result, the conditions of development of different longitudinal modes are practically the same and in the consideration of the effects of interest in the transverse development of the giant pulse we can obviously confine ourselves to the case of one longitudinal mode.

The electric field $\mathscr{E}(x, t)$ in the resonator satisfies the equation [9]

$$\frac{\partial \mathscr{E}(x, t)}{\partial t} = i\frac{c^2}{2\varepsilon_0\omega_0}\frac{\partial^2 \mathscr{E}(x, t)}{\partial x^2} - \frac{\omega_0}{2\varepsilon_0}[\varepsilon_0'' - \varepsilon''(x, t) + i\delta\varepsilon'(x)]\mathscr{E}(x, t), \tag{4}$$

where $\mathscr{E}(x, t) = U(x, t)\, e^{-i\alpha(x,t)}$, U(x, t) and α(x, t) are the "slowly" varying field amplitude and phase on the laser mirror, and the field inside the resonator has the form

$$E(x, z, t) = \mathscr{E}(x, t)\sin\frac{\omega_0 z}{c}\, e^{-i\omega_0 t},$$

where $\omega_0 = n\pi c/L$, n is the longitudinal mode number, and c is the velocity of light in the medium.

The equation for the imaginary part of the dielectric constant $\varepsilon''(x, t)$, which describes the amplification of the medium, has the form

$$\frac{\partial \varepsilon''(x, t)}{\partial t} = -\frac{2\sigma}{\hbar\omega_0}\varepsilon''(x, t)\, I(x, t), \tag{5}$$

where $I(x, t) = \frac{c\varepsilon_0}{8\pi}\frac{1}{2}\mathscr{E}\mathscr{E}^*$ is the radiation flux density (in erg/cm^2 · sec). In equation (5) we neglected the following processes, which are slow in comparison with the time of development of the giant pulse: spontaneous decay of the inverse population and change in inversion due to pumping. To solve the system of equations (4) and (5) we need to assign initial conditions.

§ 2. Initial Conditions

The amplitudes and phase of the initial field in each mode are determined by the spontaneous emission of the medium at the instant of Q-switching. The mean number $\langle n_0 \rangle$ of spontaneous photons in one mode is given by the expression

$$\langle n_0 \rangle \approx \frac{\int N_2 dV}{8\pi T_1 \Delta\nu} \left(\frac{\lambda}{a} \right)^2, \tag{6}$$

where $\int N_2 dV$ is the total number of excited atoms in the resonator, $\Delta\nu$ is the width of the spontaneous emission line of the atoms, T_1 is the lifetime of the excited atom for spontaneous transition to a lower level, $(\lambda/a)^2$ is the solid angle of directions of the wave vector of the mode. The mean-square strength of the spontaneous emission field in the mode is then equal to

$$\langle E_0^2 \rangle \approx \frac{\overline{N}_2}{T_1 \Delta\nu} \left(\frac{\lambda}{a} \right)^2 \hbar\omega_0, \tag{7}$$

where $\overline{N}_2 = \frac{1}{V} \int N_2 dV$ is the mean density of excited atoms in the resonator, and V is the volume of the resonator. In the presence of an inverse population the spontaneous emission is slightly amplified. Hence, to the expressions for $\langle n_0 \rangle$ and $\langle E_0^2 \rangle$ we need to add the gain k_0 during one transit of the resonator. For the parameters typical of a Q-switched laser, $\langle n_0 \rangle \gg 1$.

The statistics of the spontaneous emission of the medium in the absence of feedback must be similar to the statistics of equilibrium emission [10]. Hence, the probability of filling with n photons of one mode is given by the Bose–Einstein distribution, which at the considered classical limit $\langle n_0 \rangle \gg 1$ has the form (see [12, 13], for instance)

$$p(n) = \frac{1}{\langle n_0 \rangle} \exp\left(-\frac{n}{\langle n_0 \rangle} \right). \tag{8}$$

Hence, the values of the amplitudes of the initial fields in the modes must be chosen in a random manner with a considerable dispersion relative to the mean value. The phases of the initial fields in different modes are quite independent and are uniformly distributed.

The initial spatial distribution $\varepsilon''(x)$ is inhomogeneous. Usually the inverse population is a maximum on the crystal axis and is significantly (several times) less than close to the side surface [14-16]. The degree of inhomogeneity of the distribution $\varepsilon''(x)$ depends on the design of the illuminator, the treatment of the side surface of the crystal, which alters the focusing action of the crystal for the pumping radiation, and so on, but a change of 20–30% in $\varepsilon''(x)$ from the center to the edge of the crystal appears to be unavoidable [15, 16].

At the moment of Q-switching the active medium (crystal, glass) inside the resonator inevitably has an inhomogeneous distribution of refractive index, which distorts the plane-parallel resonator. Even if the medium is optically homogeneous before pumping the optical properties of the resonator during pumping are changed. Such effects have been thoroughly investigated in several experimental works [17-19]. Part of the pumping light energy either due to nonradiative transitions of excited impurity ions or to direct absorption goes towards heating of the medium. The inhomogeneous distribution of the pumping light and cooling of the sample leads to inhomogeneous heating, which distorts the optical homogeneity of the medium. In a first approximation the inhomogeneity arising during heating can be described by a small positive or negative lens inside the resonator. The sign of the lens depends on the treatment of the surface of the sample, the method of cooling, and so on [19].

§3. Linear Development of Generation. Solution of Equations

In the region of linear development of generation the gain saturation $\varepsilon''(x, t)$ can be neglected and we can regard $\varepsilon''(x, t) \equiv \varepsilon''(x)$. Then the investigation reduces to the solution of one equation (4), which is analytically possible only for some special forms of the functions $\delta\varepsilon'(x)$ and $\varepsilon''(x)$. Of practical interest are $\varepsilon'(x)$ and $\varepsilon''(x)$ which have the following form:

$$\varepsilon'(x) = \varepsilon_0 + \frac{\delta\varepsilon_0'}{\operatorname{ch}^2 px}, \quad \varepsilon''(x) = \frac{\varepsilon_m''}{\operatorname{ch}^2 px}. \tag{9}$$

Here $\varepsilon'(x)$, according to the sign of $\delta\varepsilon_0'$, is a positive or negative lens,* while $\varepsilon''(x)$ describes the distribution of the inverse population with a maximum at the center of the crystal.

We will seek the solution of equation (4) in the form

$$\mathscr{E}(x, t) = U(x)\, e^{-i\Omega t}. \tag{10}$$

We obtain the following equation for the eigenfunctions and eigenvalues:

$$\frac{d^2U}{dx^2} + k^2\left(i\varepsilon_0'' - \frac{i\varepsilon_m'' + \delta\varepsilon_0'}{\operatorname{ch}^2 px} + \frac{2\Omega\varepsilon_0}{\omega_0}\right)U = 0, \tag{11}$$

where $k = \omega_0/c$. We will consider only the discrete spectrum. The solution of (11), finite at $x = \infty$, is [21]

$$U = (1-\xi^2)^{\frac{\mu_\pm}{2}} F\left(\mu_\pm - q,\ \mu_\pm + q + 1,\ \mu_\pm + 1,\ \frac{1-\xi}{2}\right), \tag{12}$$

where F is a hypergeometric function, $\xi = \operatorname{th} px$

$$2q + 1 = \sqrt{1 - \frac{4k^2}{p^2}(\delta\varepsilon_0' + i\varepsilon_m'')}, \quad \mu_\pm = \pm i\frac{k}{p}\sqrt{i\varepsilon_0'' + \frac{2\Omega\varepsilon_0}{\omega_0}},$$

the + sign corresponds to the positive lens ($\delta\varepsilon_0' > 0$), and the − sign to a negative lens. If the solution is to be finite when $x \to \infty$, it is necessary that (then F is a polynomial of n-th degree in ξ):

$$\mu_+ - q = -n,$$

$$\mu_- - q = n + 1, \qquad n = 0, 1, 2\ldots \tag{13}$$

Relationship (13) gives the complex natural frequencies $\Omega_n = \Omega_n' + i\Omega_n''$ of the modes. In lasers using luminescence crystals and glasses $\varepsilon_m'' \approx 10^{-6}$, $\lambda \approx 10^{-4}$ cm, $p \approx 1/a \simeq 1\ \text{cm}^{-1}$, $\delta\varepsilon_0' \gtrsim 10^{-5}$. Hence, we can use the approximation

$$4\frac{k^2}{p^2}|\delta\varepsilon_0'|, \quad 4\frac{k^2}{p^2}\varepsilon_m'' \gg 1. \tag{14}$$

Then when $\delta\varepsilon_0' \gg \varepsilon_m''$ the expressions for the natural frequencies have the form

$$\frac{2\Omega_n'\varepsilon_0}{\omega_0} = -\delta\varepsilon_0' + \left(n + \frac{1}{2}\right)\frac{p}{k}\left\{\begin{matrix} 2\sqrt{\delta\varepsilon_0'} \\ \dfrac{\varepsilon_m''}{\sqrt{|\delta\varepsilon_0'|}} \end{matrix}\right\} - \left(n + \frac{1}{2}\right)^2\frac{p^2}{k^2} \tag{15}$$

and

$$\frac{2\Omega_n''\varepsilon_0}{\omega_0} = \varepsilon_m' - \varepsilon_0'' - \left(n + \frac{1}{2}\right)\frac{p}{k}\left\{\begin{matrix} \dfrac{\varepsilon_m'}{\sqrt{|\delta\varepsilon_0'|}} \\ 2\sqrt{|\delta\varepsilon_0'|} \end{matrix}\right\}, \tag{16}$$

*The case $\delta\varepsilon_0' = 0$ was considered in [20].

where the upper line in the braces corresponds to $\delta\varepsilon_0' > 0$, and the lower one to $\delta\varepsilon_0' < 0$, When $\delta\varepsilon_0' = 0$, we accordingly have

$$\frac{2\Omega_n'\varepsilon_0}{\omega_0} = \left(n+\frac{1}{2}\right)\frac{p}{k}\sqrt{2\varepsilon_m''} - \left(n+\frac{1}{2}\right)^2\frac{p^2}{k^2}, \qquad \frac{2\Omega_n''\varepsilon_0}{\omega_0} = \varepsilon_m'' - \varepsilon_0'' - \left(n+\frac{1}{2}\right)\frac{p}{k}\sqrt{2\varepsilon_m''}. \tag{17}$$

The expressions for the natural frequencies Ω have the following physical meaning. The term $(\omega_0/2\varepsilon_0)\,\delta\varepsilon_0'$ in (15) is the same frequency shift for all transverse modes by an amount determined by the change in optical length in the center of the crystal. This term is of no interest to us. The second term, linear in n, is much greater than the third term, quadratic in n, since for reasonable values $n(p/k)(n + 1/2) \ll 1$. Thus, there is a set of practically equidistant modes, which agrees with confocal resonator theory [22, 23]. The term $(\omega_0/2\varepsilon_0)(\varepsilon_m'' - \varepsilon_0'')$ in (16) describes the same rate of increase in amplitude, determined by the gain in the center of the crystal, for all modes. The second term, which is usually much smaller than the first, characterizes the reduction in gain for the mode as its number increases.

Substituting the values of μ in (12) we find expressions for the distribution of the square of the field amplitude of the dominant mode on the end face of the crystal:

$$|U_0(x)|^2 = \left(\frac{1}{\operatorname{ch} px}\right)^{b\frac{k}{p}-1}, \quad b = \begin{cases} 2\sqrt{\delta\varepsilon_0'}, & \delta\varepsilon_0' \gg \varepsilon_m'',\ \delta\varepsilon_0' > 0, \\ \sqrt{2\varepsilon_m''}, & \delta\varepsilon_0' = 0, \\ \dfrac{\varepsilon_m''}{\sqrt{|\delta\varepsilon_0'|}} & |\delta\varepsilon_0'| \gg \varepsilon_m'',\ \delta\varepsilon_0' < 0, \end{cases} \tag{18}$$

and usually $bk \gg p$.

§ 4. Field Distribution and Dependence on Initial Conditions

The field distribution in the laser at the end of linear development of the pulse is determined, generally speaking, by the superposition of the modes $\sum_n E_{0n}U_n(x)e^{-i\Omega_n\tau_d}$, where E_{0n} are the initial random complex amplitudes of the field in the modes, and τ_d is the duration of linear development of the pulse (delay time). The delay time is determined from the condition of a particular, say 20%, saturation of the gain of the medium. It follows from (5) that this requires fulfillment of the condition

$$\frac{c}{8\pi}\langle E_0^2\rangle \int_0^{\tau_d} e^{2\Omega_0''t}\,dt \approx \frac{\hbar\omega}{10\sigma}, \tag{19}$$

where $\langle E_0^2\rangle$ is the mean square intensity of the spontaneous emission in the mode, determined by expression (7), and Ω_0'' is given by expression (16). As a result we find a general expression for the time of linear development of generation

$$\tau_d \approx \frac{1}{2\Omega_0''}\ln\left(\frac{5\Omega_0''\hbar\omega}{\sigma c\langle E_0^2\rangle}\right), \tag{20}$$

which is valid for a laser with an inhomogeneous distribution of inverse population and refractive index.

If during the delay time τ_d the amplitude of the dominant mode becomes much greater than the amplitude of the following modes, then, firstly, the field distribution on the end face of the crystal by the moment of generation of the giant pulse proper will not depend on the random initial amplitudes and, secondly, the shape of the distribution will be given by expression (18) for $|U_0(x)|^2$. This is valid if

$$e^{(\Omega_0''-\Omega_1'')\tau_d} \gg 1 \quad \text{or} \quad \tau_d(\Omega_0''-\Omega_1'') \gtrsim 2. \tag{21}$$

We consider first the case $\delta\varepsilon_0' = 0$. Then condition (21) takes the form

$$\frac{\tau_d cp}{2\varepsilon_0}\sqrt{\frac{\varepsilon_m''}{2}} \gtrsim 1. \tag{22}$$

For Q-switched lasers employing luminescent crystals or glasses $\tau \approx 50$ nsec, $\varepsilon_0 \approx 3$, and $\varepsilon_m'' \approx 2 \cdot 10^{-6}$. Condition (22) is fulfilled when $p \gtrsim 4\ \text{cm}^{-1}$. Hence, inhomogeneity of the inverse population with a transverse dimension of the order of 0.5 cm leads to the giant pulse being independent of the random initial conditions for the field. In the case of inhomogeneity of the refractive index $|\delta\varepsilon_0'| \gg \varepsilon_m''$, we will have, in place of condition (22),

$$\frac{\tau_d cp}{2\varepsilon_0}\left\{\begin{matrix} \dfrac{\varepsilon_m''}{2\sqrt{\delta\varepsilon_0'}} \\ 2\sqrt{|\delta\varepsilon_0'|} \end{matrix}\right\} \gtrsim 1, \tag{23}$$

where the top line in the braces corresponds to $\delta\varepsilon_0' > 0$, and the bottom line to $\delta\varepsilon_0' < 0$. In the case of inhomogeneity of the positive-lens type $\delta\varepsilon_0'$ must be sufficiently small ($\lesssim 3 \cdot 10^{-6}$), otherwise there would be a dependence on the initial conditions. A slight inhomogeneity of the negative lens type ($|\delta\varepsilon_0'| \gtrsim 10^{-6}$), however, is sufficient for absence of dependence on the initial conditions. The physical explanation of these results is that when $\delta\varepsilon_0' > 0$ the neighboring transverse modes are grouped close to the resonator axis and, hence, are amplified almost equally in the case of an inhomogeneous distribution of $\varepsilon''(x)$. When $\delta\varepsilon_0' < 0$ the transverse modes extend into the regions with lower amplification and this leads to a difference in their gain.

Thus, inhomogeneity of the distribution of the inverse population and inhomogeneity of the refractive index of the negative-lens type lead to the field distribution on the end-face of the crystal $|\mathscr{E}(x)|^2$ being independent of the random initial conditions. In this case the field distribution by the end of the linear development of the pulse is $|\mathscr{E}(x)|^2 \simeq |U_0(x)|^2$. The size of the generation region at this moment (size of "stream") $2x_0$ is determined from the condition $|U_0(x_0)|^2 = 1/2$.

The results obtained above for an infinite layer can be applied directly to the case of a bounded layer (crystal), if the dimension of the steady field distribution $2x_0$ is less than the crystal diameter $2a$, so that diffraction effects can be neglected.

In the other limiting case, where transverse inhomogeneities of the inverse population and refractive index are absent, conditions (22) and (23) are not satisfied. Only diffraction losses on the mirror edges can discriminate the individual modes, but during the short time of linear development* ($\tau_d \lesssim 10^{-7}$ sec) the role of such discrimination for the lower modes is small. Hence, the field distribution by the end of the linear development of generation is random, but in multimode lasers, which usually generate tens of axial modes, the random distributions for different axial modes are averaged and the overall distribution is homogeneous. In such a laser the transverse development of generation and the associated lengthening of the giant pulse are absent. In single-mode Q-switched lasers with a high degree of optical inhomogeneity of the resonator and the inverse population the field distribution by the start of generation of the giant pulse will be random.

We turn now to a consideration of the transverse development of the generation region in the case where the dimension of the generation region is much less than the crystal diameter by the end of linear development.

*The case of Q-switching with clearing filters is not considered here.

§ 5. Nonlinear Transverse Development of Generation

The nonlinear generation phase, during which the giant pulse proper is generated, begins at the moment when the field distribution formed during the time of linear development attains an amplitude sufficient for gain saturation. Gain saturation occurs first in the region of the maximum amplitude $|\mathscr{E}(x, t)|^2$, and then propagates in a transverse direction as the intensity develops in the peripheral regions. Thus, the nonlinear transverse development of the generation region is the result of delay in development of the field amplitude in the edge regions of the laser.

The transverse development of generation can be considered with the aid of equations (4) and (5). The term with $\partial^2\mathscr{E}/\partial x^2$ in (4) describes the linear (diffractive) spread of the field. When $x_0^2/L\lambda \gg 1$ the contribution of this term is negligibly small and can be neglected. The results of exact calculations, given below, also show that the velocity of diffractive spread of the field is small in comparison with the rate of nonlinear transverse development of generation. In this approximation, converting to intensity I(x, t), we rewrite equations (4) and (5) in the form:

$$\frac{\partial I(x,\tau)}{\partial\tau} = \frac{\omega_0}{\varepsilon_0}\left[\varepsilon''(x)\,e^{-\frac{2\sigma}{\hbar\omega}\int\limits_0^{\tau} I(x,\tau')\,d\tau'} - \varepsilon''_0\right] I(x,\tau), \tag{24}$$

where $\tau = t - \tau_d$, τ_d is the time of linear development (delay time), and $\varepsilon''(x)$ describes the initial distribution of the inverse population. To determine the velocity of transverse development we need only investigate the transverse motion of the boundary of gain saturation.

The motion of this saturation level $\delta = \varepsilon''(x, \tau)/\varepsilon''(x)$ is given by the condition:

$$\int\limits_0^{\tau_s} I(x,\tau)\,d\tau = \frac{1}{2\sigma}\ln\frac{1}{\delta}, \tag{25}$$

where τ_s is the instant of gain saturation for level δ. We investigate the level of shallow saturation ($\delta = 0.8$), when $\varepsilon''(x, \tau < \tau_s)$ can be regarded as coinciding with $\varepsilon''(x)$. Then

$$I(x, \tau < \tau_s) = I_0(x)\,e^{\frac{\omega_0}{\varepsilon_0}[\varepsilon''(x) - \varepsilon''_0]\tau},$$

where $I_0(x) = I(x, 0)$ is the distribution of field intensity by the end of linear development. Substituting the expression for $I(x, \tau < \tau_s)$ in (25), we obtain

$$I_0(x)\,e^{\frac{\omega_0}{\varepsilon_0}[\varepsilon''(x) - \varepsilon''_0]\tau_s} \approx \frac{1}{2\sigma}\ln\frac{1}{\delta}\frac{\omega_0}{\varepsilon_0}[\varepsilon''(x) - \varepsilon''_0]. \tag{26}$$

Differentiating (26) and assuming that $\varepsilon''(x)$ is a smoother function than $I_0(x)$ $\left(\frac{1}{I_0(x)}\frac{\partial I_0(x)}{\partial x} \gg \frac{1}{\varepsilon''(x)}\frac{\partial \varepsilon''(x)}{\partial x}\right)$, we find the velocity of the saturation boundary $v = dx/d\tau_s$:

$$v = -\frac{\omega_0}{\varepsilon_0}[\varepsilon''(x) - \varepsilon''_0]\left.\frac{I_0(x)}{\frac{\partial I_0(x)}{\partial x}}\right|_{x = x(\tau_s)}. \tag{27}$$

The quantity $s = I_0(x)\left|\left[\frac{\partial I_0(x)}{\partial x}\right]^{-1}\right|_{x = x(\tau_s)}$ gives the spatial steepness of the fronts of the initial distribution $I_0(x)$. For a bell-shaped distribution with an exponential relationship $I_0(x)$ when $|x| \to a$ the spatial steepness s is constant $[I_0(x) \approx e^{-|x|/s},\ |x| \gtrsim x_0]$ and is connected with the half-width of the distribution x_0 by the approximate relationship $s \approx 0.7x_0$. However, in the presence of inhomogeneity of the refractive index in the resonator the value of s may vary along the fronts of $I_0(x)$. For instance, in a confocal resonator the value of s decreases when $|x|$ increases.

Introducing the gain and losses per unit length $\alpha(x) = (\omega_0/\varepsilon_0 c)\varepsilon''(x)$ and $\gamma = (\omega_0/\varepsilon_0 c)\varepsilon_0''$, we obtain the following expression for the velocity of transverse development of the giant pulse:

$$v = (\alpha - \gamma)cs. \tag{28}$$

The time of transverse development τ_{tr}, during which generation involves the whole crystal of diameter $2a$, can be evaluated from the expression

$$\tau_{tr} \approx \frac{a - x_0}{(\bar{\alpha} - \gamma)\,cs}, \tag{29}$$

where $\bar{\alpha} = \frac{1}{a - x_0}\int\limits_{x_0}^{a} \alpha(x)\,dx$ is the mean gain per unit length in the region of transverse development of the pulse. As an illustration we consider a laser with the following parameters: $\bar{\alpha} = 0.03\ \text{cm}^{-1}$, $\gamma = 0.02\ \text{cm}^{-1}$, $x_0 = a/4$, $a \approx x_0$. In this case the time of transverse development τ_{tr}, ≈ 10 nsec. This time is comparable with the length of the giant pulse and, hence, must be taken into consideration in the calculation of the pulse length.

Thus, the time of transverse development of the giant pulse is determined by the shape of the field distribution by the end of linear development. The obtention of the shortest light pulses necessitates reduction of the time of transverse development by an increase in the degree of homogeneity of the refractive index and inverse population of the crystals and glasses at the moment of Q-switching.

§ 6. Numerical Integration of Equations

A clearer idea of the development of the giant pulse can be obtained by numerical integration of equations (4) and (5). Moreover, the results of an exact numerical calculation demonstrate the validity of the adopted assumptions in the preceding analytical investigation of the different phases of generation of the giant pulse.

We will solve equations (4) and (5) with boundary conditions on the side surface of the active medium $\mathcal{E}(\pm a, t) \equiv 0$, which correspond to total reflection from the side surface. With such boundary conditions we neglect the effect of diffraction at the edges of the mirrors on the field distribution and losses. Neglect of the contribution of diffraction at the edges of the mirrors to the field distribution is quite rational, since, as was shown above, the main role in the spatial development of the giant pulse is played by inhomogeneities (initial and formed during generation) of the inverse population density and the refractive index. It is easy to allow for the contribution of diffraction to emission losses by introducing an additional diffraction term into ε_0'' (this however has very little effect on the solution).

In the solution of the system of equations (4) and (5) the electromagnetic field within the laser can conveniently be represented in the form of a superposition of transverse (angular) modes of the empty resonator (mirror box) with time-dependent complex amplitudes:

$$\mathcal{E}(x, t) = \sum_{k=1}^{\infty} a_k(t)\,U_k(x)\,e^{-i\Omega_k t}, \tag{30}$$

where $U_k(x)$ and Ω_k are the eigenfunctions and eigenvalues of equation (4) with $\varepsilon''(x, t) \equiv 0$, $\delta\varepsilon'(x) \equiv 0$, and zero initial conditions given by the expressions

$$U_k(x) = \frac{1}{\sqrt{a}} \begin{Bmatrix} \cos\left[\pi \frac{x}{a}\left(k - \frac{1}{2}\right)\right] \\ \sin\left(\pi \frac{x}{a} k\right) \end{Bmatrix}, \quad \Omega_k = \frac{1}{2\omega_0 \varepsilon_0} \begin{Bmatrix} \left[\frac{\pi c}{a}\left(k - \frac{1}{2}\right)\right]^2 \\ \left(\pi \frac{c}{a} k\right)^2 \end{Bmatrix}, \tag{31}$$

where the upper line corresponds to odd k, and the lower to even k. Substituting (31) in equation (4), we obtain by the usual method a system of equations for the amplitudes $a_k(t)$, which can conveniently be put in the form $a_k(t)\, e^{-i(\Omega_k - \Omega_1)t} = A_k'(t) + iA_k''(t)$. As a result, we have a system consisting of an infinite number of equations:

$$\begin{aligned} \dot{A}_k'(t) &= \frac{\omega_0}{2\varepsilon_0} \sum_{m=1}^{\infty} A_m'(t)\, \varepsilon_{km}''(t) + \frac{\omega_0}{2\varepsilon_0} \sum_{m=1}^{\infty} A_m''(t)\, \varepsilon_{km}' + (\Omega_k - \Omega_1) A_k''(t), \\ \dot{A}_k''(t) &= \frac{\omega_0}{2\varepsilon_0} \sum_{m=1}^{\infty} A_m''(t)\, \varepsilon_{km}''(t) - \frac{\omega_0}{2\varepsilon_0} \sum_{m=1}^{\infty} A_m'(t)\, \varepsilon_{km}' - (\Omega_k - \Omega_1) A_k'(t), \end{aligned} \tag{32}$$

where k = 1, 2,...; the matrix elements ε_{km}' and ε_{km}'' are given by the expressions

$$\varepsilon_{km}' = \int_{-a}^{a} U_k(x)\, \delta\varepsilon'(x)\, U_m(x)\, dx, \tag{33}$$

$$\varepsilon_{km}''(t) = \delta_{km}\left(\frac{1}{\omega_0 T_0} + \Lambda_k \frac{c}{2\omega_0 L}\right) - \int_{-a}^{a} U_k(x)\, \varepsilon''(x, t)\, U_m(x)\, dx, \tag{34}$$

and Λ_k is the diffraction losses per transit for a resonator with plane mirrors, found by Vainshtein [24], and δ_{km} is the Kronecker delta.

The initial conditions required for the solution of the system of equations (32) and (5) were determined in § 2.

Equations (32) and (5) were integrated numerically on a M-20 electronic digital computer. The results of the numerical integration give a fairly clear picture of the processes in a Q-switched laser.

The distribution of the generation field in the resonator has a certain degree of "smoothness" in the transverse direction. Physically this denotes the predominant excitation of low-order modes. At the moment of Q-switching all the modes are in the same conditions, since the mean initial amplitude $\langle E_0^2 \rangle$ does not depend on the mode order. As generation develops, however, only a few of the modes build up effectively. The decisive role in the predominant excitation of lower-order modes is played by the inhomogeneities of $\varepsilon'(x)$ and $\varepsilon''(x)$. For instance, in a filled resonator in which $\varepsilon''(x)$ changes by only 20-30% from the center to the edge of the mirror (or end face of the crystal) the field distribution has a maximum at the center, drops off appreciably towards the edge, and contains mainly lower modes. Hence, in solving the infinite system of equations (32) we can confine ourselves to a finite number k_{max} of modes, which in each case will be determined by the specific parameters of the laser (crystal diameter, length of resonator, degree of inhomogeneity of inverse population and refractive index).

Plane-Parallel Resonator. We will consider first the case where there are no inhomogeneities of refractive index of the medium in the resonator: $\delta\varepsilon'(x) \equiv 0$.

Figure 1 shows the development of the giant pulse in a ruby laser with typical parameters: L = 50 cm, $2a$ = 7 mm, $\lambda = 7 \cdot 10^{-5}$ cm, $\sigma = 4 \cdot 10^{-20}$ cm^2, $r_1 r_2 \eta^2 = 0.15$, where r_1 and r_2 are the reflection coefficients of the mirrors, η is the transparency of the crystal and shutter; the gain in one transit in the center of the crystal is $e^{\alpha L} = 12$.

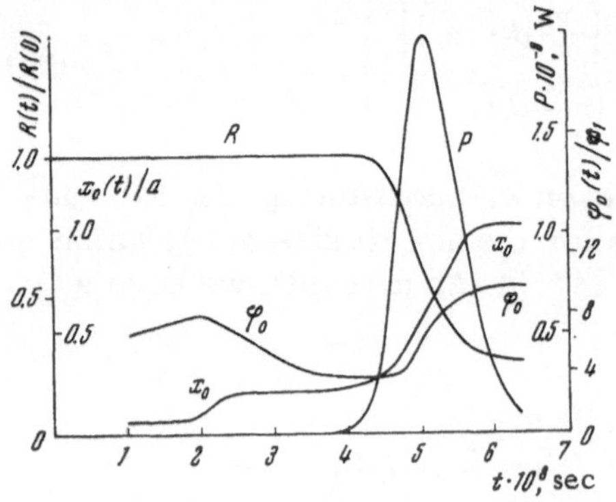

Fig. 1. Development of giant light pulse. P) Power; R) total number of active particles in resonator; $2x_0$) width of generation region ("jet") at half-height; φ_0) divergence of emission; φ_1) divergence of lowest mode.

The initial distribution of the inverse population decreased smoothly from the center to the edge to half its value at the center. The solution given was obtained for $k_{max} = 14$ modes. The termination of the infinite system at a finite number of equations was checked in every case from the nature of the change in the solution with increase in the number of modes. When the solution ceased to depend on k_{max} a complete solution of the problem was obtained. We calculated the following characteristics: power of giant pulse $P_{out} = P[(\ln r_1 r_2)/(\ln r_1 r_2 \eta^2)]$, total number of inversely populated atoms $R(t) = \int N(t)dV$, field distribution on end face of crystal, and field distribution in the far-field region. The latter, $E(t, \varphi)$, which characterizes the divergence of the giant pulse, was calculated from the formula

$$E(t, \varphi) = \sum_{n=1}^{\infty} \sqrt{\frac{2}{\pi}} \int_0^{\pi} A_n(t) \sin n\xi \cos k'\varphi\xi \, d\xi = \sum_{m=1}^{\infty} \sqrt{\frac{2}{\pi}} \left[\frac{A_{2m}}{m} \frac{\sin\left(\frac{\pi k'\varphi}{2}\right)}{1-\left(\frac{k'\varphi}{2m}\right)^2} + i \frac{A_{2m+1}}{m+\frac{1}{2}} \frac{\cos\left(\frac{\pi k'\varphi}{2}\right)}{1-\left(\frac{k'\varphi}{2m+1}\right)^2} \right], \tag{35}$$

where $k' = 4a/\lambda$.

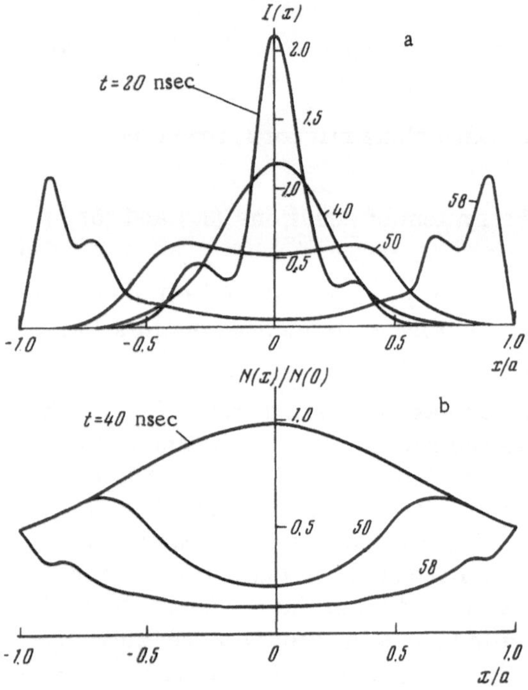

Fig. 2. Instantaneous distributions of emission intensity on end face of crystal I(x) (a) and inverse population N(x) (b).

For the complete description of the generation process Fig. 2 shows the field intensity distribution I(x) on the end face of the crystal (a) and the transverse distribution of the inverse population N(x) (b) at the most characteristic instants. Figure 3 shows the distributions of the squares of the mode amplitudes $|A_k|^2$ for all phases of development of the pulse. These indicate that the employed number of modes is quite sufficient for a self-consistent description of the generation field.

The presented solution is typical and contains all the essential factors of the spatial and temporal development of the giant pulse. At the instant of Q-switching (t = 0) the mode amplitudes and the field distribution on the end face are random, and the inverse population is a maximum at the center of the crystal. After a short time (t ≈ 10–20 nsec) the generation field takes a characteristic form (Fig. 2) with a maximum at the center and a small width – a "jet" is formed. A particular distribution of amplitude modes corresponds to the formation of the "jet" (Fig. 3). With further linear development of generation with a practically constant number of active particles R the "jet" expands to some quasi-stationary value (t ≈ 40 nsec, Fig. 1) – the normal mode of the field resonator is established. If the delay time exceeds the time of establishment of the "jet," generation is independent of the initial values

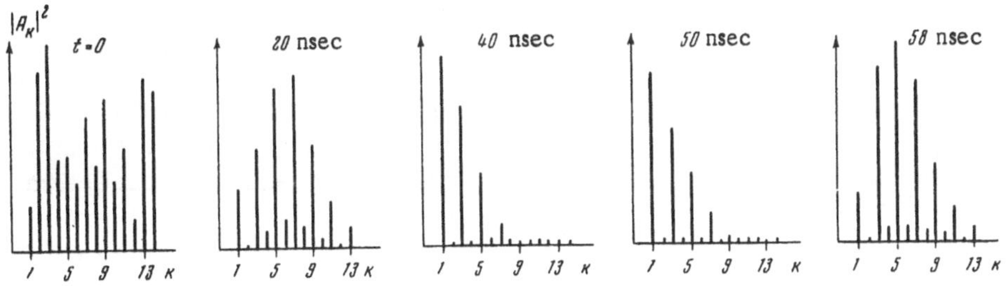

Fig. 3. Instantaneous distributions of mode intensity at different instants.

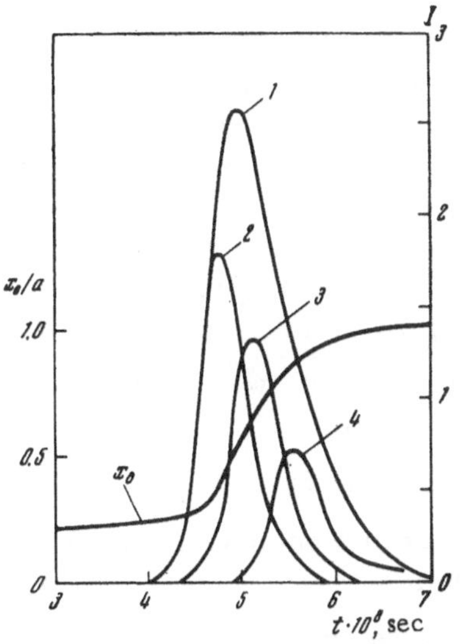

Fig. 4. Fine structure of giant pulse. 1) Pulse generated by total end face of crystal; 2,3,4) pulses generated by crystal points x = 0, 0.5a, 0.75a, respectively; x_0) halfwidth of generation region.

of the mode amplitudes. In the established "jet" (t = 40 nsec, Fig. 3) the lowest mode has the largest amplitude. At the start of generation of the giant pulse proper there is a sudden expansion of the "jet" — the field spreads towards the edges of the crystal. The divergence decreases at first, and again increases towards the end of the pulse. The decay of the pulse is characterized by an increase in the amplitude of the higher modes.

The presented solution shows that the giant pulse is emitted first by the central region of the crystal and then, during a time comparable with the pulse length, the generation region spreads in a transverse direction toward the crystal edges. Hence, the length of the pulse emitted by the entire end face of the crystal is greater than that of the pulse generated by any point of the end face. Such effects have been observed experimentally [25]. Figure 4 shows the form of the pulses generated by the entire end face of the crystal (curve 1), the central point x = 0 (curve 2), and the points x = $a/2$, $^3/_4 a$ (curves 3, 4) close to the edge of the crystal. The transverse development of the generation region is also shown. The length of the pulses generated by elements of the end face attains 5 nsec. The edge regions of the crystal radiate with a delay of 10 nsec relative to the crystal center. The time of transverse development $\tau_{tr} \approx 10$ nsec agrees with the approximate estimate from formula (29). In this case $\overline{\alpha} = 0.04$ cm^{-1}, $\gamma = 0.02$ cm^{-1}, $x_0 = 0.25a$, s = 0.7a. Formula (29) gives $\tau_{tr} \approx 7$ nsec, which is fairly close to the exact value.

In the considered case the inverse population decreased to half at the edges of the crystal. Such inhomogeneity is sufficient to make generation independent of the random initial conditions. This agrees well with the criterion (28) obtained above. In this case $\varepsilon_m'' = 10^{-6}$, $\tau_d = 4 \cdot 10^{-6}$ sec, p ≈ 3 cm^{-1}. Inequality (22) is valid in this case. However, when the parameter p decreases to 1 cm^{-1}, i.e., with reduction in the inhomogeneity of the inverse population, then, according to (22), there must be a dependence on the initial conditions. Figure 5 shows the variation of the field on the end face of the crystal and the inverse population during the generation of the giant pulse, when the initial inverse population decreases by only 20% at the crystal edges. The dependence of the field distribution on the random initial conditions, in agreement with criterion (22), is clearly seen. With such a small inhomogeneity of $\varepsilon''(x)$ the width of the generation region at the end of linear development is of the order of a. Hence, there is hardly any transverse expansion of generation, and the length of the pulse is a minimum (6 nsec). The time dependence of the pulse intensity is shown in Fig. 5c.

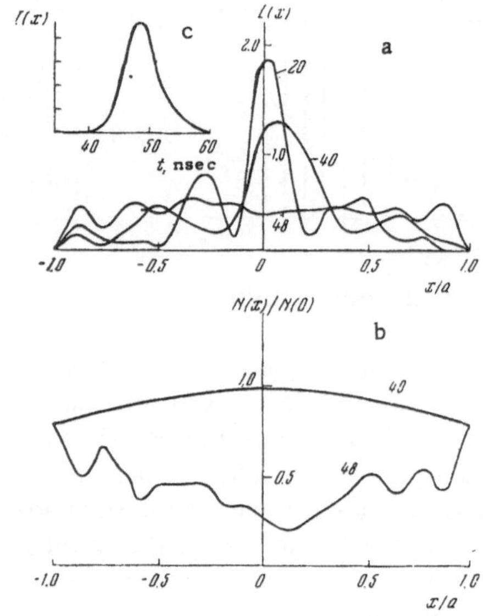

Fig. 5. Instantaneous distributions of emission intensity on end face of crystal I(x) (a), inverse population N(x) (b), and shape of giant pulse (c) in case of small inhomogeneity of initial distribution of inverse population. Figures denote time in nsec.

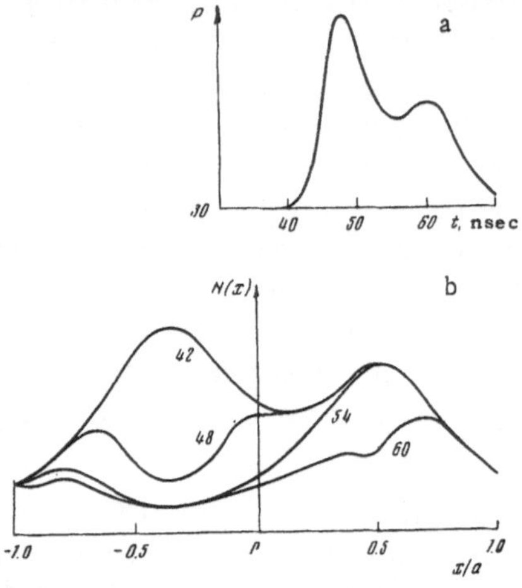

Fig. 6. Shape of pulse (a) generated in the case where there are two maxima in the initial distribution of inverse population (b).

The inhomogeneity of the inverse population can greatly alter the shape of the giant pulse. As an example, Fig. 6 shows the shape of the pulse when the initial distribution has two maxima. Generation begins first in the region of greatest amplification and then, extending in a transverse direction, attains the second inversion maximum. This gives rise to the second intensity peak. At the Third Conference on Quantum Electronics Hellwarth [7] described giant pulses with a second maxima on the trailing edge of the first pulse, but gave no explanation of this effect. The presented example shows that this effect may be due to inhomogeneity of the inverse population.

Distorted Resonator. It is of interest to investigate the dynamics of giant-pulse generation in the case where the medium inside the resonator also has an inhomogeneous distribution of refractive index, which is equivalent to distortion of the resonator. We consider the effect of a small inhomogeneity of the lens (positive or negative) or wedge type.

An inhomogeneity of the lens type can be described by putting $\varepsilon'(x)$ in the form

$$\delta\varepsilon'(x) = -\delta\varepsilon_0'(x/a)^2, \qquad (36)$$

where $\delta\varepsilon_0' > 0$ corresponds to a positive lens, and $\delta\varepsilon_0' < 0$ to a negative lens.

Figures 7 and 8 show the picture of giant-pulse development in the case of a small inhomogeneity of refractive index of the positive-lens type ($\delta\varepsilon_0' = 2 \cdot 10^{-7}$), and for definiteness all the other parameters of the laser and the initial conditions are the same as in the case of a plane-parallel resonator. The appreciable reduction of divergence to a minimum value equal to that of the fundamental mode is clearly seen. Physically this is due to the fact that the inhomogeneous distribution of inversion $\varepsilon''(x)$ with a maximum in the center is equivalent to a small negative lens distorting the wave front of the field. The introduction of a small inhomogeneity $\delta\varepsilon_0'$ of opposite sign compensates this effect and the field then approximates to a plane wave with minimum divergence. At higher values of $\delta\varepsilon_0'$ the positive-lens effect predominates and the divergence increases again. It is of interest to note that in this case the velocity of propagation of generation in the transverse direction decreases a little – the trailing edge of the pulse is prolonged, as can be seen in Fig. 7. This effect is due to an increase in the steepness, i.e., to a reduction of s on the edges of the initial distribution $I_0(x)$, due to the action of the positive lens.

Figures 9 and 10 show the development of the giant pulse in the case of a small inhomogeneity of refractive index of the negative-lens type ($\delta\varepsilon_0' = -2 \cdot 10^{-7}$) and the previous values

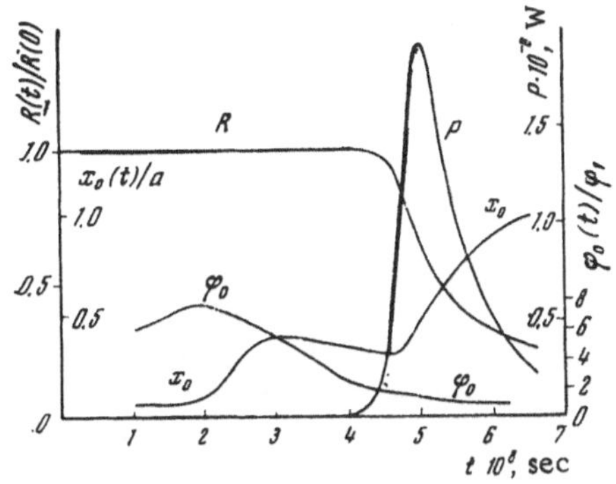

Fig. 7. Development of giant light pulse in presence of inhomogeneity of refractive index of negative-lens type in resonator.

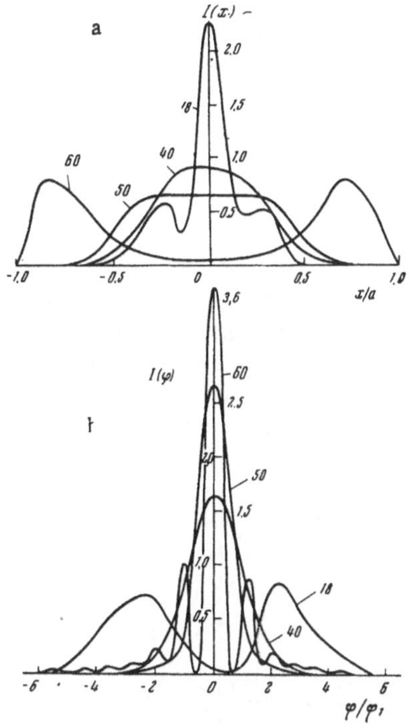

Fig. 8. Instantaneous distribution (a) of emission intensity on end face I(x) and angular distribution I(φ) (b), corresponding to the generation depicted in Fig. 7, at various times.

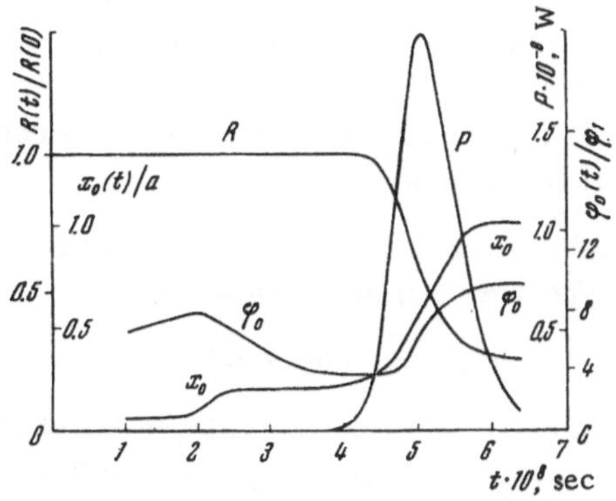

Fig. 9. Development of giant light pulse in case of inhomogeneity of refractive index of negative-lens type.

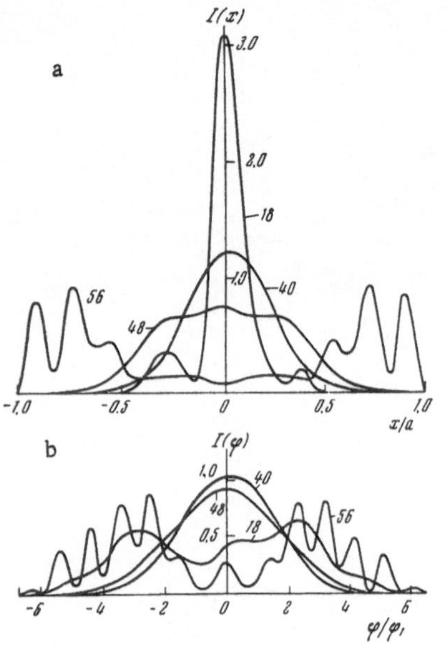

Fig. 10. Instantaneous distribution of emission intensity on end face I(x) (a) and angular distribution I(φ) (b), corresponding to generation illustrated in Fig. 9, at various times.

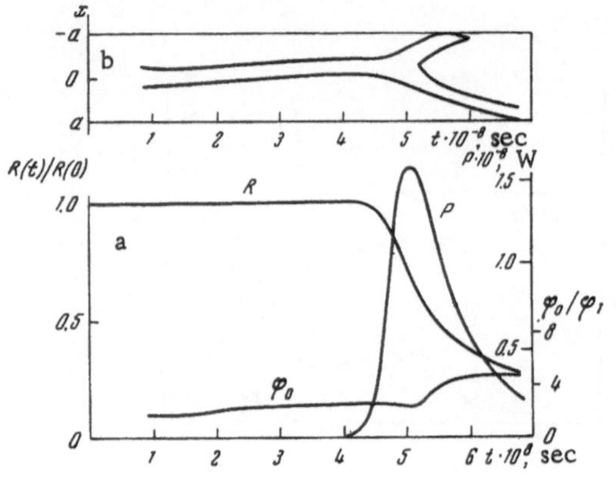

Fig. 11. Development of giant light pulse in case of wedge-type inhomogeneity of refractive index of medium in resonator (a) and motion of field distribution over end face of laser (the lines go to the half-maximum of the distribution) (b).

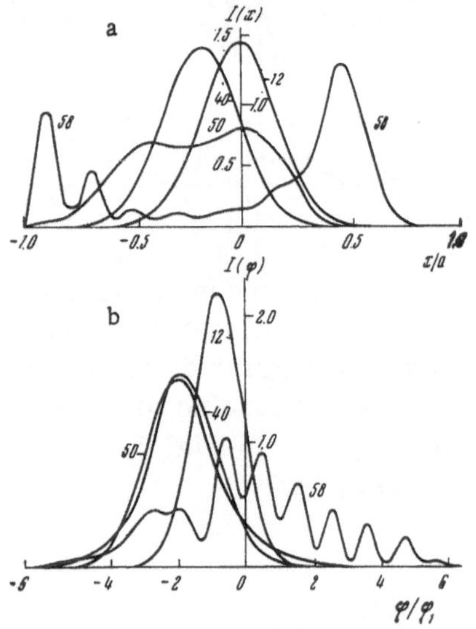

Fig. 12. Instantaneous distribution of emission intensity on end face I(x) (a) and angular distribution I(φ) (b), corresponding to generation illustrated in Fig. 11.

of the other laser parameters. The negative lens distorts the wavefront and, hence, increases appreciably the divergence of the emission. As distinct from the case $\delta\varepsilon_0' > 0$ a small negative lens does not increase the time of transverse development of generation.

The inhomogeneity of refractive index of the wedge type was assigned in the form

$$\delta\varepsilon'(x) = -\delta\varepsilon_0'(x/a). \qquad (37)$$

Figures 11 and 12 show the development of the giant pulse in this case for $\delta\varepsilon_0' = 2 \cdot 10^{-7}$ and the previous values of the other parameters. Figure 11b also shows the shift in the field distribution over the end face during generation (the solid lines correspond to the half-heights of the distribution I(x) at the given constant). It is clear that by the end of linear development of generation (t ≈ 44 nsec) the distribution is shifted towards the "open" end of the resonator. Hence, generation at the opposite end of the resonator begins much later, which leads to an increase in the time of transverse development and, hence, in the length of the giant pulse. This result agrees with the experiments carried out in (26).

Summary

In this paper we investigated the dynamics of the processes occurring in a Q-switched laser. This work was stimulated by the lack of data on the spatial and temporal development of generation, despite the obvious importance of such data in the use of giant light pulses in investigations of the nonlinear interaction of radiation and matter. From a systematic consideration of a relatively simple model of a Q-switched laser we analytically investigated two main phases of development of the giant pulse – the phase of linear development of generation, which begins with amplification of the spontaneous emission in the modes, and the phase of nonlinear transverse development, during which the giant light pulse proper is emitted. In addition, for a thorough investigation of the picture of development of the pulse as a whole the equations were numerically integrated.

Subsequent experiments [26, 27] confirmed the occurrence of transverse development of the giant pulse, while recent experiments on nonlinear amplification [28] have shown the significance of this effect in the propagation of the giant pulse in a nonlinear medium. A knowledge of the transverse development of the giant pulse would appear to be essential for the exact determination of the true strength of the light field in experiments on multiphoton processes [29]. The developed theory also leads to recommendations for the design of lasers to generate giant light pulses of minimum length and minimum divergence of emission.

Literature Cited

1. R. W. Hellwarth, Advances in Quantum Electronics, edited by J. R. Singer, Columbia University Press, New York (1961), p. 334.
2. F. J. McClung and R. W. Hellwarth, Proc. IEEE, 51:46 (1963).
3. A. M. Prokhorov, Radiotekhnika i élektronika, 8:1073 (1963).
4. A. A. Vuylsteke, J. Appl. Phys., 34:1615 (1963).
5. W. G. Wagner and B. A. Lenguel, J. Appl. Phys., 34:2040 (1963).
6. M. L. Ter-Mikaélyan and A. L. Mikaélyan, Dokl. Akad. Nauk SSSR, 155:1298 (1964) [Sov. Phys. – Dokl., 9:305 (1964)].
7. R. W. Hellwarth, Proceedings of Third International Conference on Quantum Electronics, Paris, Vol. 2, (1964), p. 1303.
8. N. G. Basov, E. M. Belenov, and V. S. Letokhov, Dokl. Akad. Nauk SSSR, 161:799 (1965) [Sov. Phys. – Dokl., 10:311 (1965)].
9. A. F. Suchkov, Zh. Éksp. Teor. Fiz., 49:1495 (1965) [Sov. Phys. – JETP, 22:1026 (1966)].
10. V. S. Letokhov and A. F. Suchkov, Zh. Éksp. Teor. Fiz., 50:1148 (1965) [Sov. Phys. – JETP, 23:764 (1966)].
11. V. S. Letokhov and A. F. Suchkov, Zh. Éksp. Teor. Fiz., 52:282 (1966) [Sov. Phys. – JETP, 25:182 (1967)].
12. V. S. Letokhov, Zh. Éksp. Teor. Fiz., 50:765 (1966) [Sov. Phys. – JETP, 23:506 (1966)]; Preprint FIAN, A-45 (1965).
13. L. Mandel, Proceedings of Third International Conference on Quantum Electronics, Paris, Vol. 1 (1964), p. 101.
14. G. E. Devlin, J. McKenna, A. D. May, and A. L. Schawlow, Appl. Optics, 1:11 (1962).
15. R. S. Congleton, W. R. Sooy, P. R. Dewhirst, and L. D. Riley, Proceedings of Third International Conference on Quantum Electronics, Paris, Vol. 2 (1964), p. 1415.
16. J. E. Geusic and H. E. D. Scovil, Proceedings of Third International Conference on Quantum Electronics, Paris, Vol. 2 (1964), p. 1211.
17. A. P. Veduta, A. M. Leontovich, and V. N. Smorchkov, Zh. Éksp. Teor. Fiz., 48:87 (1965) [Sov. Phys. – JETP, 21:59 (1965)].

18. L. J. Aplet, E. B. Jay, and W. R. Sooy, Appl. Phys. Lett., 8:71 (1966).
19. H. Welling and G. J. Bickart, J. Opt. Soc. Amer., 56:611 (1966).
20. T. J. Kuznetsova, Preprint FIAN, A-19 (1965).
21. L. D. Landau and E. M. Lifshits, Quantum Mechanics [in Russian], Fizmatgiz, Moscow (1963).
22. L. A. Vainshtein, Zh. Éksp. Teor. Fiz., 45:684 (1963) [Sov. Phys. – JETP, 18:471 (1964)].
23. G. D. Boyd and J. P. Gordon, Bell Syst. Tech. J., 40:489 (1961).
24. L. A. Vainshtein, Zh. Éksp. Teor. Fiz., 44:1050 (1963) [Sov. Phys. – JETP, 17:709 (1963)].
25. V. S. Zuev, Proceedings of First All-Union Symposium on Nonlinear Optics, Naroch' [in Russian] (1965).
26. R. V. Ambartsuman, N. G. Basov, V. S. Zuev, P. G. Kryukov, V. S. Letokhov, and O. B. Shatberashvili, Zh. Éksp. Teor. Fiz., 51:406 (1966) [Sov. Phys. – JETP, 24:272 (1967)].
27. V. V. Korobkin, A. M. Leontovich, M. I. Popova, and M. Ya. Shchelev, Pisma ZhÉTF, 4:19 (1966).
28. N. G. Basov, Address Read at Colloquium of Laboratories of Oscillations and Quantum Radiophysics [in Russian], FIAN (1965).

NONLINEAR INTERACTIONS OF ELECTROMAGNETIC WAVES IN AN ACTIVE MEDIUM AND THEIR POSSIBLE APPLICATION TO THE DEVELOPMENT OF NEW TYPES OF LASERS

B. P. Kirsanov

Introduction

Advances in laser technique now enable the production of light fields with a strength of 10^6-10^8 V/cm. In such fields the medium becomes nonlinear, i.e., the principle of superposition breaks down: Different waves interact with one another as they propagate; for instance, the propagation of waves of one frequency gives rise to waves of other frequencies, and so on. In quantum language such effects are described by multiphoton processes. Multiphoton processes are interactions of radiation with matter in which at least two photons are absorbed or emitted in each elementary act; the probability of such an act differs considerably from the product of the probabilities of individual one-photon processes. It should be noted that some multiphoton effects, such as Raman scattering, resonance fluorescence, etc., had been observed even before lasers were produced. Multiquantum processes were investigated theoretically in the early years of quantum mechanics; for instance, two-photon emission and absorption were examined as long ago as 1931 [1].

Recently there have appeared numerous theoretical and experimental investigations of multiquantum processes [2-5].

These investigations provide opportunities for the development of new types of lasers operating on the basis of the use of multiphoton effects. Such lasers can have several valuable properties which ordinary lasers lack.

The fact is that many of the potentialities of lasers are not realized, since generators of coherent optical emission based on one-photon transitions can operate only at strictly defined frequencies determined by the energy levels in the particular substance. The number of "laser" substances and, hence, of frequencies is relatively small. Smooth variation of the energy levels in one substance can be effected only in a paramagnetic maser, i.e., in the centimeter-wave region. In lasers the frequency of the generated waves is much higher and, hence, such small changes in energy levels are almost inappreciable.

Definite advances in extending the range of powerful coherent optical oscillations have been made by means of nonlinear optics. These investigations are concerned with harmonic generation and induced Raman scattering.

A particularly urgent problem is to design a powerful laser with smooth frequency tuning. Such a laser could be an ideal source for spectroscopy. It could be used to control chemical reactions (by tuning the laser to resonance with different frequencies of oscillations of the molecule and thus inducing these modes it might be possible to induce the substance to enter into chemical reactions), and so on.

At present there are three proposals for such a laser. (1) the "parametric" laser, proposed by Akhmanov and Khokhlov [3, 6], and also by Kroll [7]; (2) the laser using two-photon luminescence, proposed by Prokhorov and Selivanenko [8]; (3) the "resonance parametric laser," considered by the author of this article and Selivanenko [9]. We briefly describe the main features of these three ideas.

The operation of the parametric laser is based on a three-photon process occurring in a transparent medium. The substance virtually absorbs the coherent pumping quantum $\hbar\omega_0$ (from another laser) and then completes in succession two virtual transitions with the emission of $\hbar\omega_1$ and $\hbar\omega_2$, where $\hbar\omega_0 = \hbar\omega_1 + \hbar\omega_2$; otherwise the splitting into $\hbar\omega_1$ and $\hbar\omega_2$ is arbitrary (this means that the laser can be tuned by creating selective feedback at the different frequencies ω_1 and ω_2). Ultimately there are no changes in the medium and in place of the quantum $\hbar\omega_0$ two quanta – $\hbar\omega_1$ and $\hbar\omega_2$ – are created. From the viewpoint of a semiclassical description in terms of nonlinear susceptibilities and fields classically assigned, each two fields of the three induce a change in the third, producing a polarization response at the frequency of this field. It should be noted that significant parametric conversion requires particular phase relationships between the three interacting fields; this requires fulfillment of the "spatial synchronism" condition: $\mathbf{k}_0 = \mathbf{k}_1 + \mathbf{k}_2$, where $\mathbf{k}_\alpha$ are the wave vectors of the corresponding fields.

We note that in this method the energy of the quanta $\hbar\omega_0$ and $\hbar\omega_2$ is drawn from the coherent pumping energy $\hbar\omega_0$ from the auxiliary laser. This method has been experimentally tested [10].

In the second method induced two-quantum transitions occur in the inversely populated medium: The medium passes from the excited state 1 to the ground state 0 with the simultaneous emission of two quanta $\hbar\omega_1 + \hbar\omega_2 = \hbar\omega_{10}$ (where $\hbar\omega_{10}$ is the difference in terms of the working substance); in other respects the splitting into $\hbar\omega_1$ and $\hbar\omega_2$ is arbitrary and this allows smooth tuning. Two-quantum emission consists of two successive virtual transitions with the emission of $\hbar\omega_1$ and $\hbar\omega_2$. The probability of such a process is proportional to the product of the numbers of the quanta $\hbar\omega_1$ and $\hbar\omega_2$. We note that in this method the energy of $\hbar\omega_1$ and $\hbar\omega_2$ is drawn from the substance, in which an inverse population is produced by ordinary noncoherent pumping. A characteristic of this method is that it does not require fulfillment of the "spatial synchronism" principle, the securing of which, especially in a broad band of frequencies, entails considerable experimental difficulties. All this suggests that the second method can give a high yield of quanta $\hbar\omega_1$ and $\hbar\omega_2$ from unit volume. The probable reason why the second proposal has not yet been tested experimentally is the high threshold for quanta $\hbar\omega_1$ or $\hbar\omega_2$, which must be exceeded for startup of the two-photon laser. In the first method there is no threshold for $\hbar\omega_1$ and $\hbar\omega_2$, and there is a lower threshold for the supplied power $\hbar\omega_0$.

From the viewpoint of semiclassical treatment the two-photon laser can be described by classical fields $E(\omega_1)$ and $E(\omega_2)$ in a resonator built up by external currents which are proportional to the time derivative of the nonlinear polarizations at frequencies ω_1 and ω_2 due to the interaction of the waves $E(\omega_1)\,E(\omega_2)$ in the inversely populated medium.

The third proposal was put forward with the aim of reducing the threshold of startup of two-photon laser. As in the second method, an inverse population between levels 1 and 0 is created in the medium, and quanta $\hbar\omega_1$ and $\hbar\omega_2$ are produced, but, as in the first method, the quanta $\hbar\omega_0$ are present. Here

$$\hbar\omega_0 = \hbar\omega_1 + \hbar\omega_2 \approx \hbar\omega_{10}$$

(as in the first and second methods, in other respects the splitting into $\hbar\omega_1$ and $\hbar\omega_2$ is arbitrary, which is important for tuning). In this resonance case the quanta $\hbar\omega_1$ and $\hbar\omega_2$ can be produced either by the three-quantum process of conversion of $\hbar\omega_0$ to $\hbar\omega_1$ and $\hbar\omega_2$, or by two-quantum luminescence. The three-quantum process cannot be regarded independently of the one-quantum induced emission of $\hbar\omega_0$ and the two-quantum induced emission of $\hbar\omega_1$ and $\hbar\omega_2$; the three-quantum process emerges as the interference of these two. We note that effective three-quantum parametric conversion will occur if the "spatial synchronism" condition is fulfilled but, as distinct from the first method, this requirement in the third proposal is apparently less rigorous, since the parametric process plays more of an auxiliary role, and when the fields ω_1 and ω_2 reach a particular value the two-quantum process becomes dominant. The third method with a prescribed particular number of quanta $\hbar\omega_0$ ensures easy startup of generation of $\hbar\omega_1$ and $\hbar\omega_2$. We note that the accumulation of quanta $\hbar\omega_0$ in the resonator does not necessarily require an auxiliary laser, but can be effected by one-quantum generation in the working substance itself on the transition ω_0.

The three methods are not competitive, but rather supplement one another, since each has its drawbacks and advantages. Besides tunability, lasers based on the second and third proposals may have other useful properties. For instance, they may be used as generators of special giant pulses, as generators of submillimeter waves, as broadband amplifiers, and so on.

The present paper is devoted to the discussion of the second and third methods. The arrangement of the material is as follows. In § 1 we derive the nonlinear polarization with due regard to saturation due to three quasimonochromatic fields, and also the field equations in the "resonance parametric" and "two-photon" lasers. The operation of a "two-photon" laser is discussed in § 2. In § 3 the use of a "two-photon" laser for the generation of submillimeter waves is discussed. The operation of a "resonance parametric" laser with stable phase is investigated in § 4. In § 5 the same laser at the random-phase limit is investigated. In § 6 experimental ways of realizing the proposed schemes are discussed. In this paper nonlinear effects are treated both by semiclassical and by quantum-electrodynamic methods, which in certain cases are simpler and clearer. A brief description of this method is given in Appendix 1. As practice of recent years has shown (symposia and seminars on nonlinear optics), physicists engaged in nonlinear optics are not agreed as regards the range of applicability of the semiclassical and quantum approaches. Appendix II is devoted to a discussion of these questions.

§ 1. Derivation of Equations of Fields in "Resonance Parametric" and "Two-Photon" Lasers. Calculation of Nonlinear Polarization with Saturation Taken into Account

We consider the operation of "resonance parametric" and "two-photon" lasers in the language of classical fields and the nonlinear polarization of the active medium, calculated quantum-mechanically, due to their interaction. Such a treatment is sound practice, since we are interested in the nonlinear interaction of the substance with fields composed of large number of quanta. Such a semiclassical description, as distinct from the quantum-electrodynamic treatment, which is sometimes more convenient in other respects (see Appendices I and II), allows a simple consideration of the phase relationships, which are particularly important in the case of parametric interaction.

The obtention of a closed system of equations for the "resonance parametric" and "two-photon" lasers consists of the solution of two problems:

1) the electrodynamic problem of excitation of the resonator by an external force created by nonlinear polarization of unit volume of substance $\mathcal{P}$, acquired by the medium under the action of the field;

2) the quantum-mechanical problem of calculation of the nonlinear polarization in the presence of the resonator fields. We begin with the first problem.

As already mentioned in the "Introduction" the operation of a "resonance parametric" laser can be described by the interaction of three quasi-monochromatic fields ω_1, ω_2, and ω_3 in the substance contained in a resonator:

$$\mathscr{E}(\omega_\alpha) = E(\omega_\alpha)\, u(\omega_\alpha)\, \mathbf{j}_\alpha e^{i\omega_\alpha t}, \tag{1.1}$$

where $u(\omega_\alpha)$ are the eigenfunctions of the resonator, $\mathbf{j}_\alpha$ is the unit vector of polarization, $E(\omega_\alpha) = E^*(-\omega_\alpha)$ are the slowly (in comparison with $e^{i\omega_\alpha t}$) varying complex amplitudes, and

$$\omega_0 = \omega_1 + \omega_2 = \omega_{10}, \tag{1.2}$$

where $\hbar\omega_{10}$ is the difference in terms of the working substance. Then by known transformations from the Maxwell equations, as in the case of an ordinary laser [11], we obtain for the field amplitudes wave equations [we write the equation only for $E(\omega_1)$; the equations for $E(\omega_2)$ and $E(\omega_0)$ are similar]

$$\frac{\partial^2}{\partial t^2}\left(e^{i\omega_1 t} E(\omega_1)\right) + \frac{\omega_1}{Q_1}\frac{\partial}{\partial t}\left(E(\omega_1)\, e^{i\omega_1 t}\right) + \Omega_1^2 E(\omega_1)\, e^{i\omega_1 t} = -4\pi \frac{\partial^2}{\partial t^2}\int\limits_v \mathcal{P}_1 \mathbf{j}_1 u(\omega_1)\, dv, \tag{1.3}$$

where $\Omega_1 \approx \omega_1$ is the natural frequency of the resonator, Q_1 is the quality factor at frequency ω_1, and $\mathcal{P}_1$ is the polarization at frequency ω_1. We note that all the expressions obtained in this section and valid for a "resonance parametric" laser become the equations for a "two-photon" laser, if we put $E(\omega_0) = 0$. Generation of the field $E(\omega_0)$ in a "two-photon" laser can be regarded as an additional loss and removed by a reduction of Q_0.

We turn now to the solution of the second problem – calculation of the nonlinear polarization due to three quasi-monochromatic light fields. It should be pointed out from the very start that the polarization can be calculated only in general form by the solution of the Schrödinger equation. Calculation for specific substances entails the use of the obtained formulas, but necessitates prior knowledge of the energy levels of the substance and the corresponding wave functions. The wave functions and spectrum can be calculated in each individual case by an electronic digital computer; this is independent and difficult work. We will use our formulas only for approximate estimates and for the determination of functional relationships.

Our quantum system – a substance in the presence of classical fields – will be described by the Schrödinger equation (for simplicity of argument, and without loss of generality, we will regard the substance as one "atom" with one electron),

$$i\hbar\frac{\partial\Phi}{\partial t} = \left(\hat{H}_0 - \sum_{\pm\omega_\alpha} \hat{V}(\omega_\alpha)\, e^{i\omega_\alpha t}\right)\Phi, \tag{1.4}$$

where $\hat{H}_0$ is the Hamiltonian of the free "atom,"

$$\hat{V}(\omega_\alpha) = \boldsymbol{E}(\omega_\alpha)\,\hat{\boldsymbol{d}} = E(\omega_\alpha)(\boldsymbol{j}_\alpha \boldsymbol{d}), \tag{1.5}$$

d is the electric dipole moment operator. If the transition is forbidden, then instead of $\hat{\boldsymbol{d}}$ we must write the magnetic dipole moment μ, instead of $\boldsymbol{E}(\omega_\alpha)$ the magnetic field $\boldsymbol{H}(\omega_\alpha)$, and so on. In future we will not assume the smallness of the interaction (1.5); otherwise we cannot take into account the saturation, which is significant in strong and resonant light fields. The wave function of the atom in the classical field will be sought by representing the interaction in the form of a series

$$\Psi(t)=\sum_l a_l\varphi_l, \tag{1.6}$$

where φ_l is the eigenfunction of $\hat{H}_0$, corresponding to the energy level ε_l of the electron in the atom

$$\hat{H}_0\varphi_l=\varepsilon_l\varphi_l. \tag{1.7}$$

Substituting (1.6) in the Schrödinger equation in the representation of the interaction

$$i\hbar\frac{\partial\Psi}{\partial t}=-\left(e^{i\hat{H}_0 t/\hbar}\sum_{\pm\omega_\alpha}\hat{V}(\omega_\alpha)\,e^{-i\hat{H}_0 t/\hbar}\right)\Psi, \tag{1.8}$$

we obtain an infinite system of differential equations for a_l, which describes the behavior of the atom in the field without any restrictions:

$$i\hbar\dot{a}_0=\sum_{\pm\omega_\alpha}\left[a_1V_{01}(\omega_\alpha)\,e^{i(\omega_{01}+\omega_\alpha)t}+\sum_{l\geqslant 2}a_lV_{0l}(\omega_\alpha)\,e^{i(\omega_{0l}+\omega_\alpha)t}\right]; \tag{1.9a}$$

$$i\hbar\dot{a}_1=\sum_{\pm\omega_\alpha}\left[a_0V_{10}(\omega_\alpha)\,e^{i(\omega_{10}+\omega_\alpha)t}\sum_{l\geqslant 2}a_lV_{1l}(\omega_\alpha)\,e^{i(\omega_{1l}+\omega_\alpha)t}\right]; \tag{1.9b}$$

$$i\hbar\dot{a}_l=\sum_{\pm\omega_\alpha}[a_0V_{l0}(\omega_\alpha)\,e^{i(\omega_{l_0}+\omega_\alpha)t}+a_1V_{l1}(\omega_\alpha)\,e^{i(\omega_{l_1}+\omega_\alpha)t}+\cdots], \tag{1.9c}$$

where

$$V_{lm}(\omega_\alpha)=\mathbf{E}(\omega_\alpha)\,\mathbf{d}_{lm}=\mathbf{E}(\omega_\alpha)(\mathbf{j}_z\,\mathbf{d}_{lm}), \tag{1.10}$$

$$\omega_{lm}=\frac{1}{\hbar}(\varepsilon_l-\varepsilon_m). \tag{1.11}$$

We recall that between the levels ε_1 and ε_0 there is an inverse population and one of the acting light fields is a resonant one: $\varepsilon_1-\varepsilon_0'\approx\hbar\omega_0$; hence, in equations (1.9c) the particularly important terms are separated. All the subsequent operations are connected with simplification of the system (1.9a,b,c). The calculations in this case are generally very cumbersome and, hence, we are obliged to avoid presenting them, particularly since the method of these calculations has already been used in other investigations [12, 13]. Where possible, we will try to explain the physical sense of the mathematical approximations.

In equation (1.9c) we omit on the right hand side all the terms except the two written in explicit form: the coefficients a_l in general must be less than a_0a_1, since the population of levels ε_0, ε_1 exists, and on other levels it is small or, in any case, average for times $\Delta t\gg 1/\omega_\alpha\sim 10^{-15}$ sec. We integrate this equation over the time interval $\Delta t\gg 1/\omega_\alpha\sim 10^{-15}$ sec, but so that during this time the values of $a_0(t)$ and $a_1(t)$ do not change much. It is clear from (1.9c) that this slow change is ensured by the first term. We assume slight "detuning" of the resonance:

$$\omega_0-\omega_{10}=\Delta\neq 0, \tag{1.12}$$

but $\Delta/\omega_\alpha\ll 1$. Then the time of "slow" change of a_0, a_1 will be of the order of $1/\Delta$. We finally obtain

$$\hbar a_l=-\sum_{\pm\omega_\alpha}\left[\frac{V_{l0}(\omega_\alpha)\,e^{i(\omega_{l0}+\omega_\alpha)t}}{\omega_{l0}+\omega_\alpha}a_0+\frac{V_{l_1}(\omega_\alpha)\,e^{i(\omega_{l1}+\omega_\alpha)t}}{\omega_{l1}+\omega_\alpha}a_1\right]. \tag{1.13}$$

We substitute this expression in the first two equations (1.9a,b) and keep the terms with exponents which vary slowly with time, i.e., $e^{i\omega t}$, $\omega\to 0$. Such terms after integration will give a maximum contribution owing to the resonance denominator. We omit at present the usual polarizability on nonresonant frequencies, which is easy to allow for, but the formulas are complicated.

Introducing new symbols, we obtain a system of equations, formally analogous with the known equation in [14]:

$$i\dot{a}_0 = -K_{00}a_0 - K_{01}e^{i\Delta t}a_1,$$
$$i\dot{a}_1 = -K_{10}e^{-i\Delta t}a_0 - K_{11}a_1, \tag{1.14}$$

where we use the symbols:

$$K_{00} = \frac{1}{\hbar^2}\sum_{\pm\omega_\alpha, l}\frac{|V_{0l}(\omega_\alpha)|^2}{\omega_{l0}-\omega_\alpha}, \quad K_{11} = \frac{1}{\hbar^2}\sum_{\pm\omega_\alpha, l}\frac{|V_{1l}(\omega_\alpha)|^2}{\omega_{l1}-\omega_\alpha},$$

$$K_{01} = \frac{1}{\hbar^2}\sum_l\left[\frac{V_{0l}(\omega_1)V_{l1}(\omega_2)}{\omega_{l1}+\omega_2} + \frac{V_{0l}(\omega_2)V_{l1}(\omega_1)}{\omega_{l1}+\omega_1}\right] - V_{01}(\omega_0) = r_{01} - u_{01}, \tag{1.15}$$

$$K_{10} = \frac{1}{\hbar^2}\sum_l\left[\frac{V_{1l}(-\omega_1)V_{l0}(-\omega_2)}{\omega_{l0}-\omega_2} + \frac{V_{1l}(-\omega_2)V_{l0}(-\omega_1)}{\omega_{l0}-\omega_1}\right] - V_{01}^{+}(\omega_0) = r_{10} - u_{10},$$

where

$$\frac{1}{\hbar^2}\sum_l \ldots = r_{01}, \; V_{01} = u_{01},$$

and with accuracy to the small parameter

$$\frac{\Delta}{\omega_0} \sim \frac{\Delta}{\omega_1} \sim \frac{\Delta}{\omega_2} \ll 1$$
$$r_{01} = \overset{*}{r}_{10}, \; K_{01} = K_{10}^{*}.$$

It is more convenient to convert from coefficients a_i to the density matrix σ_{ik}:

$$\sigma_{00} = |a_0|^2; \quad \sigma_{11} = |a_1|^2; \quad \sigma_{01} = a_0 a_1^{*} e^{i\Delta t};$$
$$\sigma_{0l} = a_0 \overset{*}{a}_l; \quad \sigma_{1l} = a_1 \overset{*}{a}_l. \tag{1.16}$$

From (1.13), (1.16)

$$\sigma_{0l} = -\sum_{\mp\omega_\alpha}\frac{1}{\hbar}\left[\sigma_{00}\frac{V_{0l}(-\omega_\alpha)e^{-i(\omega_{l0}+\omega_\alpha)t}}{\omega_{l0}+\omega_\alpha} + \sigma_{01}\frac{V_{1l}(-\omega_\alpha)e^{-i(\omega_{l1}+\omega_\alpha)t}}{\omega_{l1}+\omega_\alpha}\right],$$

$$\sigma_{1l} = -\frac{1}{\hbar}\sum_{\mp\omega_\alpha}\left[\sigma_{10}\frac{V_{0l}(-\omega_\alpha)e^{-i(\omega_{l0}+\omega_\alpha)t}}{\omega_{l0}+\omega_\alpha} + \sigma_{11}\frac{V_{1l}(-\omega_\alpha)e^{-i(\omega_{l1}+\omega_\alpha)t}}{\omega_{l1}+\omega_\alpha}\right].$$

We can now replace system (1.14) by the equations for the density matrix*:

$$i\dot{\sigma}_{00} = K_{10}\sigma_{01} - K_{01}\sigma_{10} - \frac{i}{T_1}(\sigma_{00} - \sigma_{00}^0),$$
$$i\dot{\sigma}_{11} = K_{01}\sigma_{10} - K_{10}\sigma_{01} - \frac{i}{T_1}(\sigma_{11} - \sigma_{11}^0), \tag{1.17}$$
$$i\dot{\sigma}_{01} = \gamma_{10}\sigma_{10} + K_{01}(\sigma_{00} - \sigma_{11}) - \frac{i}{T_2}\sigma_{01},$$

$$\gamma_{10} = \Delta + K_{11} - K_{00}. \tag{1.18}$$

*The problem can be formulated from the very start by using the formalism of density matrix theory. We have chosen another approach in view of its greater physical clarity.

Here T_1, T_2 are the longitudinal and transverse relaxation times, which are usually introduced in such a treatment. $\sigma^0_{\alpha\alpha}$ denotes the equilibrium values of the population in the absence of light fields. Equations (1.17) describe the "slow" (in comparison with the times $1/\omega_0 \sim 1/\omega_1 \sim 1/\omega_2 \sim 10^{-15}$ sec) change in the amplitudes of the states. It is physically clear that the population can be considerably altered during the times of the spontaneous processes T_1 or during the time of transition processes in the resonator. In lasers the following condition is usually fulfilled:

$$10^{-8}\,\text{sec} \sim T_{\text{res}} \gg T_1 \gg T_2 \sim 10^{-11}\,\text{sec}. \tag{1.19}$$

The transverse relaxation time is connected with the broadening of the line due to interaction with phonons. A simple substitution shows that the third equation from (1.17) is satisfied by the expression

$$\sigma_{01} = -\int_0^t [\sigma_{00}(t') - \sigma_{11}(t^1)]\, K_{01}\, e^{\left(i\gamma_{01} + \frac{1}{T_2}\right)(t'-t)}\, dt'. \tag{1.20}$$

If the condition $t \gg T_2$ and (1.19) are satisfied, the preexponential functions, being slowly varying, can be taken out of the integrand:

$$\sigma_{01} = \frac{[\sigma_{11}(t) - \sigma_{00}(t)]}{\gamma_{10} - \dfrac{i}{T_2}} K_{01}. \tag{1.21}$$

Using this expression and (1.17) we can easily write an equation describing the inverse population:

$$\Delta\dot{N} = -\frac{4K_{01}K_{10}}{T_2\left[\gamma_{10}^2 + \dfrac{1}{T_2^2}\right]}\Delta N + \frac{1}{T_1}(\Delta N^0 - \Delta N), \tag{1.22}$$

where

$$\Delta N = \sigma_{11} - \sigma_{00}.$$

We turn our attention to one circumstance:

$$K_{01}K_{10} = (r_{01} - u_{01})(r_{10} - u_{10}) = u_{01}u_{10} + r_{01}r_{10} - (u_{01}r_{10} + u_{10}r_{01}). \tag{1.23}$$

The first term in this expression is due to one-quantum induced emission (absorption), the second to the two-quantum process, and the third term enclosed in the brackets to the parametric interaction of all three fields simultaneously. The sign of this term depends on the phase relationships of the fields, and all three processes affect the population of the "working" levels ε_0, ε_1 of the substance. From equations (1.17) and (1.21) we can calculate the density matrix and with its help find the polarization, taking saturation into account:

$$\mathcal{P} = \text{Sp}\,[\hat{\sigma} e^{i\hat{H}_0 t/\hbar}\hat{d}_l e^{-i\hat{H}_0 t/\hbar}] = \mathcal{P}_0(\omega_0) + \mathcal{P}_1(\omega_1) + \mathcal{P}_2(\omega_2). \tag{1.24}$$

We write the polarization separately:

$$\mathcal{P}_1(\omega_1) = [\sigma_{00}\mathbf{k}_{00}(\omega_1)E(\omega_1) + \sigma_{11}\mathbf{k}_{11}(\omega_1)E(\omega_1) + \sigma_{01}\mathbf{k}_{10}(\omega_1)E(\omega_1) + \sigma_{01}\mathbf{k}_{10}(\omega_1)E(-\omega_2)]\,e^{i\omega_1 t},$$

$$\mathcal{P}_2(\omega_2) = [\sigma_{00}\mathbf{k}_{00}(\omega_2)E(\omega_2) + \sigma_{11}\mathbf{k}_{11}(\omega_2)E(\omega_2) + \sigma_{01}\mathbf{k}_{10}(\omega_2)E(-\omega_1)]\,e^{i\omega_2 t}, \tag{1.25}$$

$$\mathcal{P}_0(\omega_0) = [\sigma_{00}\mathbf{k}_{00}(\omega_0)E(\omega_0) + \sigma_{11}\mathbf{k}_{11}(\omega_0)E(\omega_0) + \sigma_{01}\mathbf{d}_{10}]\,e^{i\omega_0 t}.$$

Here we use the symbols

$$\mathbf{k}_{00}(\omega_1)=\frac{1}{\hbar}\sum_{l\geqslant 2}\left[\frac{\mathbf{j}_1\mathbf{d}_{0l}\mathbf{d}_{l0}}{\omega_{l0}-\omega_1}+\frac{\mathbf{j}_1\mathbf{d}_{l0}\mathbf{d}_{0l}}{\omega_{l0}+\omega_1}\right],$$
$$\mathbf{k}_{01}(\omega_1)=\frac{1}{\hbar}\sum_{l\geqslant 2}\left[\frac{\mathbf{j}_2\mathbf{d}_{0l}\mathbf{d}_{l1}}{\omega_{l0}-\omega_2}+\frac{\mathbf{j}_2\mathbf{d}_{l1}\mathbf{d}_{0l}}{\omega_{l1}+\omega_2}\right]. \tag{1.26}$$

The vector $\mathbf{j}_\alpha$ is the unit vector of the field of frequency ω_α. The terms $\sim\sigma_{00},\sigma_{11}$ in (1.25) make a contribution only to the ordinary linear polarization and are of no interest for our treatment, particularly since we omitted such terms in the deduction of (1.14).

For our purposes we can use a simplified expression for the polarizability from (1.25), keeping only the term with σ_{10}. We write simplified expressions for the polarization:

$$\mathcal{P}_1\mathbf{j}_1=[\chi_1(\omega_1,\omega_2,\omega_0)\,\Delta N\mathcal{E}(-\omega_2)\mathcal{E}(\omega_0)+\Theta(\omega_1,\omega_2)\,\Delta N\,|\mathcal{E}(\omega_2)|^2\,\mathcal{E}(\omega_1)],$$
$$\mathcal{P}_2\mathbf{j}_2=[\chi_2(\omega_2,\omega_1,\omega_0)\,\Delta N\mathcal{E}(-\omega_1)\mathcal{E}(\omega_0)+\Theta(\omega_2,\omega_1)\,\Delta N\,|\mathcal{E}(\omega_1)|^2\,\mathcal{E}(\omega_2)], \tag{1.27}$$
$$\mathcal{P}_0\mathbf{j}_0=[\eta(\omega_0)\,\Delta N\mathcal{E}(\omega_0)+\chi_0(\omega_0,\omega_1,\omega_2)\,\Delta N\mathcal{E}(\omega_1)\mathcal{E}(\omega_2)].$$

Here

$$\chi_1(\omega_1,\omega_2,\omega_0)=\frac{1}{\hbar}\,\frac{(\mathbf{d}_{01}\mathbf{j}_0)(\mathbf{k}_{01}(\omega_1)\,\mathbf{j}_1)\left[\gamma_{10}+\frac{i}{T_2}\right]}{\gamma_{10}^2+\frac{1}{T_2^2}}, \tag{1.28}$$

$$\chi_2(\omega_2,\omega_1,\omega_0)=\chi_1(\omega_{10},\omega_2,\omega_0), \tag{1.29}$$

$$\chi_0(\omega_0,\omega_1,\omega_2)=\frac{1}{\hbar}\,\frac{(\mathbf{d}_{10}\mathbf{j}_0)(\mathbf{k}_{01}(\omega_1)\,\mathbf{j}_1)\left[\gamma_{10}+\frac{i}{T_2}\right]}{\gamma_{10}^2+\frac{1}{T^2}}. \tag{1.30}$$

We note that if $T_2\to\infty$, then

$$\chi_0(\omega_0,\omega_1,\omega_2)=\chi_1^*(\omega_1,\omega_2,\omega_0); \tag{1.31}$$

when $\gamma_{01}=0$

$$\chi_0^*(\omega_0,\omega_1,\omega_2)=-\chi_1(\omega_1,\omega_2,\omega_0). \tag{1.32}$$

We expand the other symbols:

$$\eta(\omega_0)=\frac{1}{\hbar}\,\frac{(\mathbf{d}_{01}\mathbf{j}_0)^2\left(\gamma_{10}+\frac{i}{T_2}\right)}{\gamma_{10}^2+\frac{1}{T_2^2}}, \tag{1.33}$$

$$\Theta(\omega_1,\omega_1)=\Theta(\omega_2,\omega_1)=|\mathbf{k}_{01}(\omega_1)\,\mathbf{j}_1|^2\,\frac{\gamma_{10}+\frac{i}{T_2}}{\hbar\left[\gamma_{10}^2+\frac{1}{T_2^2}\right]}. \tag{1.34}$$

For the inverse population we write equation (1.22) in more detail:

$$\frac{d}{dt}(\Delta N) = -\frac{4\Delta N}{\hbar}|E(\omega_0)|^2 u^2(\omega_0)\,\mathrm{Im}\,[\eta(\omega_0)] -$$

$$-\frac{4}{\hbar}\Delta N\,|E(\omega_1)|^2\,|E(\omega_2)|^2 u^2(\omega_1)\,u^2(\omega_2)\,\mathrm{Im}\,[\Theta(\omega_1,\omega_2)] -$$

$$-\frac{8}{\hbar}\Delta N\,\mathrm{Re}\left[\frac{\chi_0(\omega_0,\omega_1,\omega_2)}{T_2\left(\gamma_{10}+\frac{i}{T_2}\right)}E(\omega_1)\,E(\omega_2)\,E^*(\omega_0)\,u(\omega_1)\,u(\omega_2)\,u(\omega_0)\right] - \frac{1}{T_1}(\Delta N - \Delta N^0). \qquad (1.35)$$

From this equation it is easy to see the physical sense of each term: The first describes one-quantum luminescence, the second two-quantum luminescence, and the penultimate term is responsible for the parametric process.†

From (1.3) we can obtain "shortened" equations for the slow field amplitudes. For this we discard terms of the type $\ddot{E}$, $\omega\dot{E}/Q$:

$$\dot{E}(\omega_1) + \frac{1}{2}\frac{\omega_1}{Q_1}E(\omega_1) + i(\omega_1 - \Omega_1)\,E(\omega_1) =$$

$$= -2\pi\omega_1 i\,[\chi_1(\omega_1,\omega_2,\omega_0)\,E(-\omega_2)\,E(\omega_0)\int_V u^*(\omega_1)\,u^*(\omega_2)\,u(\omega_0)dv +$$

$$+\Theta(\omega_1,\omega_2)\,|E(\omega_2)|^2\,E(\omega_1)\int_V |u(\omega_1)|^2\,|u(\omega_2)|^2\,dv]\,\Delta N,$$

$$\dot{E}(\omega_2) + \frac{1}{2}\frac{\omega_2}{Q_2}E(\omega_2) + i(\omega_2 - \Omega_2)\,E(\omega_2) =$$

$$= -2\pi\omega_2 i\,[\chi_2(\omega_2,\omega_1,\omega_0)\,E(-\omega_1)\,E(\omega_0)\int_V u^*(\omega_1)\,u^*(\omega_2)\,u(\omega_0)\,dv + \qquad (1.36)$$

$$+\Theta(\omega_{,2}\,\omega_1)\,|E(\omega_1)|^2\,E(\omega_2)\int_V |u(\omega_1)|^2\,|u(\omega_2)|^2\,dv]\,\Delta N,$$

$$\dot{E}(\omega_0) + \frac{1}{2}\frac{\omega_0}{Q_0}E(\omega_0) + i(\omega_0 - \Omega_0)\,E(\omega_0) = -2\pi\omega_0 i\,[\eta(\omega_0)\,E(\omega_0)\int_V |u(\omega_0)|^2\,dv +$$

$$+\chi_0(\omega_0,\omega_1,\omega_2)\,E(\omega_1)\,E(\omega_2)\int_V u(\omega_1)\,u(\omega_2)\,u^*(\omega_0)\,dv]\,\Delta N.$$

The integrals $\int_V u^*(\omega_1)u^*(\omega_2)u(\omega_0)\,dv$ and $\int_V |u(\omega_1)|^2\,|u(\omega_2)|^2\,dv$ are constant, depending on the specific design of the laser; their value can vary from 0 to 1. For instance, in the special case of traveling waves and when the spatial synchronism condition

$$k_1 + k_2 - k_0 = 0 \qquad (1.37)$$

is fulfilled the considered integrals can be regarded as units. From equations (1.36) in this case we convert to equations for the "numbers of quanta":

$$\frac{d\,|E(\omega_1)|^2}{dt} + \frac{\omega_1}{Q_1}|E(\omega_1)|^2 = 4\pi\omega_1\,\mathrm{Im}\,[\chi_1(\omega_1,\omega_2,\omega_0)]\,E(-\omega_1)\,E(-\omega_2)\times$$

$$\times E(\omega_0)\,\Delta N + 4\pi\omega_1\,\mathrm{Im}\,[\Theta(\omega_1,\omega_2)]\,|E(\omega_1)|^2\,|E(\omega_2)|^2\,\Delta N;$$

$$\frac{d\,|E(\omega_2)|^2}{dt} + \frac{\omega_2}{Q_2}|E(\omega_2)|^2 = 4\pi\omega_2\,\mathrm{Im}\,[\chi_2(\omega_2,\omega_1,\omega_0)]\times$$

$$\times E(-\omega_1)\,E(-\omega_2)\,E(\omega_0)\,\Delta N + 4\pi\omega_2\,\mathrm{Im}\,[\Theta(\omega_2,\omega_1)\,|E(\omega_1)|^2\,|E(\omega_2)|^2\,\Delta N,$$

† In the case of the combination process $\omega_1 - \omega_2 = \omega_{10}$, $\Theta'(\omega_1,\omega_2) = \Theta(\omega_1,-\omega_2)$ must be taken instead of $\Theta(\omega_1,\omega_2)$, and $\Theta'(\omega_2,\omega_1) = \Theta^*(\omega_1,-\omega_2)$ used instead of $\Theta(\omega_2,\omega_1)$.

$$\frac{d\,|E(\omega_0)|^2}{dt} + \frac{\omega_0}{Q_0}|E(\omega_0)|^2 = 4\pi\omega_0 \operatorname{Im}[\chi_0(\omega_0,\ \omega_1,\ \omega_2)\,E(-\ \omega_1)\times$$
$$\times\,E(-\ \omega_2)\,E(\omega_0)\,\Delta N + 4\pi\omega_0 \operatorname{Im}[\eta_0(\omega_0)]\,\Delta N\,|E(\omega_0)|^2;$$
$$\frac{d\Delta N}{dt} = -\frac{4\Delta N}{\hbar}|E(\omega_0)|^2 \operatorname{Im}[\eta(\omega_0)] - \frac{4\Delta N}{\hbar}|E(\omega_1)|^2|E(\omega_2)|^2 \operatorname{Im}[\Theta(\omega_1,\ \omega_2)] -$$
$$-\frac{8\Delta N}{\hbar}\operatorname{Re}\left[\frac{\chi_0(\omega_0,\ \omega_1,\ \omega_2)}{T_2\left(\gamma_{10}+\frac{i}{T_2}\right)}E(+\ \omega_1)\,E(+\ \omega_2)\,E(-\ \omega_0)\right] - \frac{1}{T_1}(\Delta N - \Delta N^0). \qquad (1.38)$$

The term proportional to the product of the amplitudes of the three fields is responsible for the parametric process; its sign and magnitude depend on the relationship of the field phases. In concluding this section we wish to draw attention to a curious property of the resonance parametric process. For simplicity of argument we omit from expressions (1.38) the terms describing one-quantum and two-quantum luminescence, and we calculate the change in energy of all three fields:

$$\frac{\Delta W}{\Delta t} = \frac{1}{4\pi}\left[\frac{d\,|E(\omega_1)|^2}{dt} + \frac{d\,|E(\omega_2)|^2}{dt} + \frac{d\,|E(\omega_0)|^2}{dt}\right]. \qquad (1.39)$$

In formula (1.18) we introduced the quantity γ_{10}, which depends on the "detuning" Δ of the resonance and on the shift of the level of the atom due to the field. We can consider two limiting cases:

$$|\gamma_{10}T_2| \gg 1; \qquad (1.40a)$$

this case corresponds to ordinary parametric interaction in a transparent medium:

$$|\gamma_{10}T_2| \ll 1; \qquad (1.40b)$$

this case corresponds to "resonance" parametric interaction.

We recall that T_2 is the transverse relaxation time and is due to the interaction of the electron with phonons, i.e., $T_2 \sim 10^{-11}$ sec.

In the first case

$$\frac{\Delta W_1}{\Delta t} = \frac{L}{T_2\gamma_{10}^2}$$

and (1.41)

$$\frac{\Delta W_2}{\Delta t} = LT_2,$$

in the second case, where

$$L = \operatorname{Re}\left\{\frac{2}{\hbar}(\mathbf{d}_{01}\mathbf{j}_0)(\mathbf{k}_{01}(\omega_1)\,\mathbf{j}_1)\,E(-\ \omega_1)\,E(-\ \omega_2)\,E(-\ \omega_0)\right\}\Delta N.$$

We find the ratio

$$\frac{\Delta W_1}{\Delta t}\bigg/\frac{\Delta W_2}{\Delta t} = \frac{1}{(\gamma_{10}(1)\,T_2)^2} \ll 1, \qquad (1.42)$$

for $\gamma_{10}(1) - \gamma_{10}$ for case (1.40a).

We assumed that T_2 is the same in both cases and only the value of Δ in (1.40) is different. But it is clear from (1.41) that for the second case the ratio $\Delta W_2/\Delta t$ is independent of γ_{10}, and consequently inequality (1.42) is fulfilled. On the other hand, it follows from (1.38)*

*One-quantum and two-quantum processes are again ignored.

$$\frac{d\Delta N}{dt} = -\frac{2}{\hbar\omega_0}\frac{\Delta W}{\Delta t}. \tag{1.43}$$

Returning again to inequalities (1.42) and (1.43), it immediately becomes clear that in one limiting case (1.40b) the parametric interaction greatly affects the population of resonance levels, and in the other (1.40a) it has a weak effect. If we introduce the pumping of energy from one field into the two others

$$\frac{\Delta R}{\Delta t} = \frac{1}{2\pi}\frac{d}{dt}\left[|E(\omega_1)|^2 + |E(\omega_2)|^2 - |E(\omega_0)|^2\right], \tag{1.44}$$

this quantity also behaves differently for the two cases (1.40). In the case of a strong effect on the population of the levels (1.40b), pumping of energy between the fields is reduced, although the total energy of the fields changes considerably. In the other case (1.40a) pumping of energy between the fields is increased when the change in inversion is reduced (the total energy of the fields is unaltered). For certain definite phase and amplitude relationships of the fields in both cases there is no change in the inverse population of the levels [see (1.22) and (1.23)].

These interesting features are not encountered in the case of ordinary parametric interaction in transparent media.

§ 2. Two-Photon Laser

A powerful tunable laser based on two-quantum luminescence was proposed in [8]. Papers [12, 13, 15-18] are devoted to the discussion of the two-photon laser. We give a brief account of this proposal.

On transition from the excited state ε_1 to the unexcited state ε_0, the atom emits one light quantum $\varepsilon_1 - \varepsilon_0 = \hbar\omega_0$. However, there is [19, 20] a nonzero probability of two-quantum luminescence, provided there is no violation of the law of conservation of energy $\varepsilon_1 - \varepsilon_0 = \hbar\omega_1 + \hbar\omega_2$ (Fig. 1). This effect was recently discovered experimentally [21]. Such a process can also be induced. By means of perturbation theory it is easy to calculate the probability of two-quantum induced luminescence in one mode [see (A.I.16)].

$$W = \frac{2\pi}{\hbar}(n_1+1)(n_2+1)\left|\sum_m \frac{v_{1m}v_{m0}}{\varepsilon_1 - \varepsilon_m - \hbar\omega\Delta_{1,2}}\right|^2 \frac{\Delta N}{\varepsilon}, \tag{2.1}$$

where ΔN is the inverse population.

Formula (A.I.16) is obtained in the approximation $\varepsilon \ll \varepsilon_1 - \varepsilon_0$. It is clear that the accuracy of "division" of the quanta is limited by the widths of the levels ε_1 and ε_0. For the estimates required later we take ε of the order of the line width. For the R-line of ruby, for instance, $\varepsilon \sim 5\text{-}7\ \text{cm}^{-1}$ at $T \le 300°K$. From (2.1) it is easy to write the well-known condition of self-excitation for one of the types of quanta [8]:

$$\frac{2\pi}{\hbar}(n_1+1)(n_2+1)\left|\sum_m \frac{v_{1m}v_{m0}}{\varepsilon_1 - \varepsilon_m - \hbar\omega_{1,2}}\right|^2 \frac{\Delta N}{\varepsilon} \geqslant \frac{\omega_1}{Q_1} n_1. \tag{2.2}$$

Here Q_1 is the quality factor of the resonator. By introducing a field of quanta from an external source (laser) into the working substance of the laser the condition for self-excitation (2.2) can be satisfied and such a laser started up. Since the quanta $\hbar\omega_1$ and $\hbar\omega_2$ are created in pairs, there can be a self-maintaining generation regime, during which, by alteration of the frequency-selective quality factor $Q_{1,2}$ of the resonator, the laser can be tuned to the generation frequency. Numerical estimates show that the density of quanta in the beam of a megawatt laser can be sufficient and startup presents no insuperable difficulties [12, 13, 15-18]. As distinct from the parametric generator the two-quantum laser does not require fulfillment of the synchronism condition.

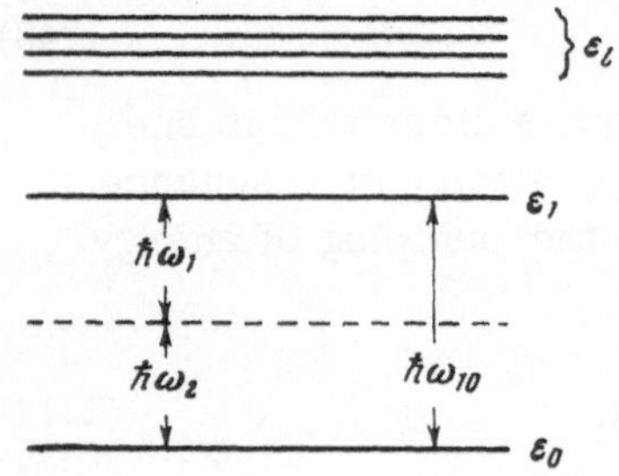

Fig. 1. Diagram of energy levels. ε_0, ε_1, ε_l, energy levels of substance.

We turn our attention to another important circumstance. According to formula (2.2), the probability of induced two-quantum luminescence increases as the square of the number of quanta, in contrast to the linear increase in the one-quantum process. This means that the two-photon laser can be used as a giant-pulse generator [17, 18, 22]. In such a laser the steepness of the leading edge will be much greater than in existing lasers, which may be important for practical application.

We now consider the degenerate regime $\omega_1 = \omega_2 = \omega$ in a "two-photon" laser by using the semiclassical results of § 1 (see also [13, 15, 16]). We find the stationary values of the polarization and ΔN from (1.22) and (1.27) by equating the derivative to zero in (1.22):

$$\Delta N_{st} = \Delta N^0 \left[1 + \frac{K_{01}K_{10}T_1}{T_2\left[\gamma_{10}^2 + \frac{1}{T_2^2}\right]}\right]^{-1}, \tag{2.3}$$

$$P(\omega)\mathbf{j} = \frac{\Theta(\omega,\omega)}{2}\Delta N_{st}\,|E(\omega)|^2 E(\omega)e^{i\omega t}. \tag{2.4}$$

From equations (1.36) we can obtain the stationary values of the amplitude and phase [we assume the field $E(\omega_0)$ equal to zero]*:

$$\frac{1}{Q}|E(\omega)|^2 = 2\pi\,\mathrm{Im}\,[\Theta(\omega,\omega)]\,\Delta N_{st}\,|E(\omega)|^4; \tag{2.5}$$

$$|E(\omega)|^2(\omega - \Omega) = -2\pi\omega\,\mathrm{Re}\,[\Theta(\omega,\omega)]\,\Delta N_{st}\,|E(\omega)|^4. \tag{2.6}$$

For the sake of simplicity we put

$$\Delta = \Omega - \omega = 0, \quad |\gamma_{01}| \ll \frac{1}{T_2}.$$

Then, using (2.3), (1.23), (1.26), (1.34), and (2.5), we obtain

$$\Delta N_{st} = \Delta N^0[1 + K_{01}K_{10}T_1T_2]^{-1}, \tag{2.7}$$
$$K_{01}K_{10} = \frac{1}{\hbar^2}(\mathbf{k}_{10}(\omega)\,\mathbf{j})^2\,|E(\omega)|^4,$$
$$\mathrm{Im}\,\Theta(\omega,\omega) = \frac{(\mathbf{k}_{10}(\omega)\,\mathbf{j})^2}{\hbar}T_2,$$

and also the values of the square of the field amplitude in three equilibrium positions:

$$|E(\omega_2)|_1^2 = 0,$$

$$|E(\omega)|_{2,3}^2 = \frac{\hbar\left(\pi Q\Delta N^0 \pm \sqrt{(\pi Q\Delta N^0)^2 - \frac{T_1}{T_2(\mathbf{k}_{10}(\omega)\,\mathbf{j})^2}}\right)}{T_1}. \tag{2.8}$$

It is easy to see that the first trivial equilibrium position is stable. $|E(\omega)|_2^2$ determines the threshold of self-excitation (with a higher initial inverse population the fields required for excitation of the system decrease). $|E(\omega)|_3^2$ give the amplitude in the steady regime.

* In the case of the degenerate regime (1.36) θ and χ_1 must be replaced by $\theta(\omega_1, \omega)/2$ and $\chi_1(\omega, \omega, \omega_0)$ in the equation for $d|E(\omega)|^2/dt$ and by $\theta(\omega_1, \omega)/4$ and $\chi(\omega, \omega, \omega_0)$ in the equation for $d\Delta N/dt$.

From equation (2.8), since $\mathrm{Im}\,|E(\omega)|^2 = 0$, we must have the inequality

$$\pi Q \Delta N^0 (\mathbf{k}_{10}(\omega)\,\mathbf{j})\, T_2 > \sqrt{T_1 T_2}$$

or, in a different form, (2.9)

$$(\pi Q \Delta N^0)^2 > \frac{T_1}{\hbar\, \mathrm{Im}\,[\Theta(\omega,\omega)]},$$

which imposes a restriction on $Q\Delta N^0$. If inequality (2.9) is not satisfied, there will be no equilibrium positions (2.3) and the laser will operate in the single-pulse regime. If inequality (2.9) is satisfied with plenty of reserve, the values of the fields in the equilibrium position are as follows:

$$|E(\omega)|_2^2 = \hbar/T_2 (\mathbf{k}_{10}(\omega)\,\mathbf{j})^2 \Delta N^0 Q\pi, \qquad (2.10)$$

$$|E(\omega)|_3^2 = 2\hbar\pi Q \Delta N^0/T_1. \qquad (2.11)$$

For our estimates we take values typical of ruby (in the cgse system):

$$T_2 = 10^{-11}, \quad Q = 10^8, \quad d_{1l} = d_{l0} = 10^{-18}, \quad \omega = 10^{15}, \quad T_1 = 3\cdot 10^{-3},$$
$$\mathbf{k}_{10}(\omega) \sim 2\,|d_{10}|^2/\hbar\omega = 2\cdot 10^{-24}.$$

In the sum for $\mathbf{k}_{10}(\omega)$ we have chosen one term [23]. Then from (2.9) we obtain

$$\Delta N \geqslant 3\cdot 10^{19}. \qquad (2.12)$$

For the estimates from (2.10) and (2.11) we will assume that $\Delta N^0 = 10^{20}$. Then

$$|E(\omega)|_2^2 = 6\cdot 10^2 \text{ cgse units} \qquad (2.13)$$

corresponding to the flux density $P_2 = 0.3$ MW/cm², and

$$|E(\omega)|_3^2 = 2\cdot 10^4 \text{ cgse units} \qquad (2.14)$$

for $P_3 = 10$ MW/cm².

Thus, the estimates (2.13) show that startup of the laser requires a field $E(\omega)$ with a flux density $P_2 = 0.3$ MW/cm², which agrees with the estimates from (2.2). The stationary value (2.14) may be stable (for a fuller discussion see [15, 16]).

§ 3. Generation of Submillimeter Waves by Two-Quantum Luminescence

As a specific example we consider in detail the use of a two-quantum laser to generate submillimeter waves [24]. The importance of developing generators of submillimeter waves ($1000 < \lambda \leq 10$) is well known. The proposed method holds out the hope of obtaining powers of tens of kilowatts in this range and of smooth tuning. The practical realization of the method requires the selection of a working substance and a laser of particular power, generating light of such frequency that the energy of its quanta is a little less (in the case of Raman scattering a little more) than the difference between two particular energy levels in the substance. This frequency difference will correspond to the energy of the submillimeter quantum. In the working substance inversion must be created between these two levels. If a laser beam is now directed into a substance prepared in this way, then two-quantum luminescence (or Raman scattering) will lead to amplification and in certain conditions to generation of an electromagnetic field on the difference frequency, i.e., in the submillimeter range.

Since in the case of condensed media, particularly luminescent crystals, the absorption of submillimeter emission has been poorly investigated, and in monatomic rarefied gases there are wide "windows" for submillimeter radiation, to illustrate the proposed method we take as a working substance the gas neon, illuminated by the light of a ruby laser through a frequency converter.

The maximum attainable useful power produced in unit volume in the proposed method can be

$$P = \frac{\Delta N \hbar\omega}{\tau_0}, \tag{3.1}$$

where τ_0 is the time of action of the exciting laser, ω is the frequency of the submillimeter radiation. When $\Delta N \sim 10^{10}$ cm^{-3}, $t_0 = 10^{-8}$ sec, $h\omega \sim 0.01$ eV $\sim 10^{-17}$ J, P $\sim 10^3$ W/cm^3, and from the whole instrument of volume 100 cm^3 it is possible to take 10^3 W.

The probability of two-quantum emission in a mode we take in the form (A.I.18) and, like (2.2), we write the self-excitation) condition for quanta $n_1 = n$. We note that the quantity $n_2 = n_0$ is now a parameter which can be fixed by an external source – the other laser.

It is convenient to replace the density of quanta of the exciting laser radiation n_0 by the radiation energy flux

$$S_0 = c\hbar\omega_0 n_0. \tag{3.2}$$

Averaging over the possible orientations of the atoms in (2.2) and introducing the quantity f_{ik} the oscillator strength,

$$\overline{|(\mathbf{j}\,\mathbf{d})|^2_{ik}} \frac{\hbar e^2}{2m\omega_{ik}} f_{ik} = \frac{\hbar r_0 c^2}{2\omega_{ik}} f_{ik}, \tag{3.3}$$

(where $r_0 = e^2/mc^2$ is the classical radius of the electron, c is the velocity of light) and keeping the first term in the sum (2.2) [23], we convert the self-excitation condition (2.2) to the form

$$\frac{2\pi^3}{\hbar^2} r_0^2 c^3 \frac{f_{ik} f_{k2}}{(\omega_{ik} - \omega_0)^2} \frac{S_0(\omega_0)}{\omega_{ik}\omega_{k2}} \hbar\omega \frac{\Delta N}{\Delta\omega} \geqslant \frac{\omega}{Q}. \tag{3.4}$$

In this formula $\hbar$ is Planck's constant, ω_{ik} are frequencies corresponding to the transitions between the levels i and k; ω_0 is the frequency of the exciting radiation; ω is the frequency of the submillimeter waves; f_{ik} are the oscillator strengths for the corresponding transitions; $S_0(\omega_0)$ is the energy flux from the exciting source; $\Delta\omega$ is the width of the luminescence line; Q is the quality factor of the resonator on submillimeter waves. The inverse population on the levels of the neon atom is produced by using a gas discharge and the addition of another inert gas – helium. The excitation energy of the helium atom is very efficiently transferred to the neon atoms. The mixture of gases used must have the same parameters as in the usual helium-neon laser; the pressure of He is 1 mm Hg and that of neon is 0.1 mm Hg. The supplied high-frequency power for maintenance of the gas discharge is 100 W. It is known that in these conditions there is an inverse population and generation of light between the groups of levels 8s-3p, 3p-2s, 2s-2p. There is an inverse population also between the groups of levels 3p-2p, which we use.

One-quantum transitions between the states 3p-2p are forbidden according to the orbital quantum number and, hence, ordinary (one-quantum) generation for these transitions does not occur. Since the exciting laser (in this particular case a ruby laser) is used in pulse operation, it is possible to work with a pulsed gas discharge, in which a high inverse population can be achieved. However, for our considered case of helium–neon we must bear in mind that the time of transfer of excitation from the helium atom to the neon atom (i.e., the time of formation of the inverse population 3p-2p) is of the order of 10^{-4} sec and the discharge time cannot be smaller. Even in a continuous discharge, however, an inverse population of 10^{10} atoms/cm^3 can be obtained. The intermediate levels which determine the intensity of two-quantum luminescence will obviously be the group of levels 2s.

We consider the transition: $3p^1[\frac{3}{2}]-4p^1[\frac{3}{2}]_1$. We write the numerical values for formula (3.4). The intermediate state must be 3d: $\omega_0 = 2\pi c \cdot 12782$ sec^{-1}, $\omega = 2\pi c \cdot 78$ sec^{-1}, $\omega_{1k} = 2\pi c \cdot 1000$ cm^{-1}, $\omega_{k2} = 2\pi c \cdot 11782$ cm^{-1}, $r_0 = 2.8 \cdot 10^{-13}$ cm, $c = 3 \cdot 10^{10}$ cm/sec. The exciting radiation of frequency ω_0 is the emission of a ruby laser transmitted through a styrene converter. The luminescence line width can be taken as of the same order of the Doppler width in gas in these conditions: $\Delta\omega = 2\pi c \cdot 0.03$ cm^{-1}. The exact values of the oscillator strengths contained in formula (3.4) are unknown at present. It is known, however, that in the case of transitions entailing no change in the principal quantum number the oscillator strengths are of the order of unity. Hence, we can put $f_{1k} \sim f_{k2} \sim 1$.

For the calculations it is more convenient to use the gain α of emission in the mode per unit length. For this we convert the inequality (3.4) to the form

$$\alpha > \frac{\omega}{Qc} = \frac{1-r}{l}. \tag{3.5}$$

Here l is the length of the resonator, r is the reflection coefficient of the mirrors. The gain in the case of the given parameters is

$$\alpha\,(\text{cm}^{-1}) = 3.6 \cdot 10^{-21} S_0\,(\text{W/cm}^2)\,\Delta N\,(\text{cm}^{-3}). \tag{3.6}$$

When $S_0 = 10^7$ W/cm^2, $\Delta N \sim 10^{10}$ cm^{-3}, we obtain $\alpha = 3.6 \cdot 10^{-6}$ cm^{-3}, and the inequality (3.5) is satisfied only when the mirrors have a very high reflection coefficient for submillimeter waves r = 0.999 and the length l = 3 m.

We consider now other facts of a fundamental nature which can prevent the generation of submillimeter radiation.

1. Reduction of the real part of the dielectric constant of the plasma.

2. Absorption of submillimeter radiation by the plasma.

As is known, the critical wavelength at which the plasma is opaque, is

$$\lambda_{cr} = \sqrt{\frac{\pi mc^2}{e^2 N_e}} = \sqrt{\frac{\pi}{r_0 N_e}}, \tag{3.7}$$

where N_e is the number of free electrons in 1 cm^3, and r_0 is the classical radius of the electron. When $N_e = 10^{12}$ cm^{-3}, we obtain λ_{cr} = 3 cm, and emission with $\lambda > \lambda_{cr}$ will strongly decay in the plasma.

It should be noted that the electron concentration significantly affects the population of the neon levels. Hence, it becomes necessary to work with $N_e > 10^{12}$ cm^{-3}. The difficulties due to reduction in λ_{cr} can be circumvented by operating in a decaying plasma, i.e., after the discharge supply is switched off. In fact, the majority of acts of excitation of the upper neon levels are due to collisions of neon with helium metastables 2^1S_0, the lifetime of which is 10^{-3}-10^{-4} sec. Free electrons disappear in the plasma in a time of the order of 10^{-5}. Thus, after the discharge is switched off there is an interval of time when there are no electrons, and the upper level still has a practically unchanged population.

We consider now the possible absorption of submillimeter radiation by the plasma. The absorption coefficients is connected with the imaginary part of the dielectric constant of the plasma at these frequencies and is given by the known formula

$$\alpha' = \frac{4\pi N_e e^2}{m\omega}\frac{\nu_{eff}}{c} = \left(\frac{\lambda}{\lambda_{cr}}\right)^2 \frac{\nu_{eff}}{c}, \tag{3.8}$$

where ν_{eff} is the number of collisions of the electron per second. An assessment of the various collision mechanisms showed that in the discharge conditions of interest to us ($N_e \sim 10^{12}$ cm^{-3}, atom concentration ~10^{17} cm^{-3}) the value of ν_{eff} can reach 10^8 sec^{-1}. In this case for $\lambda = 0.3$ mm, $\alpha' \sim 10^{-6}$ cm^{-1}. Thus, absorption by the plasma is fairly small. The above-obtained gain $\alpha = 3.6 \cdot 10^{-6}$ cm^{-1} with the selected parameters and the estimated possible losses suggest that generation of submillimeter radiation is possible. It is clear that there is a possibility of increasing the gain by increasing the power of the exciting laser or by increasing the inverse population, and so on. Nevertheless, as the estimates show, the startup of such a generator is fairly difficult from the experimental aspect. In addition, at present lasers with the power required for startup generate pulses of duration $\tau \sim 10^{-8}$ sec, and in such a time the considered laser with a low gain cannot attain the maximum generation level (for a discussion of this difficulty see § 6).

We point to another practically important feature, which can be realized not only in the given example, but also in other cases.

It is known that the frequency of emission of a ruby laser depends on the crystal temperature and varies from 14,420 cm^{-1} at 77°K to 14,400 cm^{-1} at 270°K. Hence, alteration of the crystal temperature can be used for smooth tuning of a submillimeter laser in a range of ~20 cm^{-1}.

Thus, the proposed method in this particular case enables very simple smooth tuning of a submillimeter laser.

At present in quantum radioelectronics wide use is made of conversion of the frequency of a powerful generator by Raman scattering in styrene, nitrobenzene, liquid nitrogen, etc. This opens up new opportunities for different combinations. In condensed media such a method of generation is also possible and has a definite advantage, since inversion of the medium is much greater (by seven orders) than in gases, but owing to the lack of data on the absorption of submillimeter waves and the presence of broad absorption bands of the substance, special investigations are required for each particular case.

§ 4. Theory of the "Resonance Parametric" Laser in the Stable Phase Case

As was shown above, one of the main drawbacks of the two-photon generator from the viewpoint of practical application is the difficult startup with the aid of a special auxiliary laser of fairly high power. The threshold of startup of the two-quantum laser can be reduced by using the resonance parametric interaction of three quasi-monochromatic fields. The main equations for polarization of the substance and the wave equations of the light fields in this case were obtained earlier (see § 1). We will now investigate these processes more thoroughly. The energy levels of the substance and the values of the interacting quanta are shown schematically in Fig. 2. As already mentioned in § 1, the three-quantum resonance parametric process cannot be considered separately from the one-quantum and two-quantum processes but, as distinct from these last two processes, the resonance parametric process is important only when the "spatial synchronism" condition [(1.37) et seq.] is fullfilled:

$$\mathbf{k}_0 = \mathbf{k}_1 + \mathbf{k}_2. \tag{4.1}$$

Hence, the selected working substance must be a medium in which condition (4.1) can be satisfied.

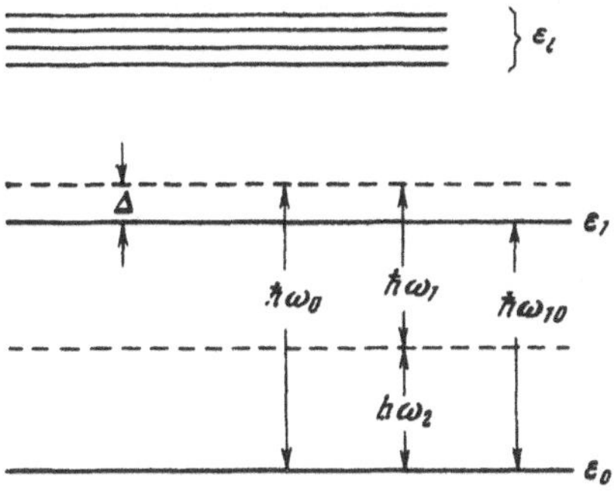

Fig. 2. Diagram of energy levels. Same notation as in Fig. 1.

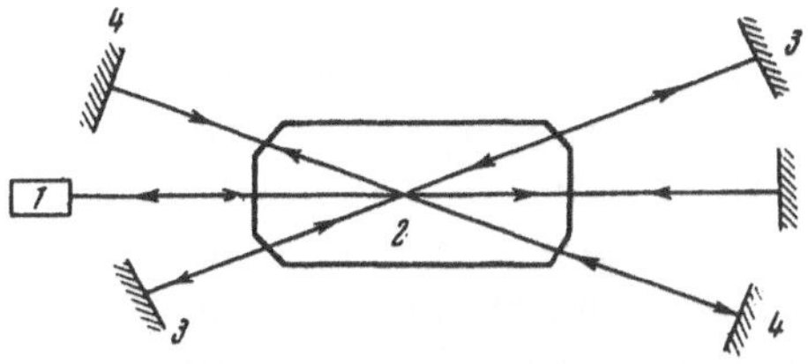

Fig. 3. Diagram of resonance parametric laser. 1) Laser generating frequency ω_0; 2) active working substance; 3) mirrors for frequency ω_1; 4) mirrors for frequency ω_2.

The setup of the laser is illustrated in Fig. 3. It operates in the following way. An inverse population is created between the levels 1 and 0 in the working substance. A field $E(\omega_0)$ is produced in the resonator either by an auxiliary laser or by one-quantum generation in the substance itself. When $E(\omega_0)$ attains the threshold value the resonance parametric process begins to act and produces quanta $\hbar\omega_1$ and $\hbar\omega_2$, which accumulate in the resonator due to the reflection of the mirrors, which are positioned to ensure satisfaction of the "spatial synchronism" condition (actually we have two or three Fabry-Perot resonators mounted at particular angles to one another). We note that the energy of the quanta $\hbar\omega_1$ and $\hbar\omega_2$ is ultimately drawn from the substance, and not from the field $\hbar\omega_0$, as distinct from the "parametric" laser based on the first method (see "Introduction"). When a sufficient number of quanta $\hbar\omega_1$, $\hbar\omega_2$ have accumulated in the resonator the probability of the induced two-quantum process exceeds the probability of the parametric process, and generation is due mainly to two-quantum luminescence with an efficiency independent of condition (4.1), which is of much less importance than in the "parametric" laser based on the first method (see "Introduction"). Thus, it is clear that the "resonance parametric" generator can actual operate in two regimes:

1. A regime in which the resonance parametric process is important and hence, satisfaction of the synchronism condition (4.1) is necessary. This is the regime of easy startup of the laser (see below) and of generation of small powers of the fields ω_1, ω_2;

2. A regime in which two-quantum luminescence is the leading process. The parametric laser can be brought gradually into this regime from the first regime (see below).

In both cases division into quanta $\hbar\omega_1$ and $\hbar\omega_2$ is arbitrary within the restriction of the law of conservation of energy $\hbar\omega_1 + \hbar\omega_2 = \hbar\omega_{10} = \hbar\omega_0$ and this allows smooth tuning. Smooth tuning can be effected in the first regime, as in the "parametric" laser based on the first method [10], by ensuring the synchronism condition for the new division of frequencies by altering the refractive index, rotating (or heating) the crystal, and adjusting the position of the mirrors in accordance with the new synchronism condition. It is difficult in practice to ensure condition (4.1) in a wide frequency range and, hence, it may be sensible to tune at the stage of the second regime, as in the two-photon laser (see § 6). In this case the synchronism condition (4.1) need not be satisfied and tuning over a large range of frequencies is possible.

We consider the operation of a "resonance parametric" laser. Since we assume that the phase of the three quasi-monochromatic fields is stable (see Appendix II), then in this case we use equations (1.36), assuming for simplicity, as at the end of § 1, that the integrals

$$\int_V u^*(\omega_1)\, u^*(\omega_2)\, u(\omega_0)\, dv, \quad \int_V |u(\omega_1)|^2\, |u(\omega_2)|^2\, dv$$

are of the order of unity, and this may occur in the case of spatial synchronism, which we assume:

$$\dot{E}(\omega_1) + \frac{1}{2}\frac{\omega_1}{Q_1} + i(\omega_1 - \Omega_1)E(\omega_1) =$$
$$= -2\pi i\omega_1[\chi_1 E(-\omega_2)E(\omega_0) + \Theta|E(\omega_2)|^2 E(\omega_1)]\Delta N,$$
$$\dot{E}(\omega_2) + \frac{1}{2}\frac{\omega_2}{Q_2} + i(\omega_2 - \Omega_2)E(\omega_2) =$$
$$= -2\pi i\omega_2[\chi_2 E(-\omega_1)E(\omega_0) + \Theta|E(\omega_1)|^2 E(\omega_2)]\Delta N,$$
$$\dot{E}(\omega_0) + \frac{1}{2}\frac{\omega_0}{Q_0} + i(\omega_0 - \Omega_0)E(\omega_0) =$$
$$= -2\pi i\omega_0[\eta_0 E(\omega_0) + \chi_0 E(\omega_1)E(\omega_2)]\Delta N, \qquad (4.2)$$
$$\Delta\dot{N} = -\frac{4\Delta N}{\hbar}|E(\omega_0)|^2 \operatorname{Im}\eta_0 - \frac{4\Delta N}{\hbar}\operatorname{Im}\Theta|E(\omega_1)|^2|E(\omega_2)|^2 -$$
$$-\frac{8\Delta N}{\hbar}\operatorname{Re}\left[E(\omega_1)E(-\omega_2)E(-\omega_0)\frac{\chi_0}{T_2\left(\gamma_{10} + \frac{i}{T_2}\right)}\right] - \frac{1}{T_1}(\Delta N - \Delta N^0).$$

The right sides of equations (4.2) contain the coefficients η, θ, χ which are connected with the polarization of the substance due to one-quantum, two-quantum, and three-quantum processes, respectively, and are calculated from formulas (1.28)-(1.34). Using equations (4.2) we first find the condition for startup of the laser. We solve the problem on the assumption that the field $E(\omega_0)$ is assigned as a parameter. Such a situation is probably the simplest in experiment and, in addition, the calculations are significantly simplified. If $E(\omega_0)$ is prescribed by the auxiliary source, and at the initial instant $E(\omega_1) = E(\omega_2) = 0$, then the first two equations (4.2) can be investigated separately from the others. We represent $E(\omega_{1,2})$ formally in the form

$$E(\omega_{1,2}) = E_0(\omega_{1,2}) + \delta E(\omega_{1,2}), \qquad (4.3)$$

where $\delta E(\omega_{1,2})$ are small deviations from $E_0(\omega_{1,2}) = 0$. Then

$$\delta\dot{E}(\omega_1) + i(\omega_1 - \Omega_1)\delta E(\omega_1) = -2\pi i\omega_1\chi_1\Delta N\delta E(-\omega_2)E(\omega_0) - \frac{1}{2}\frac{\omega_1}{Q_1}\delta E(\omega_1);$$
$$\delta\dot{E}(-\omega_2) - i(\omega_2 - \Omega_2)\delta E(-\omega_2) = -2\pi i\omega_2\chi_2\Delta N\delta E(\omega_1)E(-\omega_0) - \frac{1}{2}\frac{\omega_2}{Q_2}\delta E(-\omega_2), \qquad (4.4)$$

where terms of higher order of smallness in δE are dropped. We will seek the solution in the form

$$\delta E(\omega_\alpha) = \rho(\omega_\alpha)e^{\mu t}. \qquad (4.5)$$

Then, substituting (4.5) in (4.4), we obtain

$$\left[\mu + i(\omega_1 - \Omega_1) + \frac{1}{2}\frac{\omega_1}{Q_1}\right]\rho(\omega_1) + 2\pi\omega_1\chi_1\Delta N E(\omega_0)\rho^*(\omega_2) = 0,$$
$$\left[\mu - i(\omega_1 - \Omega_0) + \frac{1}{2}\frac{\omega_2}{Q_2}\right]\rho^*(\omega_2) + 2\pi\omega_2\chi_2^* E^*(\omega_1)\Delta N\rho(\omega_1) = 0. \qquad (4.6)$$

The condition for solvability of (4.6) is that the determinant

$$\begin{vmatrix} \mu + i(\omega_1 - \Omega_2) + \frac{1}{2}\frac{\omega_1}{Q_1} & 2\pi\omega_1\chi_1\Delta N E(\omega_0) \\ 2\pi\omega_2\chi_2^*\Delta N E^*(\omega_0) & \mu - i(\omega_2 - \Omega_2) + \frac{1}{2}\frac{\omega_2}{Q_2} \end{vmatrix} = 0. \qquad (4.7)$$

We note in (1.29) that

$$\chi_2(\omega_2, \omega_1, \omega_0) = \chi_1(\omega_1\omega_2\omega_0).$$

Solving (4.7), we find μ :

$$\mu = \frac{1}{2} i(\Delta\omega) - \frac{1}{2}\left(\frac{\omega_1}{2Q_1} + \frac{\omega_2}{2Q_2}\right) \pm \left\{\frac{1}{4}\left[i(\Delta\omega) - \left(\frac{1}{2}\frac{\omega_1}{Q_1} + \frac{1}{2}\frac{\omega_2}{Q_2}\right)^2\right] + \omega_1\omega_2 4\pi^2 |\chi_1|^2 \Delta N^2 |E(\omega_0)|^2\right\}^{1/2}, \quad (4.8)$$

where $\Delta\omega = \omega_0 - \Omega_1 - \Omega_2$.

The condition for self-excitation will be given by the inequality

$$\mathrm{Re}\mu > 0. \quad (4.9)$$

For further investigation it is convenient in equations (4.4) to convert from complex amplitudes $E(\omega_\alpha)$ to real amplitudes $A(\omega_\alpha) > 0$ and phases φ_α. In addition, for simplicity we consider only the degenerate case: $\omega_1 = \omega_2 = \omega$, $Q_1 = Q_2 = Q$. Then the complete system of equations for a resonance-parametric laser in the case of a prescribed field $E(\omega_0)$ will be

$$\frac{dA(\omega)}{dt} + \frac{\omega}{2Q} A(\omega) = 2\pi\omega\,\mathrm{Im}\,[\chi_1]\cos\Phi A(\omega)A(\omega_0)\Delta N + 2\pi\omega\,\mathrm{Im}\,[\Theta]A^3(\omega)\Delta N, \quad (4.10)$$

$$A\frac{d\Phi}{dt} = -4\pi\omega\,\mathrm{Im}\,[\chi_1]\Delta N A(\omega_0)A(\omega)\sin\Phi, \quad (4.11)$$

$$\frac{d\Delta N}{dt} = -4\frac{\Delta N}{\hbar}\mathrm{Im}\,[\eta_2]A^2(\omega_0) - \frac{4\Delta N}{\hbar}\mathrm{Im}\,[\Theta]A^4(\omega) - 4\frac{\mathrm{Im}\,[\chi_0]}{\hbar}\cos\Phi A^2(\omega)A(\omega_0)\Delta N - \frac{\Delta N - \Delta N^0}{T_1}, \quad (4.12)$$

where $\Phi = 2\varphi_1 - \varphi_0$.

We investigate the special features of the behavior of the physical system in accordance with equations (4.10-4.12). It follows directly from equation (4.11) that the phase has two stationary points $\Phi = 0, \pi$, the value $\Phi = 0$ being stable, while the value $\Phi = \pi$ is unstable and the phase during the operation of the laser tends to $\Phi = 0\ (2\pi)$; in fact, since $d\Phi/dt \sim -\sin\Phi$, then when $0 < \Phi < \pi$, $d\Phi/dt < 0$, and when $\pi < \Phi < 2\pi$, $d\Phi/dt > 0$.

We note that the sign of the term describing the paramagnetic process in the equation for the amplitude (4.10) depends on $\cos\Phi$. In the case $\Phi = 0$ this term is positive, which corresponds to an increase in the field of frequency ω due to the parametric process; when $\Phi = \pi$ this term is negative, i.e., it corresponds to a decrease in the field. Thus, the parametric process, depending on the phase Φ, can either "help" or "hinder" generation. Since the phase tends to the stable value $\Phi = 0$ we can directly conclude that the "resonance-parametric" process must "speed up" the laser. We turn to the conditions for startup of the laser. The two-quantum luminescence term in equation (4.10) can be neglected in comparison with the parametric term in the case of small fields $A(\omega)$ (in startup conditions). In addition, in the investigation of the startup threshold we can assume $\Delta N = \Delta N^0$.

It is clear that in the case of onset of fluctuation of the field $A(\omega)$ with phase $\Phi = \pi$ the field will be immediately absorbed by the system; on the other hand, the case $\Phi = 0$ leads to driving of the laser and a further increase in the field $A(\omega)$ due to parametric interaction. Thus, to determine the generation threshold we need to take the value $\Phi = 0$ (maximum gain) and equation (4.10) gives the condition for self-excitation of generation

$$4\pi\,\mathrm{Im}\,[\chi_1(\omega, \omega, \omega_0)]\Delta N A(\omega_0) > \frac{1}{Q}. \quad (4.13)$$

Using the expression for χ_1 from § 1 (1.28) and leaving one term in the sum (23), we have

$$\frac{4\pi}{\hbar^2}\left|\frac{d_{01}d_{0l}d_{l0}}{\omega_{l0} - \omega_1}\right|\Delta N A(\omega_0) T_2 > \frac{1}{Q}. \quad (4.14)$$

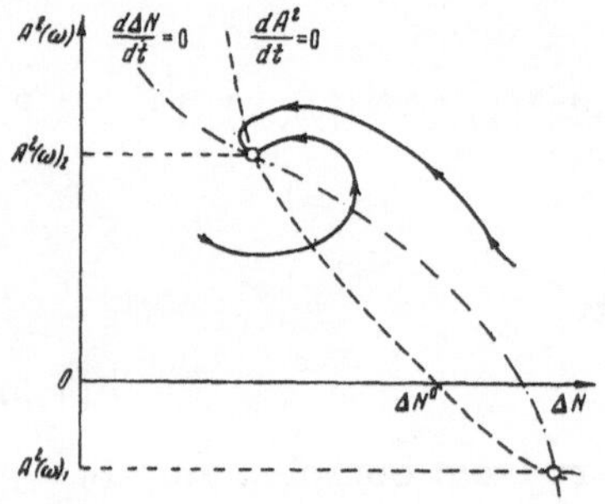

Fig. 4. Approximate picture of phase plane of resonance parametric laser. ΔN, inversion; ΔN^0, initial inversion; $A^2(\omega)$, square of field amplitude.

If we take the value of the parameters in the cgse system

$$d_{0l} = d_{1l} \sim 10^{-18}, \quad d_{01} \sim 10^{-20}, \quad T_2 \sim 10^{-11}, \quad \Delta N = 10^{19},$$
$$\omega \sim 10^{15}, \quad Q \sim 10^{18},$$

then from (4.14) we obtain the threshold pumping field

$$A(\omega_0) \geqslant 0.8 \text{ cgse units,} \tag{4.15}$$

which corresponds to a flux power

$$P = \frac{1}{2\pi} A^2(\omega_0) c \geqslant 300 \text{ W/cm}^2. \tag{4.16}$$

We turn now to an investigation of the steady-state regime. To do this we require to find the nonzero stationary points. It is clear that now we must take two-quantum luminescence into account. We showed earlier that the stationary points of the phase are $\Phi = 0, \pi$, and only the value $\Phi = 0$ is stable. Putting the derivatives and Φ equal to zero in (4.10)-(4.12), we find the stationary values of $|E(\omega)|^2$:

$$|E(\omega)|^2_{1,2} = \frac{\hbar}{T_1}\left(\pi Q \Delta N^0 - \frac{T_1 \chi_1}{\hbar \operatorname{Im} \Theta} A(\omega_0) \pm \sqrt{(\pi Q \Delta N^0)^2 - \frac{\chi_1^2 T_1^2 A^2(\omega_0)}{(\operatorname{Im} \Theta)^2 \hbar^2} - \frac{T_1}{\hbar \operatorname{Im} \Theta}}\right). \tag{4.17}$$

Since $\operatorname{Im} |E(\omega)|^2 = 0$, then, as in § 2, the condition for $Q\Delta N^0$ (the necessary condition for a stationary point) has the form

$$(\pi Q \Delta N^0)^2 \geqslant \frac{T_1}{\hbar \operatorname{Im} \Theta} + \frac{\chi^2 T_1^2 A^2(\omega_0)}{(\operatorname{Im} \Theta)^2 \cdot \hbar^2}. \tag{4.18}$$

Hence, with the values of the other parameters used above, $\Delta N^0 > 10^{20}$. When $A(\omega_0) \to 0$, inequality (4.18) is converted to condition (2.9) from § 2, and (4.17) becomes the stationary values for the "two-quantum" laser. If inequality (4.18) is fulfilled with a margin, then (4.17) is simplified:

$$|E(\omega)|^2_1 \approx -\frac{\operatorname{Im}[\chi_1] A(\omega_0)}{\operatorname{Im}[\Theta]} < 0, \tag{4.19}$$

$$|E(\omega)|^2_2 \approx \left[2\hbar Q \Delta N^0 - \frac{T_1 \operatorname{Im}[\chi_1] A(\omega_0)}{\operatorname{Im}[\Theta]}\right] T_1^{-1}. \tag{4.20}$$

With $\Delta N = 3 \cdot 10^{20}$ and the above values of the other parameters for $A(\omega_0) \sim 0.8$ cgse units (threshold) the field in the second (as can be shown) stable equilibrium position is

$$|E(\omega)|^2_2 = 1.3 \cdot 10^4 \text{ cgse units,} \tag{4.21}$$

which corresponds to a flux power

$$P = 6 \text{ MW/cm}^2.$$

An approximate picture of the phase plane in the variables ΔN, $A^2(\omega)$ is illustrated in Fig. 4.

In the conclusion of this section we note that we have considered the interaction of three "modes." In ordinary lasers multimode operation occurs and, hence, our conclusions are mainly of a quantitative nature and the real case requires additional investigation. This may be carried out in accordance with the proposed scheme with the aid of a computer. We turn now to the discussion of the other limit – random phase.

§ 5. Theory of the "Resonance Parametric" Laser at the Random-Phase Limit

Since the phase of the emission of a laser is unstable to some extent, it is worth while considering another limiting case – the random-phase limit. In Appendix II we discuss the region of applicability of the two approximations. The energy diagrams of the levels of the substance and the interacting quanta for the two cases are the same (Fig. 2). The fundamental design of the laser is also the same (see Fig. 3). The equations for the fields in the random-phase approximation can be derived from the general equations (1.38) by averaging over the phase (see Appendix II). The equations then contain terms representing the probabilities of the one-quantum, two-quantum, and three-quantum process. The last depends on condition (1.37) and in the region in which (1.37) is satisfied is proportional to the square of the inversion ΔN:

$$W = 32\pi^3\hbar \,|\chi_1(\omega_1, \omega_2, \omega_0)|^2 \frac{\omega_1\omega_2\omega_0}{\Delta\omega} \Delta N^2 \left(\int\limits_V u^*(\omega_1) u^*(\omega_2) u(\omega_0)\, dv\right)^2 \times$$
$$\times (n_1n_2 + n_2n_0 + n_1n_2) = W_3(n_1n_0 + n_2n_0 + n_1n_2)\Delta N^2. \quad (5.1)$$

In formula (5.1) the direct and reverse processes are taken into account (without the terms describing spontaneous emission).

We will investigate at what threshold fields n_0 does resonance parametric scattering begin, i.e., we consider the condition for excitation of frequency $\omega_{1,2}$. In our laser with a prescribed field ω_0

$$W_3(n_1n_0 + n_2n_0 + n_1n_2)(\Delta N)^2 \geqslant \frac{\omega_{1,2}}{Q_{1,2}} n_{1,2}. \quad (5.2)$$

When $n_1 = n_2 = 0$ at the initial instant

$$W_3n_0(\Delta N)^2 \geqslant \frac{\omega_1}{Q_1}. \quad (5.3)$$

Hence, with the values of the parameters used below we can obtain the threshold value for $n_0 \geq 10^{12}$ quanta/cm^3, which corresponds to

$$|E(\omega_0)|^2 = 2\pi \text{ cgse units } \simeq 6 \text{ cgse units}, \quad (5.4)$$

i.e., several orders lower than the value of the threshold fields required for the two-quantum laser. This means that if a substance with the required characteristics is placed in the resonator, the process of resonance parametric scattering will begin and the two channels will be filled immediately with quanta $\hbar\omega_1$ and $\hbar\omega_2$. When the threshold value for $\hbar\omega_1$ (or $\hbar\omega_2$) is obtained two-quantum emission will become the decisive factor.

We will now consider the question of steady-state operation of the laser and find the amplitude of the steady-state oscillations in the degenerate case $\omega_1 = \omega_2 = \omega$; we note that the field ω_0 is not regarded as prescribed, but arises inside the laser. We write the complete system of equations for the numbers of quanta and the inverse population:

$$\frac{dn_0}{dt} = -\frac{n_0}{\tau} + b_0n_0\Delta N + b_2(2n_0n + n^2)(\Delta N)^2,$$

$$\frac{dn}{dt} = -\frac{n}{\tau} + 2b_1n^2\Delta N + 2b_2(2n_0n + n^2)(\Delta N)^2, \quad (5.5)$$

$$\frac{d(\Delta N)}{dt} = \frac{\Delta N^0 - \Delta N}{T} - 2b_0n_0\Delta N - 2b_1n^2\Delta N - 2b_2(2n_0n + n^2)(\Delta N)^2.$$

Here $\Delta N^0/T$ characterizes the pumping of the inverse population, b_0 is the probability of the one-quantum process without the field, b_1 is the probability of the two-quantum process without the field, b_2 is the probability of scattering without the field, and τ is the decay time of the oscillations in the resonator.

We take the following values of the parameters (corresponding to ruby):

$$b_0 = 10^{-7}\ \text{sec}^{-1}, \quad b_1 = 3\cdot 10^{-23}\ \text{sec}^{-1},$$
$$b_2 = 10\ \text{sec}^{-1}, \quad \tau = 10^{-8}\ \text{sec}, \tag{5.6}$$
$$T = 3\cdot 10^{-3}\ \text{sec}, \quad \Delta N^0 = 10^{19}\ \text{cm}^{-3}.$$

We find the stationary values of $n_1 n_0 \Delta N$ by equating the derivatives in (5.5) to zero. In view of its difficulty the algebraic nonlinear system was solved on a computer with parameters (5.6). It was found that the change in parameters in a wide range (several orders) had practically no effect on the stationary value of the quantities

$$n_0 = 3\cdot 10^{16}\ \text{quanta/cm}^3, \quad n = 3\cdot 10^{14}\ \text{quanta/cm}^3, \quad \Delta N = 4\cdot 10^{14}\ \text{atoms/cm}^3. \tag{5.7}$$

Investigations by Lyapunov's method show that the obtained stationary point is stable. The values of the field ω corresponding to the equilibrium position are

$$|E(\omega)|^2 = 2\cdot 10^3\ \text{cgse units} \tag{5.8}$$

or the flux $P \approx 1\ \text{MW/cm}^2$.

It is of interest to note that, as was to be expected, at the random-phase limit the threshold for generation of the field $E(\omega)$ occurs at higher fields $E(\omega_0)$ than in the stable-phase case (§ 4).

§ 6. Discussion of Results and Experimental Approach to Their Realization

The mathematical apparatus which we have discussed can provide a basis for further development of theory. One of the important theoretical results of our work is the evaluation of the startup threshold for the "two-photon" laser. Startup occurs, as the estimates in § 2 indicate, when $|E(\omega)|^2 \sim 6\cdot 10^2$ cgse units, which corresponds to a coherent exciting power $P \sim 1\ \text{MW/cm}^2$. We found that the resonance parametric laser does not require coherent excitation at the tunable frequencies ω_1, ω_2. Startup of the laser requires only the resonance field $\omega_0 = \omega_1 + \omega_2 = \omega_{10}$, where ω_{10} is the frequency of the transition between the working levels of the substance; this field is due to one-quantum luminescence (ordinary laser) and is fairly small: In the stable-phase case $|E(\omega_0)|^2 = 0.64$ cgse units, which corresponds to a flux density $P = 300\ \text{W/cm}^2$, and at the random-phase limit $|E(\omega_0)|^2 = 6$ cgse units or a flux density $P \approx 3\ \text{kW/cm}^2$.

Using the equations we have derived we can investigate other theoretical questions; transient regimes, multimode operation, theory of laser line width and tuning, as well as other problems similar to those discussed in Akhmanov and Khokhlov's well known monograph [3]. None of these problems has yet been examined and they are all of definite interest.

We will assume, however, that at this stage the main point of interest is experimental work. It is worth while here to discuss some experimental possibilities. We begin with a laser based on two-quantum luminescence. We note that so far no one has observed the physical phenomenon of induced two-quantum luminescence. The observation of this effect requires illumination of the excited phosphor with coherent light from an extraneous source (laser).

Purely spontaneous two-quantum luminescence (without the presence of a coherent field) has been detected in a monatomic gas [21]. The experimental difficulty in observing two-quantum luminescence may be the impossibility of separating the two-quantum emission from the background of noncoherent pumping. Perhaps the property of long afterglow of the phosphor or even "constant-action" phosphors should be used. We note that the detection of spontaneous luminescence does not depend greatly on the aggregate state of the phosphor and powdered phosphors can be used. The background of noncoherent pumping for excitation of the phosphor can be got rid of, as is done in the case of semiconductors, by using another laser. To obtain maximum sensitivity of the apparatus it is essential that the "additional" luminescence quantum $\hbar\omega_2 = \hbar\omega_{10} - \hbar\omega_1$ lies within the range of sensitivity of the photomultiplier, i.e., the wavelength should be $\lambda_2 \leq 0.9\,\mu$. This imposes particular demands on the frequency of the exciting laser and the difference in energy levels of the phosphor, i.e., a "blue" phosphor must be used, and illumination is effected with a long-wave laser, which could be a CaF_2–Dy (~2.5 μ) laser or a powerful gas laser. It would be possible to conduct an experiment with two exciting lasers ω_1 and ω_2, where $\hbar\omega_1 + \hbar\omega_2 = \hbar\omega_{10} = \varepsilon_1 - \varepsilon_0$. One of the exciting lasers could be a low-power ($P \sim 10^{-3}$ W/cm^2) gas laser. The continuous generation of such a laser and the large set of discrete frequencies would facilitate its use in the experiment. By observing the amplification at frequency ω_2 it would also be possible to investigate induced two-quantum emission at the frequency ω_2 of the gas laser. Perhaps it might be easier to obtain amplification at the additional frequency ω_2 in the case of excitation with a sufficiently strong field ω_1. In this case detection of the emission ω_2 will be much simpler. The difficulties of such an experiment are illustrated by the example discussed below. Let the emission of the neodymium exciting laser $\hbar\omega_1$ ($\lambda = 1.06\,\mu$) of power 1 MW/cm^2 be incident on a 7-cm long ruby with inversion $\Delta N \sim 10^{18}$–10^{19} atoms/cm^3, and the sensitivity of the detecting instruments (photoresistors) at the additional frequency ($\hbar\omega_2 = \hbar\omega_{10} - \hbar\omega_0$) be 10^{-4} W/cm^4. With such sensitivity it would be possible to detect only the amplified induced two-quantum emission ω_2. We will consider when amplification of ω_2 is possible in such a system. The intensity of the emission ω_2 emerging from the ruby will be

$$I_2(l) \approx I_2(0)\,e^{kl},$$

where $I_2(l)$ is the intensity of emergence, $I_2(0)$ is the intensity of the emission ω_2 without amplification in the presence of fields ω_1 ($I_1 \sim 1$ MW/cm^2), k is the difference of the gain and absorption coefficient, and l is the length of the ruby. The absorption coefficient in ruby can be taken as $\alpha \gtrsim 10^{-2}$ cm^{-1} [25]. Hence, the gain must be greater than 10^{-2} cm^{-1}, and on excitation with 1 MW/cm^2 the amplification is 10^{-4} cm^{-1}, i.e., there will be no amplification in the system. In addition, if k were $\gtrsim 10^{-2}$ cm^{-1}, then for adequate gain we would require $kl \sim 1$ and $l \geq 10^2$ cm. Thus, "buildup" of the exponent requires longer lengths. This can be achieved by feedback, which increases the effective length.

In conclusion of the discussion of the two-quantum laser we show schematically how it can be tuned (Fig. 5). Tuning is possible in the steady-state operation of the laser. This is effected by altering the position of the mirror 4, thus creating selective feedback at the altered frequency ω_2. The instrument is constructed so that the other beam ω_1 will always be reflected from the mirror 3, thus ensuring feedback at ω_1.

The main requirement of a resonance parametric laser is satisfaction of the spatial synchronism condition (1.37). This entails using the anisotropy of the refractive index in the crystals. It must be borne in mind that the synchronism condition (1.37) can also be fulfilled in a semiconductor laser. The synchronism condition must be satisfied with a certain degree of precision and, hence, in rarefied gases, where dispersion is weak, synchronism can obviously be secured. The difficulty lies in the fact that optical anisotropy has been sufficiently well investigated only for a limited range of transparent crystalline substances, and only in the yellow-

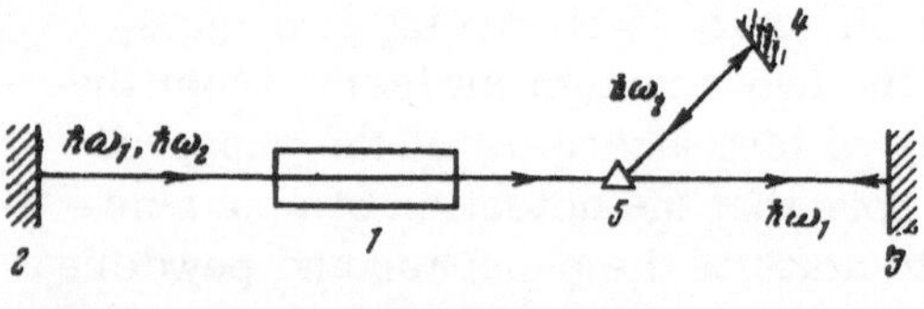

Fig. 5. Schematic of two-quantum laser. 1) Active substance; 2) fixed mirror for frequencies ω_1, ω_2; 3) mirror for frequency ω_1; 4) mirror for frequency ω_2; 5) spectral instrument.

green region of the spectrum. In our case we require to know the refractive index in the red and infrared regions of the spectrum. Here, as in the previous case, a successful choice of substance is important. Unfortunately, the synchronism condition (1.37) cannot be fulfilled in ruby. A promising substance is $LiNbO_3$, into which Cr ions could be introduced. We can conclude from general considerations that the "resonance parametric" laser will have greater power per unit volume of the working substance than an ordinary "parametric" laser. This will be due to two-quantum luminescence.

From the work we have conducted we can assert that the construction of the proposed types of lasers is a difficult, but feasible, task at the present level of technique.

I express my deep gratitude to A. S. Selivanenko for his interest and considerable help in the work for this paper.

Appendix I. Quantum-Electrodynamic Description of Multiphoton Processes

We will discuss briefly the method of describing multiphoton processes in the case where the electromagnetic fields are assumed to be quantized. Then the quantum system "substance + field" will be described by the Schrödinger equation [19] (for simplicity we will regard the substance as one "atom" with one electron)

$$i\hbar \frac{\partial \psi}{\partial t} = \hat{H}\psi, \tag{A.I.1}$$

$$\hat{H} = \hat{H}_0 + \sum_\alpha \hat{H}_\alpha + \hat{V}, \tag{A.I.2}$$

where $\hat{H}_0$ is the Hamiltonian of the "atom," $\hat{H}_\alpha$ is the energy of the free field of frequency ω_α, $\hat{V} = \sum_\alpha V_\alpha(\omega_\alpha)$ is the energy of interaction of the fields ω_α with the "atom":

$$V(\omega_\alpha) = -\frac{e\hat{\mathbf{p}}\hat{\mathbf{A}}_\alpha}{cm} + \frac{e^2\hat{A}_\alpha^2}{2c^2m}. \tag{A.I.3}$$

Here $\hat{p}$ is the momentum operator of the electron (the extension to several electrons is not difficult), $\hat{\mathbf{A}}_\alpha$ is the vector-potential of the field ω_α, e and m are the charge and mass of the electron, and c is the velocity of light.

In the dipole approximation it is sometimes more convenient to use the interaction in the form

$$\hat{V}'_\alpha(\omega_\alpha) = -\hat{\boldsymbol{d}}\hat{\boldsymbol{E}}_\alpha, \tag{A.I.4}$$

where $\hat{d}$ is the electron dipole moment operator, and $\hat{E}_\alpha$ is the electric field strength operator.

In the representation of the interaction the Schrödinger equation takes the form

$$i\hbar \frac{\partial \psi_0}{\partial t} = e^{\frac{i\hat{H}_0 t}{\hbar} + i\sum_\alpha \hat{H}_\alpha} \sum_\alpha \hat{V}_\alpha e^{\frac{-i\hat{H}_0 t}{\hbar} - i\sum_\alpha \hat{H}_\alpha} \psi_0, \tag{A.I.5}$$

where ψ_0 is the wave function in the representation of the interaction. The integral of this equation is sought in the form

$$\psi_0 = \sum_n a_n(t)\varphi_n \exp\left(-\frac{i\varepsilon_n t}{\hbar}\right), \tag{A.I.6}$$

where ε_n and φ_n are the eigenvalues and eigenfunctions, respectively, of the operator $\hat{H}_0 + \sum_\alpha \hat{H}_\alpha$.

We note that usually [13, 20] the states of the fields ω_α at the initial instant are chosen as states with a fixed number of quanta n_α. According to the uncertainty relation $\Delta n_\alpha \Delta\varphi_\alpha \sim 1$ (this relation, generally speaking, is approximate [26, 27]). In a state with fixed n_α the phases φ_α are not determined (in the calculation averaging over phases is implicitly assumed).

Thus, this variant of the calculation certainly cannot give information about the phase relationships of the fields. The probability of transition in unit time to a final state f, the energy E_f of which is in the interval dE_f, from the initial state i, with $a_i(0) = 1$, is

$$W_{if} = \frac{\partial}{\partial t}|a_f|^2 \rho(E_f)\,dE_f, \tag{A.I.7}$$

where $\rho(E_f)$ is the density of the final states.

Since the interaction $\hat{V}(\omega_\alpha)$ can be regarded as small, a_f will be sought from perturbation theory and the probability of the p-photon process is

$$W_{if}^{(p)} = \frac{2\pi}{\hbar}|K_{if}^{(p)}|^2 \rho_f \tag{A.I.7'}$$

where

$$K_{if}^{(p)} = \sum_{k_1, k_2, k_3, \ldots, k_{p-1}} \frac{V'_{ik_1} V'_{k_1 k_2} \cdots V'_{k_{p-1} f}}{(\varepsilon_i - \varepsilon_{k_1})(\varepsilon_i - \varepsilon_{k_2}) \cdots (\varepsilon_i - \varepsilon_{k_{p-1}})}. \tag{A.I.8}$$

The denominator contains the differences between the total energies of the initial and intermediate states, i.e., the p-photon process passes through p intermediate states. It can be shown that multiphoton processes are characterized by statements: (1) the law of conservation of energy holds only for the entire p-photon process as a whole and does not hold for the intermediate states; and (2) the probability of the p-photon process is proportional to the number of photons in the degree p or $|E|^{2p}$.

As an example we consider the emission of two photons

$$\hbar\omega_1 + \hbar\omega_2 = \hbar\omega_{10} \tag{A.I.9}$$

by the atom from state 1 (see Fig. 1) in the presence of fields of frequencies ω_1, ω_2:

$$K_{10}^{(2)} = \sum_l \left(\frac{V'_{1l}(\omega_1) V'_{l_0}(\omega_2)}{\varepsilon_1 - \varepsilon_l} + \frac{V'_{1l}(\omega_2) V'_{l_0}(\omega_1)}{\varepsilon_1 - \varepsilon'_l} \right), \tag{A.I.10}$$

$$\begin{aligned} \varepsilon_1 - \varepsilon_l &= \hbar\omega_{1l} - \hbar\omega_1, \\ \varepsilon_1 - \varepsilon'_l &= \hbar\omega_{1l} - \hbar\omega_2. \end{aligned} \tag{A.I.11}$$

Here ε_1 and ε_2 are the energies of the whole system (atom + field) in the initial and intermediate states. Summation is carried out over all the energy states of the system (atom + field). It is convenient to separate from $V_{1l}(\omega_\alpha)$ the "field part," i.e., the numbers of photons, for instance,

$$V'_{1l}(\omega_1) = \sqrt{n_1 + 1}\, v_{1l}(\omega_1), \tag{A.I.12}$$

where n_1 is the mean number of quanta in the mode. Then

$$K_{10}^{(2)} = \sqrt{(n_1+1)(n_2+1)}\sum_l\left(\frac{v_{1l}(\omega_1)\,v_{l_0}(\omega_2)}{\varepsilon_1-\varepsilon_l}+\frac{v_{1l}(\omega_2)\,v_{l0}(\omega_1)}{\varepsilon_1-\varepsilon_l'}\right). \tag{A.I.13}$$

From (A.I.7) the probability of the two-quantum transition is

$$W_{10}^{(2)} = \frac{2\pi}{\hbar}\int dE\,|K_{10}^{(2)}|^2\rho_E\delta(\hbar\omega_{21}-\hbar\omega_1-\hbar\omega_2). \tag{A.I.14}$$

Here

$$\rho_E = \rho_1\rho_2 = \frac{(\hbar\omega_1)^2}{(2\pi\hbar c)^3}d\Omega_1\frac{(\hbar\omega_2)^2}{(2\pi\hbar e)^3}d\Omega_2, \tag{A.I.15}$$

$d\Omega$ is an element of solid angle, $dE = d(\hbar\omega_1)d(\hbar\omega_2)$.

We are interested in the emission of two quanta of particular energy with accuracy to the small value $\varepsilon \ll \varepsilon_1 - \varepsilon_0 = \hbar\omega_0$:

$$W_{10}^{(2)} = \frac{2\pi}{\hbar}|K_{10}^{(2)}|^2\rho_1\rho_2\varepsilon \tag{A.I.16}$$

or the probability of emission of two quanta into only two modes

$$W_{10}^{(2)'} = \frac{W_2}{\rho_1\rho_2\varepsilon^2} = \frac{2\pi}{\hbar}|K_{10}^{(2)}|^2\frac{1}{\varepsilon} = \frac{2\pi}{\hbar}(n_1+1)(n_2+1)\left|\sum_l\left(\frac{v_{1l}(\omega_1)\,v_{l0}(\omega_2)}{\hbar\omega_{1l}-\hbar\omega_1}+\frac{v_{1l}(\omega_2)\,v_{l0}(\omega_1)}{\hbar\omega_{1l}-\hbar\omega_2}\right)\right|^2\frac{1}{\varepsilon}. \tag{A.I.17}$$

The expression (A.I.17) can be converted to the form

$$W_{10}^{(2)'} = \frac{(2\pi)^3}{\hbar}\frac{\omega_1\omega_2(n_1+1)(n_2+1)}{\varepsilon}\left|\sum_l\frac{(\boldsymbol{d}_{1l}\boldsymbol{e}_1(\omega_1))(\boldsymbol{d}_{l0}\boldsymbol{e}(\omega_2))}{\omega_{1l}-\omega_1}+\frac{(\boldsymbol{d}_{1l}\boldsymbol{e}(\omega_2))(\boldsymbol{d}_{l0}\boldsymbol{e}(\omega_1))}{\omega_{1l}-\omega_2}\right|^2, \tag{A.I.18}$$

where $\mathbf{e}(\omega_\alpha)$ is the unit vector of the field.

Knowing the probabilities of multiquantum processes we simply write the balance equation for the inverse population and the numbers of quanta, which, as already noted, ignore phase relationships. To take into account the phase of the fields in quantum electrodynamics we must consider the initial states of the fields with a fairly well-defined phase [28]. This can be achieved by selecting the initial states of the field in the form of "packets":

$$|\psi_{0\alpha}\rangle = \sum_{n_\alpha}C_\alpha(n_\alpha)|n_\alpha\rangle, \tag{A.I.19}$$

where $|C_\alpha(n_\alpha)|^2$ is the probability of finding $>n_\alpha$ photons $\hbar\omega_\alpha$ in the state $|\psi_{0\alpha}$. By an appropriate choice of $C_\alpha n_\alpha$ we can make $\Delta\varphi_\alpha$ sufficiently small. For the change in energy of the field ω_α we can obtain

$$\frac{d\hat{H}_\alpha}{dt} = \frac{i}{\hbar}[\hat{V}, \hat{H}_\alpha]. \tag{A.I.20}$$

From (A.I.20) we find the mean value of $d\hat{H}_\alpha/dt$ with due regard to nonlinear interaction

$$\left\langle\frac{d\hat{H}_\alpha}{dt}\right\rangle = \left\langle\psi\left|\frac{d\hat{H}_\alpha}{dt}\right|\psi\right\rangle, \tag{A.I.21}$$

where ψ is given by (1.1).

Using perturbation theory ψ is found from the Schrödinger equation (A.I.1) if the initial states of the fields are selected in the form (A.I.19). The calculation of the "parametric" interaction of three fields

$$\omega_1 + \omega_2 = \omega_3 \tag{A.I.21'}$$

requires calculation of the wave function with accuracy to the second order; since the calculations are laborious we are obliged to omit them. As a result, from (A.I.20) and the solution of ψ for the average over the time we obtain an expression, from which we write out the term corresponding to parametric interaction,

$$\left\langle \overline{\frac{d\hat{H}_1}{dt}} \right\rangle \sim \sum_{n_1 n_2 n_3} C_1(n_1) C_1^*(n_1+1) C_2(n_2) C_2^*(n_2+1) C_3(n_3) C_3^*(n_3+1) \sqrt{n_3(n_1+1)(n_2+1)}\, B_1 B_2 B_3 + \ldots \tag{A.I.22}$$

The quantities B_α are the real modes in the resonator. The standing-wave operator is

$$\hat{\mathbf{E}}_\alpha = \mathbf{B}_\alpha \hat{P}_\alpha = e(\omega_\alpha) B_\alpha \hat{P}_\alpha, \tag{A.I.23}$$

where $\hat{P}_\alpha$ are generalized momenta of the field [19, 20, 28]. By definition the mean value of the field operator $\hat{E}_\alpha$ in the state $|\psi_\alpha\rangle = \sum_\alpha c_\alpha(n_\alpha)\,|n_\alpha\rangle$:

$$E(\omega_\alpha) = \langle\psi|\,\hat{E}_\alpha\,|\psi\rangle = iB_\alpha \sum_{n_\alpha} C_\alpha^*(n_\alpha+1) C_\alpha(n_\alpha) \sqrt{n_\alpha+1}\; \times$$

$$\times \sqrt{\hbar \frac{\omega_\alpha}{2}}\, e^{i\omega_\alpha t} + \text{constant terms} = e(\omega_\alpha)\,[E_+(\omega_\alpha)\, e^{i\omega_\alpha t} + E_-(\omega_\alpha)\, e^{-i\omega_\alpha t}]. \tag{A.I.24}$$

The expression (A.I.24) gives the classical field with phase by means of "packets." Using (A.I.24), we convert (A.I.22).

$$\left\langle \overline{\frac{d\hat{H}_1}{dt}} \right\rangle \sim E_+(\omega_1) E_+(\omega_2) E_-(\omega_3)\, \overline{\exp\,[i(\omega_1+\omega_2-\omega_3)]\,t} + \text{constant terms}. \tag{A.I.25}$$

In view of (A.I.21') the exponent in (A.I.25) on averaging is equal to unity. The result (A.I.25) corresponds to the classical description of parametric processes (see § 1). Such procedures are tedious and it is much simpler to take into account phase relationships by calculating the nonlinear polarization of the quantum system in the presence of classically assigned fields (see § 1).

Appendix II. Nonlinear Interactions with Random Phase. Comparison of Quantum and Semiclassical Descriptions of Nonlinear Interactions

In this appendix we discuss the region of applicability of the quantum and semiclassical description of nonlinear interactions, and we also examine nonlinear interactions at the random-phase limit. The fact is that in the case of "parametric" interactions, such as the three-quantum process of combination of frequencies, the results obtained from the usual quantum theory of transitions differ from the results obtained by the semiclassical method (§ 1). It follows from the usual quantum theory of transitions that the probability of the parametric process depends on the square of the moduli of the field amplitudes, whereas the semiclassical approach indicates that it depends on the amplitudes [28]. This diffference requires explanation. The explanation is, as mentioned above, that the usual quantum theory of transitions ignores the role of phase relationships and corresponds to the random-phase limit.

As can be seen from (1.38), the change in field energy in the case of some nonlinear interactions depends on the phase relationships between the fields [the penultimate term on the right in the first equation of (1.38)], while in the case of others it is independent [the last term on the right in the first equation of (1.38)] and agrees with the corresponding value obtained by the usual quantum theory. We note that for the first ("parametric" [28]) interactions it depends only on the product of the squares of the moduli of the amplitudes. In preceding sections the fields were assumed to be "quasi-monochromatic." In this case the width $\Delta\omega_\alpha$ of the emission line is assumed to be sufficiently small, i.e., the time of the phase mismatch $\Delta\omega_\alpha^{-1} = \tau_{\varphi_\alpha}$ is greater than some characteristic time of nonlinear interaction τ (see below). In the case where $\tau_\varphi > \tau$, it is necessary to average equations (1.38) over the time [29]. Obviously, averaging in (1.38) has no effect on the terms responsible for two-quantum ("combination") processes, but has a significant effect on "parametric" processes. We keep only the "parametric" terms in (1.36):

$$\begin{aligned} \dot{E}_a(\omega_1) &= -2\pi\omega_1 i\chi_{1abc}E_b(-\omega_2)E_c(\omega_0)\Delta N, \\ \dot{E}_a(\omega_2) &= -2\pi\omega_2 i\chi_{2abc}E_b(-\omega_1)E_c(\omega_0)\Delta N, \\ \dot{E}_a(\omega_0) &= -2\pi\omega_0 i\chi_{0abc}E_b(\omega_1)E_c(\omega_2)\Delta N. \end{aligned} \tag{A.II.1}$$

Henceforth for simplicity we will assume that all the fields are directed in the same way, along j_1, for instance, and then the equation can be put in the form*

$$\partial E(\omega_1)/\partial t = \eta_1 E^*(\omega_2)E(\omega_0), \tag{A.II.2}$$

$$\eta_1 = -2\pi\omega_1 i\chi_{1,111}(\omega_1, \omega_2, \omega_0)\Delta N. \tag{A.II.3}$$

In view of (1.29) and (1.31) and assuming that all ω_α are of the same order, it is easy to see that

$$\eta_1 \sim \eta_2 \sim \eta_0 \sim \eta. \tag{A.II.4}$$

An important role in what follows will be played by two quantities: the characteristic time of phase mismatch and the time of nonlinear interaction. The time of phase mismatch of the wave ω_α is

$$\tau_{\varphi_\alpha} = \frac{1}{\Delta\omega_\alpha}, \tag{A.II.5}$$

where $\Delta\omega_\alpha$ is the width of the emission line.

For simplicity we will henceforth assume that all τ_{φ_α} are equal:

$$\tau_{\varphi_\alpha} = \tau_\varphi. \tag{A.II.6}$$

We introduce the characteristic time of nonlinear interaction from equations (A.II.2), assuming again for simplicity that all the fields $E(\omega_\alpha)$ are of the same order:

$$E(\omega_1) \approx E(\omega_2) \approx E(\omega_0) \approx E, \tag{A.II.7}$$

$$\tau \approx \tau_\alpha \approx \frac{E(\omega_\alpha)}{\frac{\partial E(\omega_\alpha)}{\partial t}} = \frac{E(\omega_\alpha)}{\eta_\alpha E(\omega_\beta)E(\omega_\delta)} = \frac{1}{\eta E}. \tag{A.II.8}$$

Here we have used (A.II.4).

In the random-phase case, as will be seen from what follows, nonlinear interaction will be determined by another time:

$$\tau' = \frac{\tau^2}{\tau_\varphi} = \frac{1}{\eta^2 E^2 \tau_\varphi}. \tag{A.II.9}$$

---

*Henceforth we will write all the equations for ω_1, since the equations for ω_2 and ω_0 are similar.

Below we will calculate the mean change in amplitude and energy of the fields in accordance with equation (A.II.2) during a time $T \leq \tau$. In regard to the relationships of τ_φ and τ we consider only two limiting cases:

$$\tau_\varphi \gg \tau \gtrsim T, \tag{A.II.10}$$

$$\tau_\varphi \ll T \lesssim \tau. \tag{A.II.11}$$

Case (A.II.10): Phase of All Fields Stable

Case (A.II.10) was considered in § 1. In view of (A.II.10) during time T the phases of the field vary slowly, and there is no need to average over the phase. From (A.II.2) the change in field energy is

$$\frac{\partial |E(\omega_1)|^2}{\partial t} = 2\mathrm{Re}\{\eta_1 E^*(\omega_2) E^*(\omega_1) E(\omega_0)\} \sim \frac{|E|^2}{\tau} + \frac{|E|^2}{\tau}\frac{T}{\tau} + \ldots \tag{A.II.12}$$

Thus, the energy transfer is determined by τ.

Case (A.II.11): Phase of All Fields Random

In this case all three $E(\omega_\alpha)$ are random functions of time which can be put in the form

$$E(\omega_\alpha) = E_0(\omega_\alpha) + \delta E(\omega_\alpha), \tag{A.II.13}$$

where $E_0(\omega_\alpha)$ are statistically independent random functions, while $\delta E(\omega_\alpha)$ take into account the correlations between the fields due to nonlinear interaction. Henceforth we assume that $E_0(\omega_\alpha)$ are stationary functions with sufficiently "good" properties.

Owing to the randomness of the fields we will be interested in the mean change in time of the field energies in a time of the order of T. According to ergodic theorems this mean is equal to the mean over the ensemble of random functions $E_0(\omega_\alpha)$, which we will denote by angular brackets. We introduce the correlation functions of the fields $E_0(\omega_\alpha) = E_0(\omega_\alpha, t)$:

$$\langle E_0^*(\omega_\alpha, t) E_0^*(\omega_\beta t')\rangle = \delta_{\alpha\beta} K_\alpha(t - t'), \tag{A.II.14}$$

where

$$K_\alpha(t - t') = I(\omega_\alpha) \int_{-\infty}^{\infty} \varphi_\alpha(\Delta\omega)\, e^{i\Delta\omega(t-t')}\, d\Delta\omega, \tag{A.II.15}$$

where

$$I(\omega_\alpha) = \langle |E(\omega_\alpha)|^2\rangle, \tag{A.II.16}$$

and $\varphi_\alpha(\Delta\omega)$ is the spectral density of the field energy, measured from the center of the line.

Here

$$\int_{-\infty}^{\infty} \varphi_\alpha(\Delta\omega)\, d\Delta\omega = 1. \tag{A.II.17}$$

In view of (A.II.17) $\varphi_\alpha(\Delta\omega)$ can be approximated:

$$\varphi_\alpha(\Delta\omega) = \begin{cases} \dfrac{1}{\Delta\omega_\alpha} = \tau_{\varphi_\alpha} & \text{for } |\Delta\omega| \leqslant \dfrac{1}{2\tau_{\varphi_\alpha}}, \\ 0 & \text{for } |\Delta\omega| > \dfrac{1}{2\tau_{\varphi_\alpha}}. \end{cases} \tag{A.II.18}$$

It is clear from (A.II.15) and (A.II.16) that $K_\alpha(t-t')$ differs significantly from zero when $|t-t'| \le \tau_{\varphi_\alpha}$. In the zero approximation in δE the mean change in field energy by virtue of (A.II.14) is zero:

$$\left\langle \frac{\partial |E(\omega_1)|^2}{\partial t} \right\rangle = 2\mathrm{Re}\{\eta_1 \langle E_0^*(\omega_1) E_0^*(\omega_2) E_0(\omega_0)\rangle\} = 0. \quad \text{(A.II.19)}$$

In the first approximation in δE

$$\left\langle \frac{\partial |E(\omega_1)|^2}{\partial t} \right\rangle = 2\mathrm{Re}\{\eta_1 [\langle \delta E^*(\omega_1) E_0^*(\omega_2) E_0(\omega_0)\rangle + \langle E_0^*(\omega_1) \delta E^*(\omega_2) E(\omega_0)\rangle + \langle E_0^*(\omega_1) E_0^*(\omega_0) \delta E(\omega_0)\rangle]\}. \quad \text{(A.II.20)}$$

Integrating equations of the type (A.II.2) we find $\delta E(\omega_\alpha)$:

$$\delta E(\omega_1) = \eta_1 \int_{-\infty}^{t} e^{\varepsilon t} E_0^*(\omega_\alpha) E_0(\omega_0)\, dt. \quad \text{(A.II.21)}$$

Here for convenience we have put $-\infty$ as the lower limit of integration and have introduced $e^{\varepsilon t}$, $\varepsilon \to 0$ to eliminate divergence at the lower limit. We substitute (A.II.21) in (A.II.20), expressing† the triple and quadruple correlations in pairs from formulas [30]:

$$\langle ABCD\rangle = \langle AB\rangle\langle CD\rangle + \langle AC\rangle\langle BD\rangle + \langle AD\rangle\langle BC\rangle. \quad \text{(A.II.22)}$$

In first order in δE we obtain

$$\left\langle \frac{\partial |E(\omega_1)|^2}{\partial t} \right\rangle = 2\mathrm{Re}\Big\{\eta_1\eta_1^* \int_{-\infty}^{t} e^{\varepsilon t'} K_0(t'-t) K_2(t-t')\, dt' + \eta_1\eta_1^* \int_{-\infty}^{t} e^{\varepsilon t'} K_1(t-t') K_0(t'-t)\, dt' + \eta_1\eta_0 \int_{-\infty}^{t} e^{\varepsilon t'} K_1(t-t') K_2(t-t')\, dt'\Big\}. \quad \text{(A.II.23)}$$

Substituting (A.II.15) in (A.II.23), we obtain

$$\left\langle \frac{\partial |E(\omega_1)|^2}{\partial t} \right\rangle = 2\mathrm{Re}\{\eta_1\eta_2^* |E_0(\omega_2)|^2 |E_0(\omega_0)|^2 + \eta_1\eta_2^* |E_0(\omega_1)|^2 |E_0(\omega_0)|^2 + \eta_1\eta_0 |E_0(\omega_1)|^2 |E_0(\omega_2)|^2\}\tau_\varphi. \quad \text{(A.II.24)}$$

Using (A.II.15), (A.II.16)-(A.II.18), we obtain

$$\left\langle \frac{\partial |E(\omega_1)|^2}{\partial t} \right\rangle = 2\eta_1\eta_1^*\omega_2\omega_0 \left[\frac{|E_0(\omega_2)|^2}{\omega_2\omega_0} |E_0(\omega_0)|^2 + \frac{|E_0(\omega_1)|^2 |E_0(\omega_0)|^2}{\omega_1\omega_2} \pm \frac{|E(\omega_1)|^2 |E(\omega_2)|^2}{\omega_1\omega_2}\right]\tau_\varphi. \quad \text{(A.II.25)}$$

The ± sign in (A.II.25) was obtained owing to the difference in the properties of $\chi_0(\omega_0, \omega_1, \omega_2)$ in (1.31), (1.32) in the case of small and great "detuning" of the resonance γ_{10}, respectively [see (1.40)].

Similarly we obtain

$$\left\langle \frac{\partial |E(\omega_2)|^2}{\partial t} \right\rangle, \quad \left\langle \frac{\partial |E(\omega_0)|^2}{\partial t} \right\rangle, \quad \left\langle \frac{\partial \Delta N}{\partial t} \right\rangle.$$

As can be seen from (A.II.25), the time of nonlinear interaction for wave ω_1 is not equal to τ but to τ'. Thus, it is easy to find the time of energy transfer. In this case it is inversely

† In view of the statistical properties of the fields $E_0(\omega_\alpha)$ (A.II.15) in our case we can use these expressions.

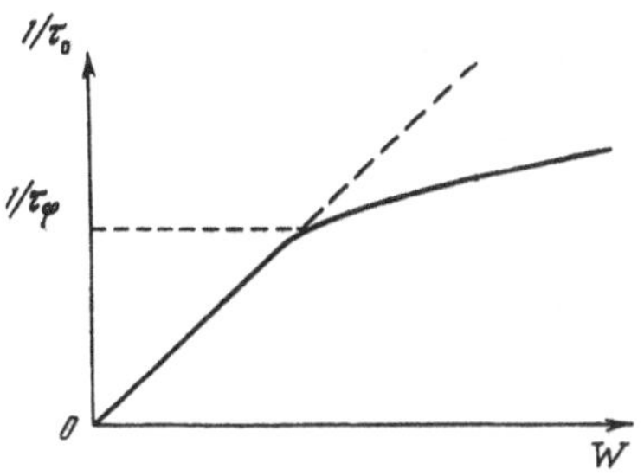

Fig. 6. Time of energy transfer τ_0 as function of emission density W.

proportional to the square of the amplitude of the fields. We introduce the numbers of quanta:

$$n_\alpha = \frac{|E(\omega_\alpha)|^2}{4\pi\hbar\omega_\alpha}. \tag{A.II.26}$$

Then from (A.II.25) with the aid of (A.II.26) we obtain

$$\frac{dn_1}{dt} = W(n_1 n_0 + n_2 n_0 \pm n_1 n_2), \tag{A.II.27}$$

$$W = 8\pi\eta_1\eta_2^* \frac{\omega_2\omega_0}{\omega_0}\hbar\tau_\varphi = 32\pi^3\hbar\,|\chi_{1.111}(\omega_1, \omega_2, \omega_0)|^2 \frac{\omega_1\omega_2\omega_0}{\Delta\omega}(\Delta N)^2. \tag{A.II.28}$$

In view of (A.II.3) it is easy to see that (A.II.28) is the same as the expression for the change in the number of quanta in the mode, obtained from the usual quantum perturbation theory with accuracy to the spontaneous term [9].

Comparison of Results of (A.II.10) and (A.II.11)

As (A.II.12) shows, in the case of (A.II.10) the time of change of energy of the field ω_α is equal to τ. In the case of (A.II.11), which is realized when all the phases are random (the time of phase mismatch τ_φ is small), or when the fields are small and τ is large, inequality (A.II.11) again holds; in addition, inequality (A.II.11) can be satisfied at low η (weak nonlinearity or low frequencies). In case (A.II.11) the amplitude on the average does not change in time T, while the transfer time is determined by τ'. The latter is

$$\tau' = \tau\left(\frac{\tau}{\tau_\varphi}\right) = \frac{1}{\eta^2 E^2 \tau_\varphi} \gg \tau. \tag{A.II.29}$$

Figure 6 shows the energy transfer time τ_0 as a function of the emission density W. As Fig. 6 shows, at low E, W, $\tau_0 = \tau' \sim 1/W$ [case (A.II.11)]; at large W, $\tau_0 = \tau' \sim 1/W^{\frac{1}{2}}$ [case (A.II.10)]. We consider the region of applicability of approximations (A.II.10) and (A.II.11). For this we calculate

$$\tau = \frac{1}{\eta E} \approx \frac{1}{2\pi\omega\chi_{abc} E \Delta N}, \tag{A.II.30}$$

and τ'. The estimate we obtain for χ_{abc}, for instance, in the case of a "transparent medium" from (1.28) by putting in cgse units $d \sim 10^{-18}$, $\omega \sim 10^{15}$, $\Delta N \sim 10^{23}$ is:

$$\chi_{abc} \sim \frac{d^3 \Delta N}{\hbar^2\omega^2} \sim 10^{-7}, \tag{A.II.31}$$

$$\tau \approx \frac{\hbar^2\omega}{2\pi d^3 \Delta N E} = \frac{10^{-9}}{E}. \tag{A.II.32}$$

It should be noted that the estimate (A.II.32) relates to a "quadratic" medium with good nonlinearity; In a weakly nonlinear medium, for instance, if one transition is forbidden in (1.28), then the estimate of χ_{abc} will be three orders less and, hence, τ will be three orders more:

$$\tau = \frac{10^{-6}}{E}. \tag{A.II.33}$$

Example. A field of power 100 kW/cm^2, which corresponds to E ~ 20 cgse units, i.e., from (A.II.32) $\tau \sim 5 \cdot 10^{-11}$ sec. Accordingly, approach (A.II.10) is valid for fields with a time of phase mismatch $\tau_\varphi \geq 10^{-11}$ sec, i.e., practically always, but for fields E = 0.01 cgse units this approach is not applicable when the time of phase mismatch is $\tau_\varphi \leq 10^{-7}$ sec.

We note that we consider all the means in time $T \leq \tau$. If we consider that the mean in time T is such that $1/\omega \leq T \leq \tau_{\varphi}$, there will be no difference between approaches (A.II.12) and (A.II.13).

In conclusion we discuss the region of applicability of the usual quantum approach and the derived probability method of obtaining the rate of equations (which is fairly simple and clear) in the case of multiphoton processes.

A comparison of the results obtained in Appendices I, II and in § 1 indicates that the probability method

1) is equivalent to the semiclassical method for the evaluation of the intensity of "combination" (two-quantum) processes;

2) describes all the nonlinear processes in the random-phase approximation;

3) as distinct from the semiclassical description (where complex generalized fluctuation-dissipation theorems are required), simply describes spontaneous emission and the noise of nonlinear devices;

4) is not suitable for the description of "parametric" interaction with stable phase. Besides this, the region of applicability of the balance equations is limited by the same factors as in the one-quantum case [28].

Literature Cited

1. M. G. Goeppert-Mayer, Ann. Phys., 9:273 (1931).
2. A. M. Bonch-Bruevich and V. A. Khodovoi, Usp. Fiz. Nauk, 85:3 (1965) [Sov. Phys. – Usp., 8:1 (1965)].
3. S. A. Akhmanov and R. V. Khokhlov, Problems of Nonlinear Optics [in Russian], Moscow (1964).
4. N. Bloembergen, Nonlinear Optics, W. A. Benjamin, New York (1965).
5. V. M. Fain and E. G. Yashchin, Zh. Éksp. Teor. Fiz., 46:695 (1964) [Sov. Phys. – JETP, 19:474 (1964)].
6. S. A. Akhmanov and R. V. Khokhlov, Zh. Éksp. Teor. Fiz., 43:351 (1962) [Sov. Phys. – JETP, 16:61 (1963)].
7. N. Kroll, Phys. Rev., 127:1207 (1962).
8. A. M. Prokhorov and A. S. Selivanenko, Author's Certificate B 459, December 24, 1963.
9. B. P. Kirsanov and A. S. Selivanenko, Opt. Spektrosk., 23:455 (1967) [Opt. Spectrosc., 23:242 (1967)].
10. S. A. Akhmanov, A. I. Kovrigin, A. S. Peskarskas, V. V. Fadeev, and R. V. Khokhlov, ZhÉTF Pis. Red., 2:300 (1965).
11. W. E. Lamb, Phys. Rev., 134:1431 (1963).
12. Yu. G. Khronopulo, Izv. Vuzov. Radiofizika, 7:674 (1964).
13. I. V. Voropaev and A. N. Oraevskii (in press).
14. L. D. Landau and E. M. Lifshits, Quantum Mechanics [in Russian], Fizmatgiz, Moscow (1963).
15. T. N. Il'inova, Vestn. MGU, seriya fiz. astron., 1:79; 5:39 (1966).
16. G. L. Gurevich and Yu. G. Khronopulo, Izv. vuzov. Radiofizika, 8:493 (1965).
17. P. P. Sorokin and N. N. Braslau, IBM J., 8:177 (1964).
18. P. L. Garwin, IBM J., 8:338 (1964).
19. W. Heitler, The Quantum Theory of Radiation, Clarendon Press, Oxford (1954).

20. A. I. Akhiezer and V. B. Berestetskii, Quantum Electrodynamics [in Russian], Gostekhizdat, Moscow (1953).
21. M. Lipeles, P. Nowick, and N. Tolk, Phys. Rev. Lett., 15:690 (1965).
22. A. S. Selivanenko, Opt. Spektrosk., 21:100 (1966) [Opt. Spectrosc., 21:54 (1966)].
23. F. V. Bunkin, Zh. Éksp. Teor. Fiz., 50:1685 (1966) [Sov. Phys. – JETP, 23:1121 (1966)].
24. S. G. Rautian and A. S. Selivanenko (in press).
25. A. P. Veduta, Candidate's Dissertation, FIAN (1967).
26. V. M. Fain, Zh. Éksp. Teor. Fiz. (in press).
27. E. D. Trifonov and A. S. Troshin, Vestn. LGU. Fizika, khimiya, No. 4, p. 69 (1966).
28. V. M. Fain and Ya. I. Khanin, Quantum Radiophysics [in Russian], Izd. Sovetskoe Radio, Moscow (1965).
29. B. P. Kirsanov, A. S. Selivanenko, and V. N. Tsitovich, Paper read at Symposium on Nonlinear Optics in Novosibirsk [in Russian] (1966).
30. V. S. Pugachev, Theory of Random Functions [in Russian], Fizmatgiz, Moscow (1962).